The Holocene and Anthropocene Environmental History of Mexico

Nuria Torrescano-Valle • Gerald A. Islebe
Priyadarsi D. Roy
Editors

The Holocene and Anthropocene Environmental History of Mexico

A Paleoecological Approach on Mesoamerica

Springer

Editors
Nuria Torrescano-Valle
Departamento Conservación
de la Biodiversidad
El Colegio de la Frontera Sur
Unidad Chetumal
Chetumal, Quintana Roo, Mexico

Gerald A. Islebe
Departamento Conservación de la
Biodiversidad
El Colegio de la Frontera Sur Unidad
Chetumal
Chetumal, Quintana Roo, Mexico

Priyadarsi D. Roy
Instituto de Geología
Universidad Nacional Autónoma de México
Ciudad Universitaria
Mexico City, Mexico

ISBN 978-3-030-31718-8 ISBN 978-3-030-31719-5 (eBook)
https://doi.org/10.1007/978-3-030-31719-5

This Springer imprint is published by the registered company Springer Nature Switzerland AG
The registered company address is: Gewerbestrasse 11, 6330 Cham, Switzerland

Preface

Analyzing the Holocene environmental history of Mexico and Mesoamerica is a complex task. It has been researched and explored by numerous scientists with the application of multidisciplinary methodologies and proxies over the last several decades. The idea of this book was developed some years ago with the principal objective of putting together most of the information on environmental changes occurred over the Holocene in this important part of the world with diverse ecosystems. We consider that this book with synthesis of the Holocene environmental histories is of wide interest to scientists, students, and the general public as it covers the geographical diversity of this region and incorporates different approaches to understand environmental and climate changes. The ample geographical and thematic ranges made it unmanageable to cover all the aspects of environmental and climate changes. The chapters in this book, however, provide the readers with an overview of research realized in Mexico to foster new research collaborations in the fields of paleoclimate and climate change in the near future.

The editors of this book thank all the authors for their cooperation and positive spirit. To Springer, we are thankful for this opportunity and patience (thanks to Joao Pildervasser and colleagues!). We also acknowledge the reviewers for their valuable time, dedication, and important contributions: Dr. Hermann Behling (University of Goettingen), Dr. Henry Hooghiemstra (University of Amsterdam), Dr. Kees Nooren (University of Utrecht), and Dr. Josué Polanco-Martínez (Basque Centre for Climate Change).

Chetumal, Quintana Roo, Mexico — Nuria Torrescano-Valle
Chetumal, Quintana Roo, Mexico — Gerald A. Islebe
Mexico City, Mexico — Priyadarsi D. Roy

Contents

Contributors

Juan David Acevedo-Acosta Instituto Politécnico Nacional, Centro Interdisciplinario de Ciencias Marinas (CICIMAR), La Paz, Mexico

Mirtha Albañez-Lucero Instituto Politécnico Nacional, Centro Interdisciplinario de Ciencias Marinas (CICIMAR), La Paz, Mexico

Alejandro A. Aragón-Moreno Departamento Conservación de la Biodiversidad, El Colegio de la Frontera Sur Unidad Chetumal, Chetumal, Quintana Roo, Mexico

José L. Arce Instituto de Geología, Departamento de Procesos Litosféricos, Universidad Nacional Autónoma de México, Coyoacán, Mexico, Mexico

Dulce M. Bocanegra-Ramírez Faculty of Biology, Universidad Michoacana de San Nicolás de Hidalgo., Edificio R. Gral. Francisco J. Múgica. CU. Felicitas del Río., Morelia, Mexico

Margarita Caballero Instituto de Geofísica, Universidad Nacional Autónoma de México, Ciudad de, Mexico, Mexico

Nancy Cabanillas-Terán Catedras Consejo Nacional de Ciencia y Tecnología, El Colegio de la Frontera Sur Unidad Chetumal, Chetumal, Quintana Roo, Mexico

Alicia Carrillo-Bastos División de Estudios de Posgrado e Investigación, Tecnológico Nacional de México/I. T. Chetumal, Chetumal, Quintana Roo, Mexico

Valerio Castro-López Centro de Investigaciones en Geografía Ambiental, Universidad Nacional Autónoma de México, Morelia, Michoacán, Mexico

Claudia M. Chávez-Lara Organic Geochemistry Unit, School of Chemistry, School of Earth Sciences, Cabot Institute for the Environment, University of Bristol, Bristol, UK

Olivia de Los Ángeles Flores-Castillo Instituto Politécnico Nacional, Centro Interdisciplinario de Ciencias Marinas (CICIMAR), La Paz, Mexico

Ana P. Del Castillo-Batista Departamento de Ecología y Recursos Naturales, Centro Universitario de la Costa Sur, Universidad de Guadalajara, Autlán de Navarro, Jalisco, Mexico

Gabriela Domínguez-Vázquez Faculty of Biology, Universidad Michoacana de San Nicolás de Hidalgo., Edificio R. Gral. Francisco J. Múgica. CU. Felicitas del Río., Morelia, Mexico

Diana Cecilia Escobedo-Urías Instituto Politécnico Nacional, Centro de Interdisciplinario de Investigación para el Desarrollo Integral Regional (CIIDIR–Unidad Sinaloa), Guasave, Mexico

Betty B. Faust Centro de Investigación Científica de Yucatán, Mérida, Yucatán, Mexico

Blanca L. Figueroa-Rangel Departamento de Ecología y Recursos Naturales, Centro Universitario de la Costa Sur, Universidad de Guadalajara, Autlán de Navarro, Jalisco, Mexico

William J. Folan Centro de Investigaciones Históricas y Sociales, Universidad Autónoma de Campeche, Campeche, Mexico

Anabel Ford ISBER/MesoAmerican Research Center, University of California, Santa Barbara, CA, USA

Helga Z. Geovannini-Acuña Centre College, Mérida, Yucatán, Mexico

Joel D. Gunn University of North Carolina at Greensboro, Greensboro, NC, USA

Sherman W. Horn III Grand Valley State University, Allendale, MI, USA

Gerald A. Islebe Departamento Conservación de la Biodiversidad, El Colegio de la Frontera Sur Unidad Chetumal, Chetumal, Quintana Roo, Mexico

Socorro Lozano-García Instituto de Geología, Universidad Nacional Autónoma de México, Ciudad de Mexico, Mexico

José L. Macías Instituto de Geofísica, Unidad Michoacán, Universidad Nacional Autónoma de Mexico, Antigua Carretera a Pátzcuaro, Morelia, Michoacán, Mexico

Aída Martínez-López Instituto Politécnico Nacional, Centro Interdisciplinario de Ciencias Marinas (CICIMAR), La Paz, Mexico

Miguel Olvera-Vargas Departamento de Ecología y Recursos Naturales, Centro Universitario de la Costa Sur, Universidad de Guadalajara, Autlán de Navarro, Jalisco, Mexico

Beatriz Ortega-Guerrero Instituto de Geofísica, Universidad Nacional Autónoma de México, Ciudad de Mexico, Mexico

Ligia Pérez-Cruz Paleomagnetismo y Paleoambientes, Instituto de Geofísica, UNAM, Mexico City, Mexico

Jesús David Quiroz-Jiménez Área Académica de Ciencias de la Tierra y Materiales, Universidad Autónoma del Estado de Hidalgo, Carr. Pachuca-Tulancingo, Mineral de la Reforma, Mexico

Pablo J. Ramírez-Barajas Departamento Conservación de la Biodiversidad, El Colegio de la Frontera Sur Unidad Chetumal, Chetumal, Quintana Roo, Mexico

División de Estudios de Posgrado e Investigación, Tecnológico Nacional de México/ I.T. Zona Maya, Othon P. Blanco, Quintana Roo, Mexico

Priyadarsi D. Roy Instituto de Geología, Universidad Nacional Autónoma de México, Ciudad Universitaria, Mexico City, Mexico

Romeo Saldívar-Lucio Centro de Investigación Científica y de Educación Superior (CICESE-ULP), La Paz, Mexico

José Luis Sánchez-Zavala Instituto de Geología, Universidad Nacional Autónoma de México, Ciudad Universitaria, Mexico City, Mexico

Alfred H. Siemens University of British Columbia, Vancouver, BC, Canada

Susana Sosa-Najera Instituto de Geología, Universidad Nacional Autónoma de México, Ciudad de Mexico, México

Nuria Torrescano-Valle Departamento Conservación de la Biodiversidad, El Colegio de la Frontera Sur Unidad Chetumal, Chetumal, Quintana Roo, Mexico

Mirna Valdez-Hernández Departamento Conservación de la Biodiversidad, El Colegio de la Frontera Sur Unidad Chetumal, Chetumal, Quintana Roo, Mexico

Alejandro A. Vela-Pelaez Departamento Conservación de la Biodiversidad, El Colegio de la Frontera Sur Unidad Chetumal, Chetumal, Quintana Roo, Mexico

Alejandro Velázquez Centro de Investigaciones en Geografía Ambiental, Universidad Nacional Autónoma de México, Morelia, Michoacán, Mexico

Gerardo Verdugo-Díaz Instituto Politécnico Nacional, Centro Interdisciplinario de Ciencias Marinas (CICIMAR), La Paz, Mexico

Chapter 1
Introduction: The Holocene and Anthropocene Environmental History of Mexico

Nuria Torrescano-Valle, Gerald A. Islebe, and Priyadarsi D. Roy

Abstract The Holocene spans the last 11,700 years of Earth's history, and the paleoecology and paleoclimate dynamics of Mexico over this interval were complex. This is apparent when considering the region's ecological and physiographic diversity, as well as human impact since the Late Holocene. The geography of Mexico varies from desert to high mountain systems, indicating conspicuous precipitation and temperature variability on latitudinal and altitudinal scales. More than one-third of Mexico's territory is classified as arid or subarid. This area receives more than 60% of its total annual precipitation during the warm season through the North American Monsoon (NAM) and tropical storm systems, and significant winter precipitation occurs only in its northwestern margin. The Anthropocene is the latest part of the Holocene and it is marked by an era of substantial human activity. Climate change during this interval is not exclusively driven by natural processes, as humans have been influencing shifts in the global climate. Several climate projections predict that anthropogenic global warming will cause further enhancement in aridity in this drought-prone region by increasing the mean temperature and reducing the average annual rainfall in the near future. Other areas of Mexico are projected to receive less annual precipitation or the same amount but in shorter time periods. Ultimately, the knowledge of Holocene environmental change can provide society with clues for conservation, management, and adaptation of Mexico's diverse environments.

N. Torrescano-Valle (✉) · G. A. Islebe
Departamento Conservación de la Biodiversidad, El Colegio de la Frontera Sur Unidad Chetumal, Chetumal, Mexico
e-mail: ntorresca@ecosur.mx

P. D. Roy
Instituto de Geología, Universidad Nacional Autónoma de México, Ciudad Universitaria, Mexico City, Mexico

N. Torrescano-Valle et al. (eds.), *The Holocene and Anthropocene Environmental History of Mexico*, https://doi.org/10.1007/978-3-030-31719-5_1

The Holocene spans the last 11,700 years of Earth's history, and the paleoecology and paleoclimate dynamics of Mexico over this interval were complex. This is apparent when considering the region's ecological and physiographic diversity, as well as human impact since the Late Holocene. The geography of Mexico varies from desert to high mountain systems, indicating conspicuous precipitation and temperature variability on latitudinal and altitudinal scales. More than one-third of Mexico's territory is classified as arid or subarid. This area receives more than 60% of its total annual precipitation during the warm season through the North American Monsoon (NAM) and tropical storm systems, and significant winter precipitation occurs only in its northwestern margin. The Anthropocene is the latest part of the Holocene and is marked by an era of substantial human activity. During this interval, the climate change is not exclusively driven by natural processes, and humans have influenced shifts in the global climate. Several climate projections predict that anthropogenic global warming will cause further enhancement in aridity in this drought-prone region by increasing the mean temperature and reducing the average annual rainfall in the near future. Other areas of Mexico are projected to receive less annual precipitation or the same amount but in shorter time periods. Ultimately, the knowledge of Holocene environmental change can provide society with clues for conservation, management, and adaptation of Mexico's diverse environments in the near future.

Paleoclimate records from the region include a variety of proxies, biological, physical, and chemical, each coming from studies with different ecosystems and goals. Across the geosciences, a general trend toward multiproxy paleoenvironmental reconstructions can be seen from research and literature reviews carried out in the last decade. These studies generally aim to describe the ecological and hydrological responses to several global events, such as the events of 8.2 ka, 5.5–5 ka, and 4.2 ka. Studies have also recorded several protracted droughts during the later part of the Holocene (i.e., Meghalayan stage), responses to changes in meteorological patterns like El-Niño Southern Oscillation (ENSO) and the migration of the Intertropical Convergence Zone (ITCZ), and hydrological and ecological changes during the Medieval Climate Anomaly (MCA) (~AD 800–1350) and the Little Ice Age (LIA) (~AD 1350–1850). All these changes point to common extra-regional drivers. There are also several regional or local paleoclimate signals that were likely caused by a combination of local and regional drivers. The climate of Mexico has also been influenced by volcanic activity. In this book, the authors aim to disentangle paleoclimate signals and their paleoecological relevance from different perspectives in order to present the readers with a comprehensive view of past environmental changes. Obviously, there are gaps in knowledge and academic debates, which we see as opportunities for new and future research.

The Holocene environmental history of Mexico is also of relevance considering that many plant species were domesticated in this region, corn (*Zea mays*) being one of the best-known examples. Cultivation of this crop started as early as ca. 6000 cal yr. BP, suggesting that humans had and still have a significant impact on the landscape. Mexico's natural landscape has been influenced by indigenous cultures across the region since the Late Holocene, with the most prominent and enigmatic ones being the ancient Maya and Aztec. After the Spanish conquest, indigenous land-use practices changed rapidly from the eighteenth century onward, subsequently altering the landscape even more.

In this book, we aim to share knowledge on Mexico's Holocene environmental history from the perspectives of paleoecology and paleoclimatology with a common aim to analyze physical and ecological dynamics. Our current understanding of present-day climate change depends strongly on understanding past climate processes, and this book aims to contribute sound knowledge on environmental change. Most of the contributions include a revision of past studies, offering novel reviews of the different disciplines concerned with Holocene environmental change.

The book is divided into 12 chapters, with each chapter providing its own analysis and synthesis of a specific geographical region of Mexico and Mesoamerica. The chapters are ordered based on geographical areas, with chapters on northern Mexico listed first, followed by chapters on the Holocene histories of the central and southern Mexico. Several chapters apply the new terminology for the division of the Holocene (Walker et al. 2019), while others take a more classic approach using terms like Early, Middle, and Late Holocene, as both approaches are valid.

Chapter 2 presents an analysis of Late Holocene environmental changes in the Gulf of California. The authors focus on interannual to centennial- to millennial-scale changes on sedimentation processes, phytoplankton records, and the relationship between hydrological processes and natural and anthropogenic factors. Both the external and internal forcing mechanisms act as climatic drivers in this important region of Mexico.

Chapter 3 deals with the Holocene paleoclimate changes of semiarid to arid subtropical Mexico. The North American Monsoon (NAM) provides precipitation to a large geographical area of this region. Considered in conjunction with ENSO and ITCZ as a climate driving force, this study also provides new knowledge on the heterogeneity of hydroclimate changes by considering other sources like tropical storms and their influences across different parts of this drought-prone region.

Chapter 4 deals with paleoecological reconstruction from the cloud forests on the Pacific side of Mexico. Two major global events of the Late Holocene were detected in the pollen record, the Little Ice Age (~AD 1350–1850) and the Medieval Climate Anomaly (~AD 800–1350). Dry conditions during the LIA were detected from ~AD 1653 to 1734. The warmest period was contemporary with the MCA, dating to between 800 and 1200 AD. The authors detected specific vegetation responses to these precipitation events in the fossil pollen spectra.

In Chap. 5, the authors present a pollen record from the coastal area of the Gulf of Mexico. Land-use practices have transformed most of the original wetlands in this region. However, until now, few pollen studies have been concerned with Holocene sea-level change in this area. The Laguna Catarina core from Veracruz suggests a drop in sea level around 5000 years ago, changing the inland processes that then caused changes in the ionic conditions of the estuarine ecosystems to make *Potamogeton* the dominant aquatic plant in the wetlands.

In Chap. 6, the authors provide an extensive revision to known paleoecological dynamics from central Mexico. Among other recorded events, glacial advances between 8.5 and 7.5 ka cal BP were correlated to the 8.2 ka event. Paleolimnological data from several sites of the Trans-Mexican Volcanic Belt show a series of protracted droughts occurred after 6 ka cal BP. During the Meghalayan stage (4.2 ka cal BP to present), the paleoecological records show a clear influence of human activities.

Chapter 7 presents a novel approach to understanding Holocene paleoecology, plant ecology, landscape ecology, and vegetation reconstruction. The authors apply chorological models to reconstruct past landscape dynamics of central Mexico and unravel ecological processes on various geographical scales.

Chapter 8 analyzes volcanic activities in Mexico during the Holocene. Eruptions from stratovolcanoes are discussed, and data indicate that most volcanic activity was situated in the south. The authors identify 153 eruptions during the Holocene. They also suggest that volcanic activity occurred more during the Late Holocene than the Early Holocene.

Chapter 9 deals with human influence versus natural climate variability in southeastern Mexico. Major regional climate drivers like ITCZ and ENSO have influenced the hydroclimate of this region. In addition, humans have transformed the landscape since the Late Holocene as agriculture began by the Middle Preclassic, around 900 BCE. Several proxy records provide mixed signals of climate and human activity, and the authors aim to disentangle them.

In Chap. 10, the paleoecological records of southern Mexico are discussed. The southern and southeastern parts of Mexico feature heterogeneous climates, landscapes, biodiversity, and human settlements. The ancient Maya inhabited this diverse region and played a key role in shaping the landscape. The authors of this chapter develop a geospatial model to understand precipitation changes throughout the last 2000 years.

The authors of Chap. 11 analyze the historical ecology of the Champoton/Candelaria area in Campeche, southeastern Mexico. A detailed analysis involving many aspects of the watershed provides information about the landscape and ecological conditions as well as the connection of ancient societies with the natural system.

Chapter 12 discusses the use of Laser Imaging Detection and Ranging (Lidar) to understand ancient and contemporary Maya forest landscapes at El Pilar, Belize. Interpretations of preindustrial land use depend on archaeological survey to locate and document settlements. Yet in the case of the Maya area, where the landscape itself was domesticated, land-use studies must also include the forest to capture human impacts on the environment through time. This analysis shows that Lidar is destined to become indispensable for ancient Maya settlement research and to improve our understanding of the forested Maya lowlands.

Paleoecological studies have shown the negative effects of protracted droughts on ancient cultures of Mexico and elsewhere. The chapters of this book clearly show an urgent need to understand hydroclimate dynamics at different scales. By understanding these diachronic, multi-scalar dynamics, the negative impact of unsustainable human activities in relation to projected climate changes can be better understood (IPCC 2018).

References

IPCC (2018) Summary for policymakers. In: Global Warming of 1.5°C. An IPCC Special Report on the impacts of global warming of 1.5°C above pre-industrial levels and related global greenhouse gas emission pathways, in the context of strengthening the global response to the threat of climate change, sustainable development, and efforts to eradicate poverty [Masson-Delmotte, V., P. Zhai, H.-O. Pörtner, D. Roberts, J. Skea, P.R. Shukla, A. Pirani, W. Moufouma-Okia, C. Péan, R. Pidcock, S. Connors, J.B.R. Matthews, Y. Chen, X. Zhou, M.I. Gomis, E. Lonnoy, T. Maycock, M. Tignor, and T. Waterfield (eds.)]

Walker M, Gibbar P, Head MJ, et al. (2019) Formal subdivision of the Holocene series/epoch: a summary. J Geol Soc India 93(2):135–141

Chapter 2
Paleoclimate of the Gulf of California (Northwestern Mexico) During the Last 2000 Years

Aída Martínez-López, Olivia de Los Ángeles Flores-Castillo, Romeo Saldívar-Lucio, Diana Cecilia Escobedo-Urías, Gerardo Verdugo-Díaz, Ligia Pérez-Cruz, Mirtha Albañez-Lucero, and Juan David Acevedo-Acosta

Abstract This selective 2000 years, paleoclimate review expounds on a range of key issues appertaining to Mexico's Gulf of California (GC). Many of these issues are unresolved and in some cases are controversial in nature. This chapter will explore the following areas: (1) significant climate variation involving differing timescales ranging from interannual, to centennial, to millennial; (2) processes responsible for the sedimentation cycle; (3) assessing the fidelity of Alfonso Basin's siliceous phytoplankton record through linkage with integrated water–vertical settling studies, in that paleoclimate proxies are one of the most relevant and important tools utilized in reconstructions of (paleo)temperature and primary production; and (4) examine some global surface hydrological and climate processes and the connections to anthropogenic changes that can structure marine records recovered from marginal environments. Additionally, brief highlights from several areas under investigation will be presented, which have the potential to further an understanding of the role that temporal changes in export production may have played on carbon sequestration and how the resultant effects might have significantly affected climate variations.

A. Martínez-López (✉) · O. de Los Ángeles Flores-Castillo
G. Verdugo-Díaz · M. Albañez-Lucero · J. D. Acevedo-Acosta
Instituto Politécnico Nacional, Centro Interdisciplinario de Ciencias Marinas (CICIMAR), La Paz, Mexico
e-mail: amartin@ipn.mx

R. Saldívar-Lucio
Centro de Investigación Científica y de Educación Superior (CICESE-ULP), La Paz, Mexico

D. C. Escobedo-Urías
Instituto Politécnico Nacional, Centro de Interdisciplinario de Investigación para el Desarrollo Integral Regional (CIIDIR–Unidad Sinaloa), Guasave, Mexico

L. Pérez-Cruz
Paleomagnetismo y Paleoambientes, Instituto de Geofísica, UNAM, Mexico City, Mexico

N. Torrescano-Valle et al. (eds.), *The Holocene and Anthropocene Environmental History of Mexico*, https://doi.org/10.1007/978-3-030-31719-5_2

Keywords Siliceous phytoplankton · Climate variations · Laminated sediments · Hydrological history · Artificial neural networks

Introduction

Earth's climate is a dynamic system with a semi-defined and semi-understood set of modulations, which are being subjected to anthropogenic perturbations that could potentially induce alterations to these dynamics with unforeseen results. The development of effective responsive strategies perforce needs to be based on, first, observing the system's present variability and, second, identifying differential changes. Historical archives and instrumental data series can document these changes though the latter's effective implementation commenced during the nineteenth and twentieth centuries.

The application of modern inferential and derivative analysis techniques to field databases allows the discernment of long-term patterns from observed planetary climatic history. The last few decades of the twentieth century have seen an awareness generated in modern society that is conceptually defined by the "global warming" meme. This encompasses the major changes attributed to the anthropogenic activities, such as food production deforestation, global mercantilism based on hydrocarbon combustion, and land usage reallocation through infrastructure development (Garcia et al. 2016). It is necessary to delineate mechanisms and processes that manifest as climate change in order to understand the global change dynamics. The fundamental scientific challenge is underlain by several profound social implications such as the decadal-scale societal vulnerabilities and the potential loss of human and physical assets through extreme weather events such as floods and droughts. Ultimately, mankind's long-term survival is dependent on essential, supportive, and viable ecosystems. Oceanic ecosystems are one example of where climate variability can modify productivity, which is closely linked to the marine fishery resources, an important source of needed protein for humanity (Cheung et al. 2010).

Global warming has become the impetus for climatic forecasting research and the subsequent development of predictive process unfoldment models, which could identify the necessity of mitigatory actions, thereby allowing mankind the opportunity to develop a sustainable response to global conditions and reprogram economic and social activities. However, these models are still in the developmental and validation stages, and therefore, a high degree of uncertainty exists as to how the components of the Earth's climate system will respond to perturbations on different timescales. The ocean, an essential climatic component, has a reaction period that runs from centuries to millennia, while the atmosphere and ecosystems have relatively shorter response cycles, that is, years to centuries (e.g., IPCC 2007). Therefore, the modeling results are circumscribed by the limited span of instrumental databases as well as a short-term societal orientation impairing a consensual ability to recognize, integrate, and economically justify a present-term response to a long-term prognostication.

According to the World Meteorological Organization, a span of 30 years (i.e., a generation cycle for man) is the minimum database coverage necessary to establish a standardized climatic average or baseline for isolating, identifying, and comparing any climatic deviations (Fairbridge 1987). However, to investigate the climate prior to the historical–instrumental era, a few centuries at most, on centennial or longer timescales, it becomes necessary to use the permanent records derived from natural processes governed by climate such as the high-resolution marine laminated sediments (Baumgartner et al. 1989).

Different components (e.g., planktonic remains, biomolecules, isotopic ratios, mineral grains) of the high-resolution marine records can be considered as extensions of the instrumental databases. However, they must be studied to demonstrate the fidelity of their response to the climate variability. The remains of planktonic organisms and/or their biomolecules represent short-lived species that can show dramatic changes in their distribution due to current transport or nutritive environments by rapidly expanding or contracting their range. Nonetheless, the degree of modification that the remains underwent during their settling has to be evaluated, after which they can function as sensitive indicators of ecosystem changes (Hays et al. 2005) and subsequently as indicators of climate variability. Thus, through sedimentary preservation, they function as separate, independent channels that have recorded natural processes governed by the climate. Indeed, their wide distribution throughout the world's oceans has been utilized to identify different oceanographic scenarios (Poelchau 1976; Takahashi and Blackwelder 1992; Onodera and Takahashi 2005). It is in this context that the high-resolution records of marine sediments from the silled basins and continent slopes of the Gulf of California (northwestern Mexico) are particularly valuable as a proxy, one tool to be used in reconstructing the last 2000 years of climate history.

The Gulf of California (GC), located on the Pacific Ocean's Northern Hemisphere, subtropical eastern shore, has been identified during the last decades as a highly sensitive and responsive region to the recent global warming (IPCC 2014). However, this area's climate variability during the last 2000 years is barely known, and consequently, our ability to evaluate contemporary changes and separate natural variability from anthropogenic forced global warming is limited. Here, the scant amount of literature detailing the paleoclimate of GC over the last 2000 years will be compiled and reviewed. Additionally, uncertainties and critical gaps in knowledge are identified for future studies to resolve.

Regional Settings

The GC is a dynamic narrow subtropical marginal sea that was formed at about 4.5 million years ago, during the Tertiary Period of the Cenozoic Era (e.g., Brusca 1977; Gastil et al. 1983). It is located in northwestern Mexico's semiarid region and is flanked on the east by the mainland's Sierra Madre Occidental mountain range and on the west by the Baja California Peninsula (Roden 1964). Northerly winter winds blowing down the gulf result in its net annual evaporative budget (Beron-Vera and Ripa 2000), but it is summer insolation in the northern region that forms Gulf Water, which subsequently sinks and flows southward (Bray 1988). Different physical processes drive the gulf's complex circulation, and it can be divided into four different regions dependent on which process dominates (Santa María del Ángel et al. 1994a; Thunell 1998). The winds and tides dominate the north, tidal mixing and internal waves are present in the central midriff islands, wind-derived coastal upwelling occurs along the eastern continental margin, and the mesoscale gyres and coastal jets dominate the central and southern areas (Pegau et al. 2002). A ~200 km wide southern entrance communicates with the Pacific Ocean, as well as capturing storm-generated southerly oceanic waves and transferring their energy into the gulf's interior (Martínez and Allen 2004; Baumgartner and Christensen 1985).

The GC's high primary productivity has been recognized since early studies at the beginning of the twentieth century (Gilbert and Allen 1943). However, it was only two decades later that radioactive carbon assimilation was used to estimate productivity. Productivity, driven by climate variability, is one feature that has generated substantial interest since the 1980s.

The first estimations of surface productivity ranged from 18.8 to 67 mg C m^{-3} d^{-1}, and the maximum integrated value for the euphotic zone was 0.95 g C m^{-2} d^{-1} (Zeitzschel 1969). In general, the annual primary productivity of the GC exceeds 300 g C m^{-2} y^{-1} (Enríquez-Andrade et al. 2005). This is comparable with other productive upwelling areas such as the Bay of Bengal, Baja California's western coast, and Africa's northern Atlantic coast (Steemann and Jensen 1957; Douglas et al. 2007). These values are two to three times higher than those reported for the Atlantic and Pacific Oceans' open waters of similar latitudes (Steemann and Jensen 1957). Temporally, the winter–early spring period has the gulf's highest seasonal production values, a result of the prevailing northwesterly winds. Higher values are registered along the eastern shore where they are associated with upwelling and in the northern "upper gulf" zone where they are associated with vertical mixing. Lower values occur along the peninsula (Álvarez-Borrego and Lara-Lara 1991; Santa María del Ángel et al. 1994a, b).

In the shallow region from the Ángel de la Guarda and Tiburón islands northward, tidal mixing generates moderate turbulence. In spite of these conditions, the primary productivity values range between 1 and 4 g C m^{-2} d^{-1}. The southern region has comparable integrated productivity values, due to the deeper euphotic zone (Santa María del Ángel et al. 1994b).

The effects of large-scale events such as the El Niño on primary productivity and chlorophyll concentration were recognized from the analysis of Coastal Zone Color Scanning and NOAA SeaWIFS satellite pigment concentration images (Santa María del Ángel et al. 1994b). The sedimentary records from the area around the Guaymas Basin show higher productivity (Baumgartner et al. 1985; Lara-Lara et al. 1984) contrary to the photogrammetric studies (Santa María del Ángel et al. 1994a, b; Thunell 1998; Smoak et al. 1999). This contradiction of the satellite imagery is possibly due to unobserved deep nutricline "shadow" assemblage productivity. During the AD 1982–1984 El Niño event, the pigment concentrations were lowest in the southern gulf with weaker vertical mixing, while the northern and central areas showed minimal or no response to the El Niño, probably due to excessive turbulence. Intense water mixing can cause excessive turbulence resulting in a consequential reduction of the euphotic zone, thereby limiting the photosynthetic process.

The hydrographic conditions such as water column stability can directly or indirectly influence the abundance and type of phytoplankton community and, therefore, their productive capacity (Gaxiola-Castro et al. 1995). The El Niño conditions of the winter of AD 1991 had moderate turbulence concurrent with observed high values of chlorophyll and integrated primary productivity in the south-central gulf. Some of the studies carried out subsequently, support the idea that climate variability affects the primary production and chlorophyll concentrations throughout the gulf; however, this still remains a controversial hypothesis (Karhu et al. 2004; Herrera-Cervantes et al. 2010; Páez-Osuna et al. 2016 and references therein). This probably results from limited or no data and/or limited temporal representation of in situ carbon assimilation (Mercado-Santana et al. 2017). Notwithstanding, there are some characteristic features that are recognized as a response to interannual variability. A decrease in the upwelling and a lower west coast correlation to the ENSO events than are found for the eastern and the northern gulf regions are some that stand out (Herrera-Cervantes et al. 2010).

Processes Responsible for the Sedimentation Cycle

Silica is an indirect reflection of the efficiency of the gulf's "biological pump" which exerts an important control over the CO_2 concentration of surface waters, through assimilation and respiration by autotrophic and heterotrophic microorganisms, and the eventual sedimentary burial (Honjo et al. 2014). The efficiency of this biological pump to draw carbon (phytoplankton biomass) down into the deeper waters is linked to the nutrient supply, quantity, and type required to build autotrophic microorganism biomass (Basu and Mackey 2018; Tréguer et al. 2018). Diatoms together with other primary producers such as silicoflagellates and coccolithophorids play a major role in the GC's oceanic carbon cycle by the downward exportation and delivery of carbon to the ocean floor (Thunell et al. 1996; Ziveri and Thunell 2000; Sancetta 1995; Martínez-López et al. 2012).

The quantity of organic carbon and skeletal material of the primary producers that reaches the sedimentary record is controlled by the physical mechanisms that modulate the euphotic zone production and settling in such manner that certain oceanographic–atmospheric events can be associated with their fluctuations (Takahashi and Blackwelder 1992).

Therefore, it is important to examine the past two millennia's climate and environmental changes from the perspective of the geological record, in addition to merely using the recent or historical period, to assess and quantify environmental change attributed to natural or to anthropogenic effects, minimizing the possibility of erroneous conclusions.

The deposition of alternating layers of sediment during the gulf's sedimentation cycle was first described by Revelle (1939), and since then, several other deposition mechanisms have been proposed. Two early proposals disagreed in the identification of which was the main constituent: fluvial discharge or exported phytoplankton production. Other proposals suggest that the principal component was being interlaced with the secondary. The light-colored layers' biogenic composition can be derived either from siliceous or calcareous detritus, while the dark layers have various combinations of terrigenous clastic detrital materials, depending upon the regional lithography.

Marine sediments have been studied to find non-bioturbated high-resolution records, such as Alfonso Basin where a sill creates anoxic conditions or Guaymas Basin whose slopes intersect the oxygen minimum layer of the water column. The Guaymas-type sediments, ubiquitous along central gulf's eastern open water basins, have well-defined, straight laminae with a thickness of 1–3 mm. These are alternating light–dark-layered sediments, varves, with a colored paired lamina representing 1 year of depositional history (De Geer 1921; Calvert 1966; Baumgartner et al. 1991; Thunell et al. 1993). They are particularly valuable in untangling the subannual climatological events by reconstructing the "proxy" climate history. Although both sides of the gulf respond synchronously to the changing climatic regime, the biological and sedimentological responses are not identical on each side (Donegan 1981). Therefore, uncertainties exist regarding what a light-dark lamina pair represents. A Guaymas Basin pair is clearly seen to be equivalent to a yearly productive output, while an Alfonso Basin pair might reflect an alternative formational period. Opposing arguments suggest that the Alfonso Basin could be a non-annual depositional system (Pérez-Cruz and Urrutia-Fucugauchi 2010), while others affirm that the couplets do represent annual accumulations (Staines-Urías et al. 2009; González-Yajimovich 2004).

Another controversy is associated with depositional process for terrigenous material, as to whether it is of fluvial, pluvial, eolian, or a mixed origin. These uncertainties are still not completely resolved. The lateral continuity of individual laminae at each site can give some clue, as to whether it represents a high- or low-energy depositional environment and its temporal stability; slope slump can negate the quantity and quality of information that can be extracted from lamina. Thus, it is important to map the lateral extent of sediment continuity through a spaced series of cores that should indicate the extent of the regional area influenced by climatic and oceanographic forcing (Baumgartner 1988). There are several areas in the GC that are presently under investigation in order to solve the questions raised above, and each area may have a different answer. One of the most studied areas is the Guaymas Basin located off the discharge zone of five major rivers that flow through the state of Sonora (Dean 2006), part of an extensive climatic zone, the "Sonoran Desert," a semiarid regime that extends from Phoenix, Arizona, USA, southward to central coastal Sinaloa.

The laminations are divided into two types by formational process, which results in differences in their appearance, thickness, and composition. The formational dynamics are different, even though both have terrigenous detrital clastic sediment. The dark lamina's formation dynamics have terrigenous contributions from the summer–fall rainy season fluvial component and an eolian depositional component derived from the summer convective storms (Baumgartner et al. 1991). It is alternated with the light diatomaceous-rich lamina. This contains eolian terrigenous material transported partly by winds during late fall/spring when considerable dust is simultaneously transported from the desert regions to the north and east and concurrently deposited at both sides of the central gulf (Fig. 2.1).

The calcareous component is produced mainly in the summer with the Sonoran Desert's low-pressure zone shifting northward to the gulf's head, resulting in light southerly and cross-gulf winds (Bordoni et al. 2004), thus allowing the intrusion of oligotrophic subtropical waters that support a different tropical phytoplankton assemblage. By the midsummer, the tropical waters from the Mexican Countercurrent have penetrated into the central gulf establishing a deep thermocline at the base of a layer of warm water (>28 °C) with a recorded thickness of up to 50 m that limits the nutrient replenishment of the photic zone, except at the peninsular margin (Álvarez-Borrego and Lara-Lara 1991; Bray and Robles 1991; Douglas et al. 2007), as well as eolic delivery of lithogenous particles to the gulf (Segovia-Zavala et al. 2009).

The Alfonso Basin-type sediments are developed in the protected southern peninsula margin of La Paz Bay. These sediments do not show distinct, uniform banding but rather are less resolved, being composed of wavy, thin lamina (<0.5 mm thickness) with indistinct boundaries, discontinuities, and up to 80% terrigenous

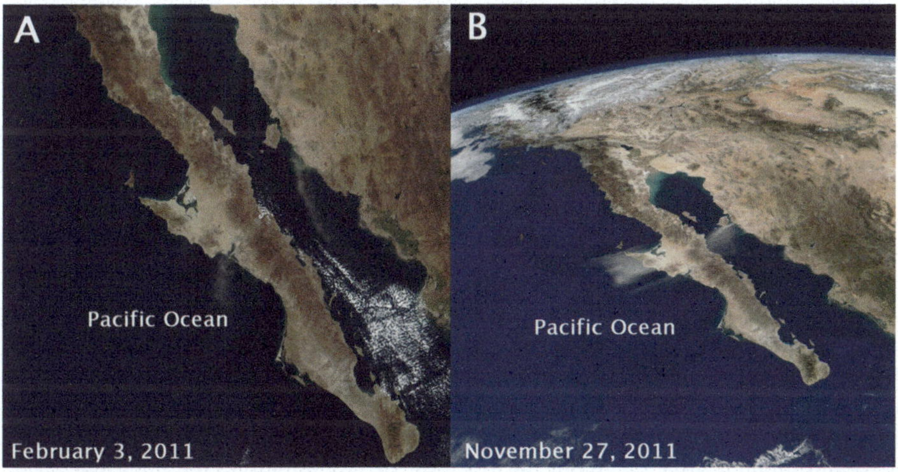

Fig. 2.1 Moderate resolution satellite images showing the dust plumes (**a**) and dust storms (**b**). The strong northeasterly winds stir up dust storms on the mainland and the peninsula. (Source: Spectroradiometer (MODIS) on NASA's Aqua satellite data of February 3 and November 272,011)

mud (Baba et al. 1991; González-Yajimovich et al. 2005). The terrigenous fraction is composed primarily of siliceous–aluminous detrital sediments sourced from the Comondu volcanic tuff sequences comprising the mountain ranges surrounding the bay (Pérez-Cruz and Urrutia-Fucugauchi 2009). Although this basin does not receive a continuous discharge of fluvial sediments, it does receive lithogenic particles from the hemipelagic rain and exceptional precipitation events, that is, hurricanes, that bring huge pluvial runoffs channeled through the numerous small arroyos and two ephemeral rivers that terminate in the San Juan and El Coyote fan deltas (Nava-Sánchez 1997; Molina–Cruz et al. 2002), which González-Yajimovich et al. (2007) have linked turbidities or flood layers in the sedimentary record to specific events. Silverberg et al. (2007) have observed remarkable outliers of high lithogenic fluxes in a sediment trap anchored at 360 m depth at the center of the basin. Regardless of these cases, the highest lithogenic fluxes have been observed during the winter–spring season in the sediment trap time series of this basin. This evidence established that the eolian pathway delivers the terrigenous particles and that it occurs during conditions of low temperature and strong northerly winds (Silverberg et al. 2014).

Hydrological Processes Linked to Climate and Anthropogenic Changes

The damming and diversion of the Colorado River has had a marked impact on the upper gulf's delta, resulting in a drastic reduction in freshwater and sediment discharge (Brusca 2015), both with significant biological influences (Kowalewski et al. 2000). However, the marginal sediment accumulation rate wasn't dramatically affected. Therefore, the hypothesis of Calvert (1966) suggesting that fluvial sediment discharge controlled varve formation became a suspect as the damming of river resulted in the retention of sediment behind the dams and reduced pluvial discharge into the ocean. The eolian dust was then alternatively proposed as a source for this unaccounted depositional material (Baumgartner et al. 1991). Three different entrainment mechanisms cumulatively contribute to the load. The winter northerly and the summer southeastern winds, the summer and fall tropical depressions and hurricanes, and the summer and fall strong adiabatic winds associated with convective thunderstorm formation (locally known as chubascos; Molina-Cruz 1988; Molina-Cruz et al. 1999) represent these mechanisms.

The Sonoran Desert receives little precipitation, so dust storms are expected. One would also expect that increased precipitation and vegetation encountered in the Tropics to the south would neutralize the eolian transportation mechanism. However, that is not absolutely the case as the clastic material is still susceptible to aero entrainment after receiving up to 6 mm of precipitation (Baumgartner et al. 1991). On the other hand, it is possible to recognize extreme precipitation events (floods) in the bedded layers of the sedimentary record even after the dam constructions.

Reservoir management and dam protection often see a remedial release of massive amounts water in preparation for prognosticated precipitation. Sediment deposits on the continental shelf near to the Fuerte River mouth, for example, show a depositional bed from the Hurricane Lidia in AD 1981 (Barbara et al. 2016). For the central region of Sinaloa state, there is however little compiled documentation of fluvial discharge. Some relevant information is provided in this contribution so that the deposition cycle conserved in the gulf sedimentary record can be correctly analyzed in future studies.

The central region of the state of Sinaloa has two watersheds, that is, the Culiacán River Basin and the San Lorenzo River Basin. The former covers an area of ~19,151 km^2, and it receives an average annual rainfall of ca. 707 mm. The Culiacán River has two tributaries, that is, the Humaya and Tamazula Rivers. Both unite in the city of Culiacán and discharge in the central zone of the Altata–Ensenada System of the Pabellón in the GC (Fig. 2.2). Its average annual runoff is 3.1 million m^3 (CONAGUA 2016). The Adolfo López Mateos Dam ("Humaya" or "El Varejonal") was built in AD 1963, at the confluence of the Humaya River and the Badiraguato River (storage capacity, 3160 million m^3). The Sanalona Dam (in operation since AD 1948) was built across the Tamazula River, and it has a capacity of 845 million m^3 (Fig. 2.2, INEGI 1995).

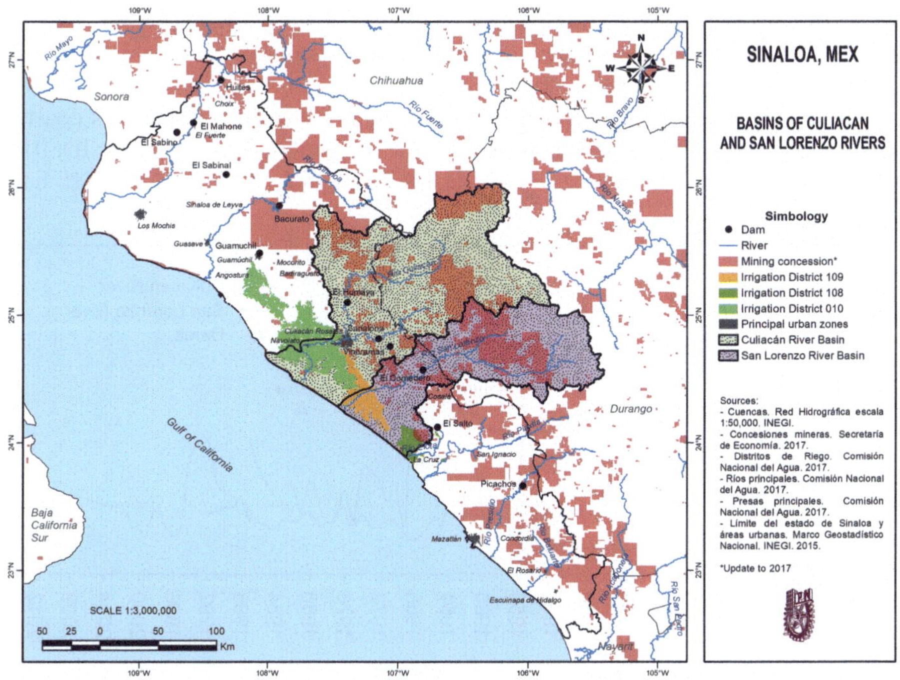

Fig. 2.2 Map showing the Culiacán and San Lorenzo River Basins in the mainland México (Sinaloa), and it includes the mining concession and irrigation district areas

The San Lorenzo River Basin has an area of ~12,013 km^2, and it receives an average annual rainfall of ~636 mm. The main stream of the San Lorenzo River originates in the state of Durango, and it covers a total length of 158 km, until it empties into the Bay of Ceuta in northern Sinaloa. Its average annual runoff is 1665 million m^3 (CONAGUA 2016). Its flow was reduced by construction of the José López Portillo Dam "El Comedero" in AD 1981, 46 km below the headwaters (Fig. 2.3).

Mining was one of the productive economic activities of great importance in the central Sinaloa and dates to the sixteenth century (AD 1565). More than 53 haciendas at the end of the nineteenth century operated 403 functioning gold and silver mines whose production was the highest recorded. By the mid-twentieth century, their importance had declined with agriculture supplanting mining as the state's principal economic activity, and this led to the development of irrigation districts. Currently, mining shows an upward trend, and there are active mines statewide (Fig. 2.2). According to the Mexican Geological Service, this region has economic deposits of aluminum, silver, gold, iron, and copper (SGM, 2016). It must be noted that the difference between mineralization and a deposit is a financial distinction dependent on the resource's market price. Mining activity has inherent detrimental results, such as the introduction of industrial chemical reagents from mill spillages, slurry pipeline spillages, or mill tailings in dewatering impoundments reaching the coastal zone and even beyond the continental margin by percolation, runoff, or eolian transport. Real anthropogenic effects must be included in the financial evaluation. However, as to date, no financial evaluation as to the replacement costs of Earth's ecosystems has occurred due to the unconsciously accepted historical assumption that undeveloped land is worthless.

The lower parts of both the watershed basins are located on Sinaloa's coastal apron, an agriculture zone composed of the irrigation districts (ID), ID 010 Culiacán–Humaya–San Lorenzo (212,123 ha), ID 108 Elota–Piaxtla (19,716 ha),

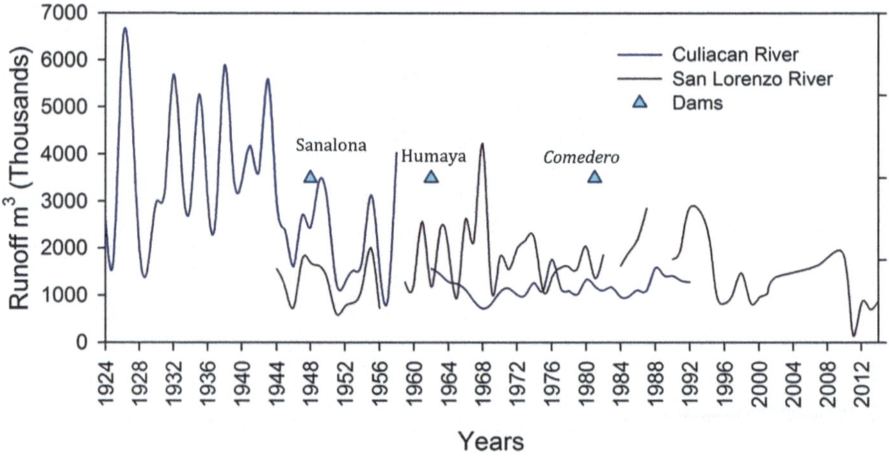

Fig. 2.3 Time series of annual runoff between AD 1924 and 2016 of the Culiacán River and San Lorenzo River. (Source: BANDAS database (CONAGUA). The triangles mark the initiations of dam operations in the central region of the Sinaloa state)

and ID 109 Río San Lorenzo (63,033 ha). This region has utilized modern agriculture techniques since AD 1933 (Fig. 2.2, INEGI 1995). Presently, the acreage under cultivation has increased, and the usage of large-scale monoculture production techniques leaves large tracts of tilled land vulnerable to erosion. The study area, a semiarid region, with low summer rainfall makes the irrigation essential for agriculture. The Sanalona Dam (on Tamazula River) was completed in AD 1948, and it allowed for large-scale agricultural activity in Sinaloa (Ortega-Noriega 1999). Today, the agricultural land occupies more than 30% of the state's territory. River damming harnessed the hydrological resources while simultaneously decreasing the flow as well as runoff variability. Figure 2.3 shows the values measured at different hydrometric stations, that is, Culiacán River's Sud-Pacífico Bridge (10018) station (107° 24′ 15" N and 24° 48′ 20" W) and San Lorenzo River's Santa Cruz (10040) station (106° 57′ 10" N and 24° 29′ 05" W).

According to the National Risk Atlas (SEGOB), Sinaloa is periodically subjected to diverse and recurring hydrometeorological phenomena, for example, a hurricane every 1.5 years, a drought every 7 years with an average duration of 3 years, and floods as quasi-annual events. Due to the lack of instrumental data, a timeline of hydrometeorological phenomena was constructed to document climatic events from an exhaustive bibliographic review of references obtained from historical records of the government agencies, scientific papers, professional and postgraduate theses, and journalistic reports. The compiled and verified information was plotted on a timeline spanning from the Colonial Period to the present day for central Sinaloa (AD 1533–2016, Fig. 2.4).

These results uncovered 42 years of flooding (22 by hurricanes), 13 droughts covering a total of 50 years, and 10 years of frosts. The most notorious floods occurred in AD 1917, AD 1946, and AD 2013, and the last one was associated with the Hurricane Manuel (Fig. 2.4; Torres-Torres et al. 1996; Salas-Salinas and Jiménez-Espinosa 2004; CENAPRED 2014; CONAGUA 2014; Palafox-Ávila et al.

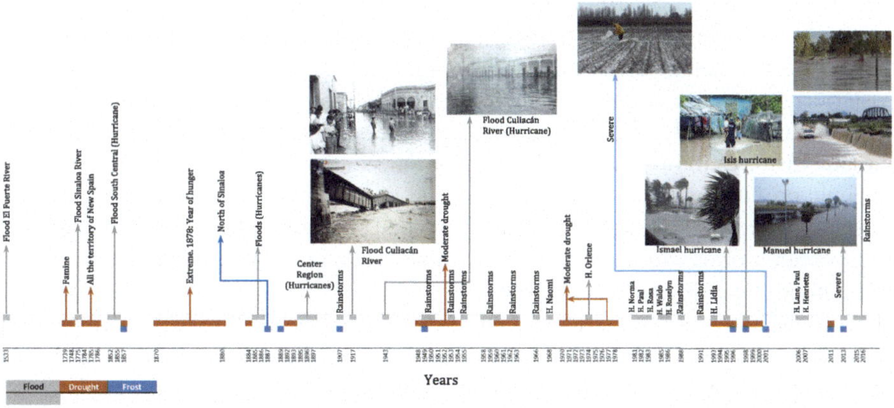

Fig. 2.4 Timeline of events in the hydrological history of the central region of Sinaloa state (México) between AD 1533 and 2016. Red horizontal bar, droughts; grey horizontal bar, floods; blue horizontal bar, frost

2014). During the Colonial Period, torrential rains in northern Sinaloa, though outside the study area, resulted in physical isolation of the study area from the rest of the country, by extensive flooding of the Fuerte River (AD 1533) and the Sinaloa River (AD 1775) triggering severe and extensive regional repercussions (Torres-Torres et al. 1996; Palafox-Ávila et al. 2014). Out of the 13 documented droughts, 3 were recorded during the Colonial Period (Fig. 2.4). The drought of AD 1739 resulted in regional famine, while that of AD 1784–1786 extended over the entire territory of New Spain (Torres-Torres et al. 1996). The most extensive drought was documented between AD 1870 and 1880, and it was characterized as an "extreme drought." By AD 1878, strong repercussions were felt at a national level due to diminished food production, and it received the sobriquet of "the year of hunger" in Mexico (García-Jiménez et al. 2002; García-Acosta et al. 2003; Torres-Torres et al. 1996; Domínguez 2016). The twentieth century has experienced two prolonged drought cycles, although of moderate severity, during AD 1948–1954 (7 years) and during AD 1970–1978 (9 years) (Torres-Torres et al. 1996; CENAPRED 2014; CONAGUA 2014). The first regional frost was recorded in AD 1857, with two additional events which occurred during the same century (Escobar-Ohmstede 2004). Four frost events were recorded during the twentieth century (Matías-Ramírez et al. 2001) with three of them occurring between AD 2001 and 2016. The frost of AD 2011 was of such severity that more than 800,000 ha of cultivation was lost, with a strong regional economic impact.

Integrated Water–Vertical Settling Studies in the Alfonso Basin

The investigative technique of sediment trap studies was initiated in the 1970s to document the biogeochemical transformations of setting particles (e.g., Soutar and Crill 1977; Soutar et al. 1977; Knauer et al. 1979). These studies have generated important information on the transformations that the settling material undergoes, including the remains of planktonic organisms before sediment interment. Posteriorly, it was recognized that an interpretational baseline was needed that would simultaneously allow three linkages, the sediment records of oceanic and climatic histories to observational data about the sources of lithogenic material and biogenic production to studies of settling and burial processes (Honjo et al. 2008). Presently, there are few regional-scale efforts that comprehensively seek to integrate water column observations with particle fluxes and burial of settled material in the ocean floor (Venrick et al. 2003, 2008).

Since the early 1980s, our lack of understanding of the settling particle dynamics in the GC became obvious due to the inferences that were being derived from the sedimentary sequences (e.g., Schrader and Baumgartner 1983; Schrader et al. 1986; Barron and Bukry 2007 and references therein). Sediment trap experiments were initiated during AD 1991–1996, allowing settling planktonic debris to be compared with the sedimentary record (Thunell et al. 1994; Ziveri and Thunell 2000). These provided, for the first time, valuable information on the influence of large-scale El Niño events inside the gulf (Sancetta 1995). A monitoring program of the Alfonso Basin conducted by an interdisciplinary research group from the Instituto Politécnico Nacional–CICIMAR commenced in AD 2002 and had its the findings summarized in Silverberg et al. (2014). The program's study on the siliceous component (diatoms and silicoflagellate) is the first initiative to form integrated linkages between production, settling, and sediment accumulation (Martínez-López et al. 2012, 2016). The results uncovered that the patterns observed in the sedimentary record closely coincide with the phytoplankton annual cycle (Martínez-López et al. 2012), suggesting that the Alfonso Basin is a favorable settling environment (settling factor, 0.45% computed by comparing annualized gross sedimentation or sinking flux with the silicoflagellate standing stock). However, the silicoflagellate *Octactis pulch*ra, the proxy for high productivity, suffered selective dissolution.

Recent results from phytoplankton's other siliceous component, that is diatoms, show around three times lower settling factor (0.14) than that obtained for silicoflagellates (Fig. 2.5, Martínez-López et al. 2016). Their quantitative diminution accounted for 76% of the standing stock at sediment trap depth, suggesting that the diatom assemblages are modified during their settling by dissolution of delicate species such as *Guinardia striata*, *Leptocylindrus danicus*, and *Pseudo-nitzschia* sp. The assemblages of both silicoflagellates and diatoms are modified, and hence, longer time series data studies are needed to confirm the recurrence of this alteration in this basin sedimentary record. Likewise, further research should include other areas of the GC to examine if the climate recording fidelity and coherence exist between sites inside the gulf.

Climate Variations: Interannual, Centennial, and Millennial Scales

Three timescale magnitudes of quasi-oscillatory climate variations collectively known as natural modes of variation are identifiable. The interannual, centennial, and millennial reference frames have identifiable mode characteristics that can be measured to give an indication of their strength at any time, and they are usually correlated across large regions through "teleconnections." The effects of these teleconnections have been observed in many of the variables and components of the Earth's climate system (Feliks et al. 2010, 2013). The El Niño–Southern Oscillation, for example, dominates the interannual scale, and it involves the interaction of seasonal cycles with the internal modes of variation of the ocean–atmosphere system in the tropical Pacific Ocean (Philander 1990). This pattern is mainly characterized by a quasiperiodic warming–cooling of the eastern tropical Pacific sea surface temperature, and each phase can last for 1–2 years.

The interannual scale of variation is the best documented in the GC (v.gr., Baumgartner and Christensen 1985; Lavín and Marinone 2003). Its effects on the seasonal cycle (summer–winter) appear with an irregular periodicity, for example,

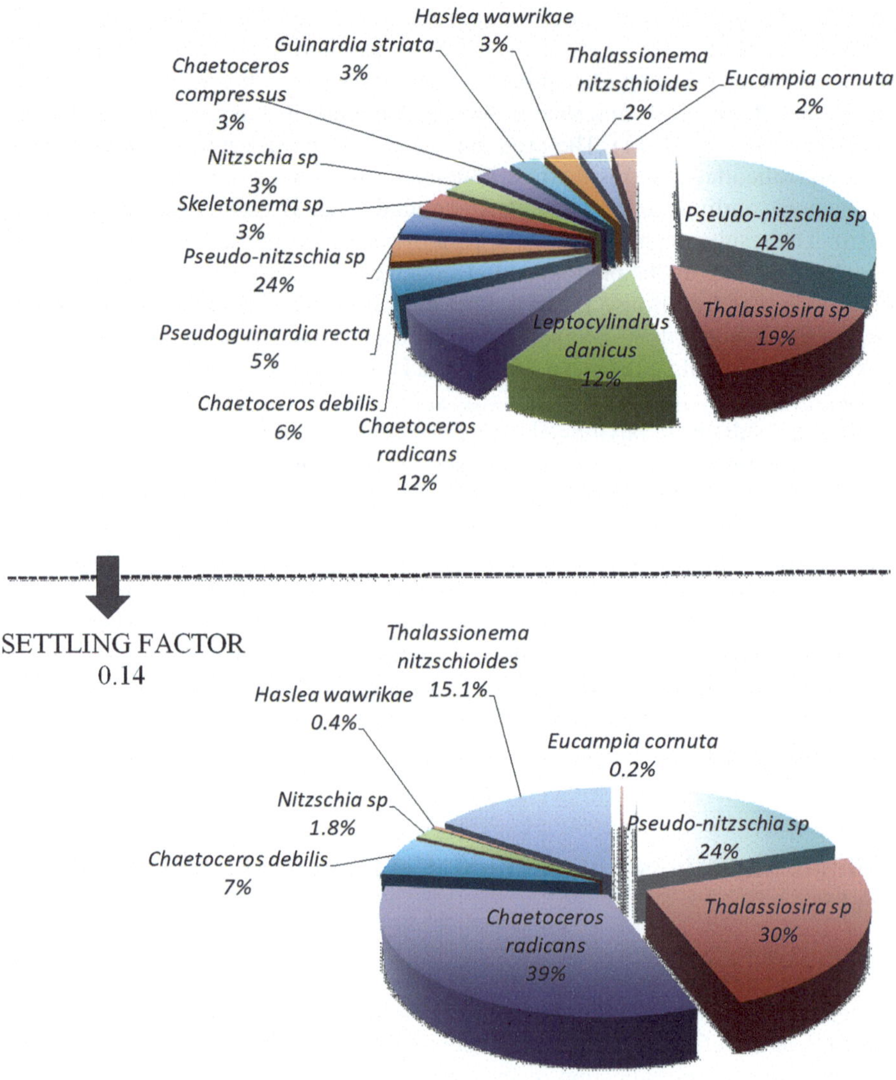

Fig. 2.5 Changes in diatom species assemblages at sediment trap depth (360 m; ≈ 60 m from bottom) in the deepest part (420 m) of the Alfonso Basin (24° 35' N, 110° 36' W), southern Gulf of California. The settling burial factor (0.14) is indicated as an arrow. Dotted line represents the base of the euphotic zone (depth of 0.1% of superficial irradiance)

2–8 years in the physical variables such as temperature and 5 years for biological variables such as changes in species distribution (Fiedler 2002; Lluch-Belda et al. 2005). The occurrence of "El Niño" gradually exerts its influence in the southern region of the gulf (Murray 1982; Murray and Schrader 1983; Schrader and Baumgartner 1983) and at times extending even to the central region (Pérez-Cruz and Molina-Cruz 1988) being characterized by changes in the composition and abundance of the siliceous phytoplankton (Baumgartner 1988).

Decadal–interdecadal climate variations in the extensive North Pacific basin – of which the GC is a part – are represented by diverse climatic patterns. These include the Pacific Decadal Oscillation and the North Pacific Gyre Oscillation (Mantua et al. 1997; Di Lorenzo et al. 2008). Periodicities of 60, 30, 18 and 11 years have also been reported, both in oceanic and atmospheric variables (Minobe 1999; Macdonald and Case 2005; Saldívar-Lucio et al. 2015, 2016). These periods are part of the climatic variability of the GC, and they appear to be the influence exerted by a consortium of factors such as the large-scale atmospheric circulation, lunar nodal tides and solar activity impinging on dominant wind patterns, exemplified by the intensity and location of atmospheric pressure centers and strength of the atmospheric circulation cells (Minobe 1999; Lluch-Belda et al. 2009; Saldívar-Lucio et al. 2016). The 60-year periodicity is one of the most significant of the natural patterns. Its strong signature has been detected in various variables such as the average global temperature (Courtillot et al. 2013), atmospheric circulation (Mazzarella 2007, 2008), global sea ice coverage (Gervais 2016), sea level change (Chen et al. 2014), and Earth's rotational speed (Mazzarella 2007).

Alterations in Earth's rotational velocity could help to explain the relationship encountered between the large-scale (global or basin) climate processes and longer (~60 years) period fluctuations. This connection is mapped onto smaller-scale oceanographic processes, such as the upwelling (Pugh 1987; Wahr 1988; Mazzarella 2007; Munk and Bills 2007). A proposed mechanism to explain the direction shift that the dominant wind undergoes is through referencing the meridional–zonal transitions affected by changing the rotational displacement speed of the landmass (Klyashtorin and Lyubushin 2007; Mazzarella 2007). Periods of planetary rotational acceleration lead to the intensification of the zonal domain circulation in the upper atmosphere, where the frictional coupling between the atmospheric layers forces a change in directional flow. This results in the southerly wind component being dominate on oceanic surface waters, thereby stimulating the upwelling activity (Klyashtorin and Lyubushin 2007; Mazzarella 2007). Contrarily, the deceleration of the Earth's rotational velocity would cause a weakening in the meridional circulation in the lower atmosphere, thereby reducing the intensity of upwelling-favorable wind (Saldívar-Lucio et al. 2016).

Atmospheric forcing isn't the sole process influencing the GC's physical environment; in this basin, different water masses interact, that is, deep water of the Pacific (>1000 m), intermediate water of the Pacific (~900 m), subsurface subtropical water (150–500 m), and basin-produced GC water that mixes with tropical surface water (0–150 m) (Baumgartner and Christensen 1985; Lavín and Marinone 2003). These interactions occur on different timescales in response to large-scale climate variability and therefore are of paramount importance to the GC's physical environment (Álvarez-Arellano and Molina-Cruz 1984).

The Gulf of California's southwestern region is influenced by the El Niño–Southern Oscillation, the Pacific Decadal Oscillation (Mantua et al. 1997), and the North Pacific Gyre Oscillation (Di Lorenzo et al. 2009). Alfonso Basin's primary production of euphotic zone signal is registered in the trap, and even though it is biased by preferential dissolution, the predominant oceanographic changes are still captured due to resistant siliceous forms. Changes in the composition of silicoflagellate microalgae indicated that there were changes in ocean circulation responding to climatic variability during AD 2002–2008. The presence of *Dictyocha messanensis* f. *spinosa* is concurrent, with the presence of warm equatorial ocean waters. Therefore, the presence of subtropical waters can be inferred through the presence of *Dictyocha messanensis* f. *messanensis* species, while the incursions of temperate and cold waters can be deduced from an assemblage comprised of *Dictyocha epiodon*, *Distephanus speculum*, and *Dictyocha pentagona* (Martínez-López et al. 2012). Likewise, an attenuation of the El Niño warm episodes (AD 2002–2003 and AD 2005–2006) was observed as well as the amplification of the cold event (La Niña) of 2008, which occurred nestled in decadal background conditions that favored temperate and cold-water incursions as well as higher primary production (Martínez-López et al. 2012).

Changes in distribution of the large-scale air masses of the instrumental era imply that meridional (zonal) component of the wind intensifies or weakens with a periodicity of 30 years. Alternation of the wind components at this scale has also been referred to as "climatic epochs" (Klyashtorin and Lyubushin 2007). This mechanism was originally described for the Atlantic Ocean, but the dominant winds over the North Pacific basin also include similar alternation (King et al. 1998; Beamish et al. 1999). It is accompanied by change in the atmospheric pressure and shifts of warming–cooling surface waters in phases (Mazzarella 2007). Although the presence of a 30-year signal is not always specified, there are several works that provide evidence of this recurrent interdecadal scale in the ocean–atmosphere interactions with effects on salmon (Mantua et al. 1997), anchovy, and sardine population (Lluch-Belda et al. 1992a, b; Lindegren et al. 2013), and others (McGowan 1990; McGowan et al. 1996).

The average 18.6-year cycle provides another pattern of climate variability, and it ranges from 17 to 20 years (Dean et al. 2004). It might be related to the lunar nodal tide cycle (Pugh 1987). Change in the declination of moon's orbital plane leads to regional differences in the gravitational attraction on water and air masses (Baart et al. 2012). The interaction between synodic lunar cycle and nodal cycle promotes a tidal response and the positional changes of atmospheric pressure centers. They, in turn, influence the medium sea level on a regional scale and consequently the thermocline depth (Rebert et al. 1985; Pugh 1987; Pizarro and Montecinos 2004). The predictive models of upwelling activity of Pineda (1995) demonstrated that the parameters such as pressure centers, sea level, and thermocline are determinants in the magnitude of Ekman's transport (Rebert et al. 1985; Trenary and Han 2012). The nodal cycle modulates all the components of the tides (e.g., diurnal, semidiurnal), and the Gulf of California's tides oscillate in parallel with those of the Pacific Ocean (Ripa and Velázquez 1993). Due to the amplificatory effects of different components, the tidal changes are stronger in the northern gulf, achieving height up to 6–7 m higher than the southern region. These result in strong tidal currents and mixing processes, subsequently break up the stratification, and generate upwelling (Lavín and Marinone 2003). It is implicitly recognized that the most intense tidal mixing are the areas of higher biological productivity (Álvarez-Borrego and Lara-Lara 1991; Lluch-Cota 2000).

Another periodicity is the 11-year cycle, and its presence is noted in physical and biological variables of the gulf. This may result from the solar activity inducing thermal differences, which in turn modify the pressure gradients and the associated wind patterns (Beer et al. 2000; Chen et al. 2002). The equatorial region is the principal recipient of solar energy or irradiance ($kW * m^2$), and this energy is then redistributed to the rest of the planet through different mechanisms (Cornejo-Garrido and Stone 1977; Lohmann et al. 2004). The teleconnections are an example of how the 11-year signal could be transmitted from the equator, in the form of heat, to high latitudes, and in other forms of energy (e.g., wind). In the Pacific Ocean, the energy dissipation has been documented between the El Niño–Southern Oscillation and variations in the Alaska and California Currents (Wooster and Fluharty 1985). Although the GC constantly interacts with the California Current and is influenced by changes in the atmospheric pressure centers in the North Pacific, the transmission of 11-year signal to the interior of the GC is more likely due to its strong interaction with waters of the equatorial region, particularly during the summer (Lavín et al. 2014).

The abovementioned periodicities are similar to the cycles revealed in the depositional time series of diatom mat laminae (11 years, 22–24 years, and 50 years) extracted from the Holocene laminated sediment (Pike and Kemp 1997). The importance of different scales of climatic variation with effects on the distribution and abundance of marine organisms of the GC and neighboring waters has been pointed out by several authors (e.g., Baumgartner et al. 1992; Lluch-Belda et al. 2003, 2009; Salvadeo et al. 2011; Lluch-Cota et al. 2017). Interestingly, in the 1990s, the Sun's 50-year coronal mass ejection cycle was proposed to be a possible climate change forcer (Anderson 1992). It is believed that the 50-year sediment diatom period is linked to this solar cycle, though no clear mechanism is known to transmit the Sun's energy output to the lower atmosphere. This cycle has also been recorded in the dynamics of fish population related to an alternating ocean–atmosphere circulation in the northern Pacific Ocean that favors either the California Current or North Equatorial Current (Pike and Kemp 1997). A coherent pattern with a periodicity from 120 to 140 year evinces a changing ecosystem structure that has been deduced from the time series sequence of AD 1730–1980, based on the analysis of the fish scales (Holmgren-Urba and Baumgartner 1993). Accordingly, the large-scale teleconnections need to be explored with the time series from the sedimentary records.

Centennial–Scale Variability

The recovery of a 152 m long (250 kyr) marine laminated sediment sequence in the GC by the Deep Sea Drilling Project in AD 1978 opens the possibility of studying the complete Holocene period. The paleoclimate reconstructions are focused on the multi-millennial timescales (e.g., Sancetta 1995; Pride et al. 1999; Álvarez et al. 2012). Nonetheless, it is still debated as to whether the late Holocene climate was relatively stable and represents a warm period (Barron et al. 2003; Pérez-Cruz 2006) or whether there has been an increase in climate variability over the last two millennia (Douglas et al. 2002; Goñi et al. 2001; Barron et al. 2004; González-Yajimovich et al. 2005; Barron et al. 2005).

There are very few works with focus on this scale, and they recognized the changes over the last two millennia (Fig. 2.6). A warmer interval coincides with the

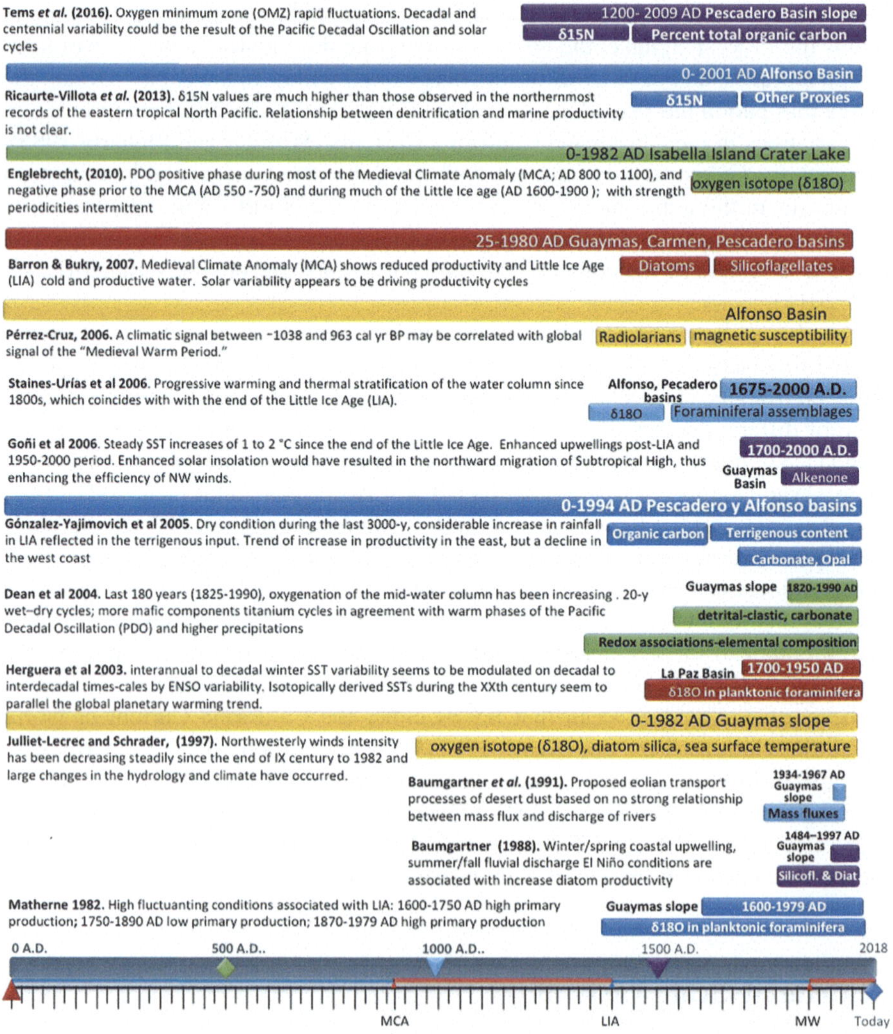

Fig. 2.6 Compilation of paleoclimatic studies from the high-resolution laminated sediments of the Gulf of California representing the last two millennia. Information about the studied period, the most relevant results, and the applied proxies are included. MCA Medieval Climate Anomaly, LIA Little Ice Age, MW modern warming

Medieval Climate Anomaly (MCA), and it is considered a period of reduced productivity (Pérez-Cruz 2013; Barron and Bukry 2007). High levels of sunspot activity imply reduced productivity and warm sea surface temperature (Barron and Bukry 2007). The MCA was succeeded by the Little Ice Age (LIA) and the coeval gulf's cooling (Douglas et al. 2002; Barron and Bukry 2007) associated with enhanced solar insolation and wind-derived coastal upwelling causing higher production (Juillet-Leclerc and Schrader 1987; Barron et al. 2003; Goni et al. 2006). The LIA is, however, characterized by an important spatial and temporal variability expression, particularly visible at a regional scale (e.g., Pages 2k Consortium 2013). It has been attributed to a combination of natural external and internal forcing (solar activity and large volcanic eruptions).

Climate Modeling

Climate modeling is still fraught with a high degree of uncertainty due to an incomplete understanding of the underpinning foundation (decadal or multi-decadal climatic patterns), with their inherent variability and multicomponent interactive triggering mechanisms. Another difficulty is due to the paucity of extended instrument time series that would allow for the recognition and mapping of signals for the prediction and anticipation of a specific manifestation. The integration of paleoclimatic studies into climate models is essential in that they extend the functional database. Better decadal timescale predictions are particularly relevant, and the dependability of this scale for climate predictions will enormously influence social and economic policy decisions.

Nowadays, the artificial intelligence (AI) models such as the Artificial Neural Networks (ANN), a computational model based on biological neural networks, have a logical unit of data input and auto-learning of self-adaptable parameter modification with expectations of simplifying problem abstraction and calculation. The ANN methodology has been used for the climate modeling at regional and global levels, as well as for the climatic reconstruction of the global temperature. The use of this methodology is justified in that it has a high level of flexibility, its learning ability as well as having the facility to establish nonlinear relationships that are encountered in the climatic, atmospheric, oceanic, and other natural processes (Pasini 2009; Liu et al. 2010). The self-organizing map (SOM) is a neural network, and it uses unsupervised rule learning to perform the cluster analysis and to represent probability densities and project a high-dimensional space onto a smaller one. It self-discovers the common features, regularities, correlations, or categories in the input data and incorporates them into its structure during the learning process (Albañez-Lucero 2010).

The SOM structure is based on two layers including an input data ($x = [x_1, x_2, \ldots, x_i]^T$) layer connected to a bidimensional output layer through weighting vectors ($w = [w_{j1}, w_{j2}, \ldots, w_{ji}]^T$). The aim of SOM is to estimate the w_j values that preserve the topological properties of the input data in the output layer by maintaining the relative distance between the input patterns. Before starting the main process, the SOM initializes the weight vector and utilizes a random number generator in a linear or sequential manner, causing the SOM to converge to a "good solution" more quickly. The SOM process consists of an iterative process, when the unit indicates values closest to inputs, known as the best matching unit (BMU). It is declared as a "winner" unit, and this process is computed as follows (Hollmen 1996; Carniel et al. 2009):

$$|x - w_c| = \min_i |x - w_i|$$

where w_c is the "winner" unit, and x_i and w_i were defined previously. The BMU determines the spatial location of the neighborhood nodes and its topology, which is defined by a neighborhood function $h_{ci}(t)$ (gaussiana, sigmoidal, step, etc.; Haykin 1998; Hollmen 1996), where:

$$h_{ci}(t) = h(d,t) \cdot \alpha$$

The SOM rule for the weighting vector according to the discrete time (iteration) involves the weight vector $w_j(t)$ of the node and a time. The updated weight vector $w_j(t + 1)$ at time $t + 1$ is defined by the following equation (Kohonen 2001; Haykin 1998):

$$w_j(t+1) = w_j(t) + \alpha(t) h_{ci}(t)\left(x - w_j(t)\right)$$

It allows the weight vectors to take their places according to the SOM structure through a learning rate and the neighborhood function. The learning rate $\langle(t)$ is a function that decreases with time. The rate starts with a value of 0.1 and should be reduced gradually according to the parameter alpha in these equations:

$$\alpha(t) = \alpha(0)(1.0 - t / rlen)$$

and an inverse function:

$$\alpha(t) = C\alpha(0) / (C + t)$$

where C may be, for example, $C = rlen/100$, and $rlen$ is the number of iterations (Kohonen et al. 1996).

As a first approximation, we explore the use the SOM neural network as a modeling tool to carry out the learning process and generate a matrix that represents the relationships among the observed biological and environmental oceanic data. Four representative silicoflagellate species (*Octactis pulchra*, *Dictyocha messanensis* forma *spinosa*, *Dictyocha messanensis* forma *messanensis*, *Dictyocha fibula* var. *robusta*) were used along with depth, sigma-t, salinity, and temperature, to determine the variables with which they exhibited close relationships by the accuracy and preservation of the algorithm topology. Therefore, the data was standardized to avoid bias (Vesanto et al. 2000; Kohonen 2001). Once the data was topologically sorted and represented by a U-matrix (*unified distance matrix*; Ultsch and Siemon 1990), the quality of the representation is determined by the accuracy and preservation of the topology. The procedure was repeated until the output satisfied the desired criteria for representation. The criteria were based on the biological relationship among the environmental variables and the species in question previously established for the region (Martínez-López et al. 2016).

We chose a 10×10 matrix as the optimal size to characterize the relationship among the variables, and it demonstrated a quantization and topological error rate smaller than other potential analyses. The database consisted of 196 samples with the relative abundances of four species previously mentioned and four environmental variables. The results generated from this analysis, represented by a U-matrix (Fig. 2.7), indicate that four species have a relationship with the environmental variables, except for the temperature.

Even the species *Octactis pulchra* and *Dictyocha messanensis* forma *spinosa* could be found in very similar habitats, and sometimes, *Dictyocha messanensis* forma *messanensis* and *Dictyocha fibula* var. *robusta* could be found in the same habitats, although the latter in lesser quantity.

After analyzing the species separately with the four environmental variables mentioned above, the best result, with the least quantization error, was obtained in

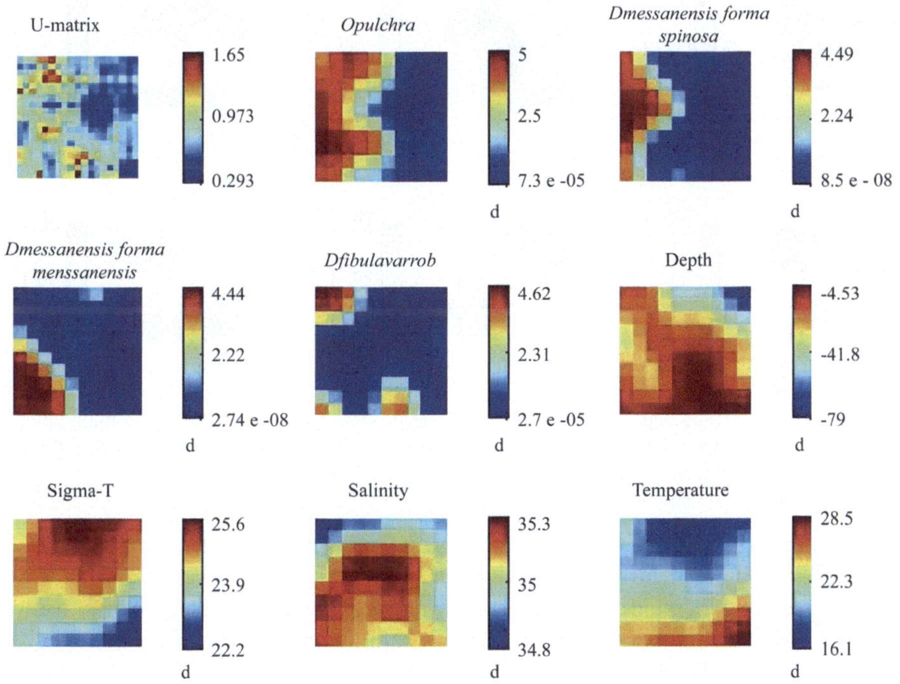

Fig. 2.7 U-matrix obtained as an output of the SOM modeling process representing the relationship between the relative abundances of species and depth, sigma, salinity, and temperature; d, Euclidean distances between the neighboring neurons

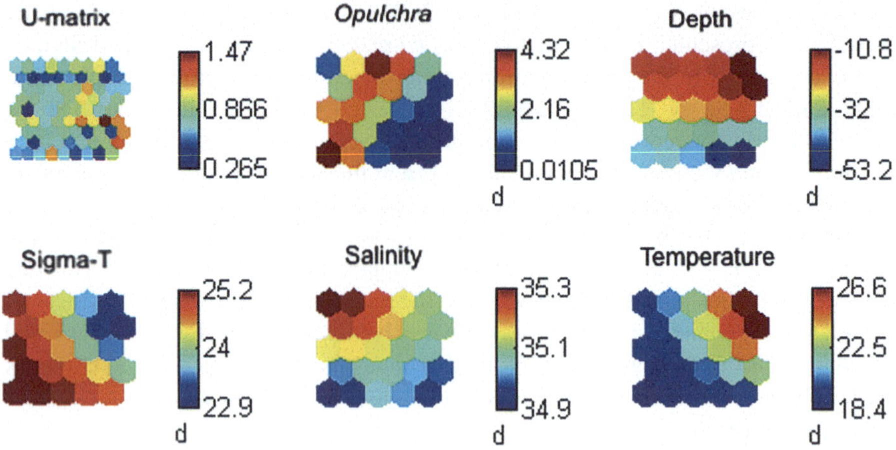

Fig. 2.8 U-matrix obtained as an output of the SOM modeling process representing the relationship between the relative abundances of *Octactis pulchra* and depth, sigma, salinity, and temperature; d, Euclidean distances between the neighboring neurons

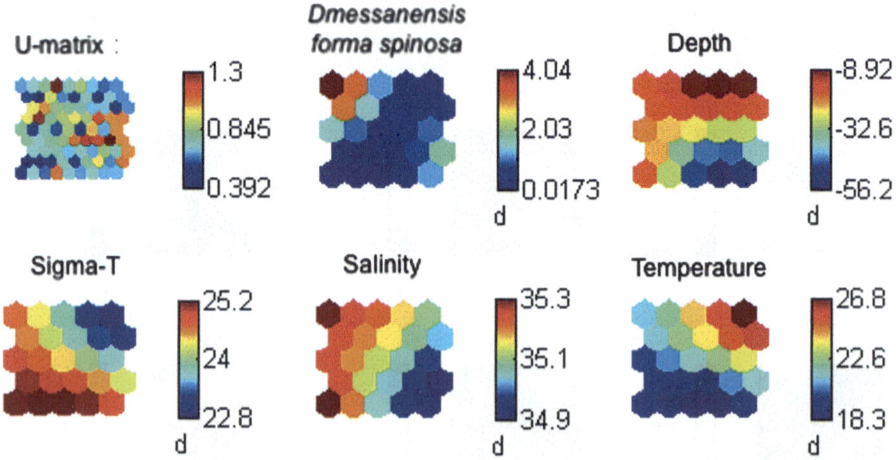

Fig. 2.9 U-matrix obtained as an output of the SOM modeling process representing the relationship between the relative abundances of *Dictyocha messanensis* forma *spinosa* and depth, sigma, salinity, and temperature; d, Euclidean distances between the neighboring neurons

a U-matrix 5×5. The analysis of the species *Octactis pulchra* shows greater association with depth even with sigma-t and reverse with the salinity (Fig. 2.8).

The results generated in the analysis of species *Dictyocha messanensis* forma *spinosa* indicate that there is a greater relationship with salinity than with the other variables (Fig. 2.9).

The analysis of *Dictyocha messanensis* forma *messanensis* together with the other species (Fig. 2.10) indicated this relationship, when tested alone to

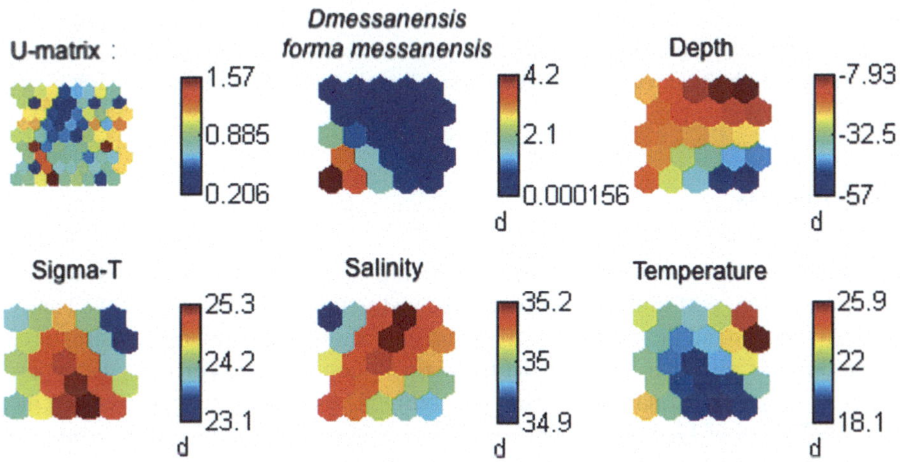

Fig. 2.10 U-matrix obtained as an output of the SOM modeling process representing the relationship between the relative abundances of *Dictyocha messanensis* forma *messanensis* and depth, sigma, salinity, and temperature; d, Euclidean distances between the neighboring neurons

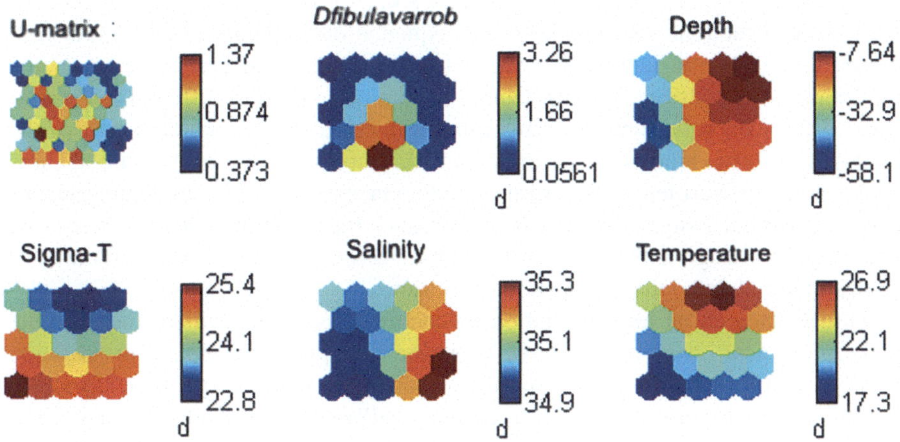

Fig. 2.11 U-matrix obtained as an output of the SOM modeling process representing the relationship between the relative abundances of *Dictyocha fibula* var. *robusta* and depth, sigma, salinity, and temperature; d, Euclidean distances between the neighboring neurons

environmental variables. The results indicate an inverse association with the depth and with the salinity (Fig. 2.10).

The species *Dictyocha fibula* var. *robusta* is positively associated with sigma-t and salinity variables, and it is inversely associated with the temperature (Fig. 2.11).

Results of the first stage of analysis show that the SOM algorithm turns out to be an adequate method, and it indicates that the silicoflagellate species was influenced by all the four environmental variables used in this study. The study also indicated that the variables with the most influence were sigma-t and salinity, while the temperature was not a variable that directly influenced this species. It even came to present a reverse influence. Therefore, the climate prognostication of sigma-t and salinity is possible through the development of the silicoflagellate-based transfer function.

Conclusions

The high-resolution marine sedimentary records (and proxies contained within) occupy a central part in the paleoclimate reconstructions. We, however, have barely begun to understand the major processes and the factors that regulate their structures in the Gulf of California. Their preindustrial revolution formation gives us the opportunity to untangle the natural climate variability. The conceptual framework of the last 2000 years of historic climate variability as being important and necessary to resolving present day problems was non-existent. Thus, the marine records serve as an external fulcrum to apply a lever of integrated understanding to augment our ability to evaluate the contemporary changes within the context of natural variability isolated from the anthropogenic forced global warming.

Presently, there is a lack of consensus for some of the key issues in the paleoclimatic reconstructions such as the exact sequence or combination of natural processes that form the laminated sediments. The unresolved identification of the mechanisms responsible for dark lamina formation has led to different interpretations. The basin-wide lamina stratigraphy is missing for the southern part of the Gulf of California hindering a twentieth-century chronology reconstruction. There is also a lack of verified fidelity of the sedimentary records with contemporaneous data of potential climate proxies. It is necessary for a comprehensive effort to be undertaken to integrate the water column observations, particle fluxes, and burial of settled material data into an integrated data suite.

Future research on high-resolution paleoclimate proxies needs to focus on the many prominent unresolved aspects. However, the apparent temporal dissociation between instrumental databases from those obtained from the geological records represents perhaps one of the greatest current challenges in the study of climate. The integrated and coherent records of the contemporary climatic variability observations meshed with those obtained from paleo-archives can be used to generate robust reconstructions of the past climate conditions and obtain models to anticipate the future.

Acknowledgments This study was funded by the grants of Dirección de Estudios de Posgrado e Investigación, Instituto Politécnico Nacional (grants SIP 20144020 and 20151586). AML, DCEU, GVD, and MAL were supported by the Comisión de Operación y Fomento de Actividades Académicas (COFAA) and Estímulos al Desempeño en Investigación (EDI) fellowships from the Instituto Politécnico Nacional of Mexico. ODLAFC and JDAA received fellowships from Programa Institucional de Formación de Investigadores of the Instituto Politécnico Nacional and CONACYT.

References

Albañez-Lucero MO (2010) Distribución espacial del mero (Epinephelus morio) en el Banco de Campeche, Mexico. Ph.D. Dissertation, IPN–CICIMAR

Álvarez-Arellano A, Molina-Cruz A (1984) Aspectos paleoceanográficos cuaternarios del Golfo de California, evidenciados por conjuntos de radiolarios. Anales del Instituto de Ciencias del Mar y Limnología 475:1–43

Álvarez-Borrego S, Lara-Lara JR (1991) The physical environment and primary productivity of the Gulf of California. In: Dauphin JP, Simoneit BRT (eds) The Gulf and peninsular province of the Californias, vol 47. American Association Petrology Geology Memoir, pp 555–567

Álvarez MC, Pérez-Cruz L, Hernández-Contreras R (2012) Coccolithophore and silicoflagellate records in Middle–Late Holocene sediments from La Paz Basin (Gulf of California): Paleoclimatic implications. Stratigraphy 9:169–181

Anderson RY (1992) Possible connection between surface winds, solar activity and the Earth's magnetic field. Nature 358:51–53

Baart F, van Gelder P, de Ronde J et al (2012) The effect of the 18.6 year lunar nodal cycle on regional sea level rise estimates. J Coast Res 28:511–516

Baba J, Peterson CD, Schrader HJ (1991) Modem fine-grained sediments in the Gulf of California. In: Dauphin JP, Simoneit BRT (eds) The Gulf and peninsular province of the Californias, vol 47. AAPG Memoir, pp 569–587

Barbara L, Schmidt S, Urrutia-Fucugauchi J et al (2016) Fuerte River floods, an overlooked source of terrigenous sediment to the Gulf of California. Cont Shelf Res 128:1–9

Barron AJ, Bukry D (2007) Solar forcing of Gulf of California climate during the past 2000 years suggested by diatoms and silicoflagellates. Mar Micropaleontol 62:115–139

Barron JA, Bukry D, Bischoff JL (2004) High resolution paleoceanography of the Guaymas Basin, Gulf of California during the past 15,000 years. Mar Micropaleontol 50:185–207

Barron JA, Bukry D, Dean WE (2005) Paleoceanographic history of the Guaymas basin, Gulf of California, during the past 15,000 years based on diatoms, silicoflagellates, and biogenic sediments. Mar Micropaleontol 56:81–102

Barron JA, Bukry D, Bischoff JL (2003) A 2000-yr-long record of climate from the Gulf of California. In: West GJ, Blomquist NL (eds) Proceedings of the 19th PACLIM Workshop, March 3–6, 2002, Technical Report 71 of the Interagency Ecological Program for the San Francisco Estuary, Asilomar, CA, pp 11–21

Basu S, Mackey KRM (2018) Phytoplankton as key mediators of the biological carbon pump: their responses to a changing climate. Sustainability 10:869

Baumgartner TR (1988) High resolution paleoclimatology from the varved sediments of the Gulf of California. Ph.D. Dissertation, Oregon State University

Baumgartner TR, Christensen N (1985) Coupling of the Gulf of California to large-scale interannual climatic variability. J Mar Res 43:825–848

Baumgartner, TR, Ferreira–Bartrina V, Schrader H, Soutar A (1985) A 20-year varve record of siliceous phytoplankton variability in the central Gulf of California. Marine Geology 64:113–129

Baumgartner TR, Ferreira-Bartrina V, Moreno-Hentz P (1991) Varve formation in the central Gulf of California: a reconsideration of the origin of the dark laminae from the 20th century varve record. In: Dauphin JP, BRT S (eds) The Gulf and peninsular province of the Californias, vol 47. American Association Petrology Geology Memori, pp 617–635

Baumgartner TR, Soutar A, Ferreira-Bartrina V (1992) Reconstruction of the history of Pacific sardine and northern anchovy populations over the past two millennia from sediments of the Santa Barbara Basin, California, Calif. Coop. Oceanic Fish. Invest. Rep. 33, 24–40

Baumgartner TR, Michaelsen J, Thompson LG et al (1989) The recording of interannual climatic change by high-resolution natural systems: tree-rings, coral bands, glacial ice layers, and marine varves. In: Peterson DH (ed) Aspects of climate variability in the Pacific and the Western Americas, vol 55. American Geophysical Union, Geophysical Monograph, pp 1–14

Beamish RJ, Noakes DJ, McFarlane GA et al (1999) The regime concept and natural trends in the production of Pacific salmon. Can J Fish Aquat Sci 56:516–526
Beer J, Mende W, Stellmacher R (2000) The role of the sun in climate forcing. Quat Sci Rev 19(1–5):403–415
Beron-Vera FJ, Ripa P (2000) Three-dimensional aspects of the seasonal heat balance in the Gulf of California. J Geophys Res 105:11441–11457
Bordoni S, Ciesielski PE, Johnson RH et al (2004) The low-level circulation of the North American monsoon as revealed by QuikSCAT. Geophys Res Lett 31:L10109
Bray NA (1988) Water mass formation in the Gulf of California. J Geophys Res 93:9223–9240
Bray NA, Robles JM (1991) Physical oceanography of the Gulf of California. In: The Gulf and peninsular province of the Californias, vol 47. Memoir – American Association of Petroleum Geologists, pp 511–553
Brusca CR (1977) A handbook to the common intertidal invertebrates of the Gulf of California. The University of Arizona Press, Tucson
Brusca RC (2015) Sea of Cortez. In: Dimmitt MA, Comus PW, Brewer LM (eds) A natural history of the Sonoran Desert, 2nd edn. University of California and Arizona-Sonora Desert Museum Press, Arizona, pp 24–26
Calvert SE (1966) Accumulation of diatomaceous silica in the sediments of the Gulf of California. Bull Geol Soc Am 77:569–596
Carniel R, Barbui L, Malisan P (2009) Improvement of HVSR technique by self-organizing map (SOM) analysis. Soil Dyn Earthq Eng 29:1097–1101
CENAPRED (2014) Diagnóstico de Peligros e Identificación de Riesgos de Desastres en México. Atlas Nacional de Riesgos de la República Mexicana. SEGOB, México
Chen J, Carlson BE, Del Genio AD (2002) Evidence for strengthening of the tropical general circulation in the 1990s. Science 295:838–841
Chen X, Feng Y, Huang NE (2014) Global sea level trend during 1993–2012. Glob Planet Chang 112:26–32
Cheung WWL, Lam VWY, Sarmiento JL et al (2010) Large-scale redistribution of maximum fisheries catch potential in the global ocean under climate change. Glob Chang Biol 16:24–35
CONAGUA (2014) Plan Operativo de Inundaciones de la Ciudad de Culiacán, Sinaloa Organismo de Cuenca Pacífico Norte. SEMARNAT, México
CONAGUA (2016) Estadísticas del Agua en México. México: Secretaría de Medio Ambiente y Recursos Naturales, México
Cornejo-Garrido AG, Stone PH (1977) On the heat balance of the Walker circulation. J Atmos Sci 34(8):1155–1162
Courtillot V, Le Mouël JL, Kossobokov V et al (2013) Multi–decadal trends of global surface temperature: a broken line with alternating ~30 yr linear segments? Atmos Clim Sci 3:364–371
Dean WE (2006) The geochemical record of the last 17,000 years in the Guaymas Basin, Gulf of California. Chem Geol 232:87–98
Dean W, Pride C, Thunell R (2004) Geochemical cycles in sediments deposited on the slopes of the Guaymas and Carmen Basins of the Gulf of California over the last 180 years. Quat Sci Rev 23:1817–1833
De Geer G (1921) Correlation of late glacial annual clay–varves in North America with the Swedish time scale. Geologiska Föreningens i Stockholm Förhandlingar (GFF) 43:70–73
Di Lorenzo E, Schneider N, Cobb KM et al (2008) North Pacific Gyre Oscillation links ocean climate and ecosystem change. Geophys Res Lett 35:L08607
Di Lorenzo E, Fiechter J, Schneider N et al (2009) Nutrient and salinity decadal variations in the Central and Eastern North Pacific. Geophys Res Lett 36:L14601
Domínguez J (2016) Revisión histórica de las sequías en México: de la explicación divina a la incorporación de la ciencia. Tecnología y Ciencias del Agua VII(5):77–93
Donegan DP (1981) Modern and ancient marine rRhythmites from the sea of Cortez and California continental borderland: a sedimentological study. M.Sc. Dissertation, Oregon State University, Corvallis,

Douglas RG, González-Yajimovich O, Ledesma-Vázquez J, Staines-Urias F (2007) Climate forcing, primary production and the distribution of Holocene biogenic sediments in the Gulf of California. Quat Sci Rev 26:115–129
Douglas RG, Gorsline D, Grippo A, Granados I, González-Yajimovich O (2002) Holocene Ocean-Climate Variations in Alfonso Basin, Gulf of California, Mexico. Proceedings of the eighteenth annual PACLIM Workshop Asilomar Conference Grounds, California, pp 7–20
Enríquez-Andrade R, Anaya-Reyna G, Barrera-Guevara JC et al (2005) An analysis of critical areas for biodiversity conservation in the Gulf of California region. Ocean Coast Manag 48:31–50
Escobar-Ohmstede A (2004) Desastres agrícolas en México: Catálogo histórico II Siglo XIX (1822–1900). Centro de Investigaciones y Estudios Superiores en Antropologia Social, Fondo de Cultura Económica, México
Fairbridge RW (1987) Climatic variation in the historical record. In: Oliver JE, Fairbridge RW (eds) The encyclopedia of climatology. Van Nostrand Reinhold, New York, pp 305–323
Feliks Y, Ghil M, Robertson A (2010) Oscillatory climate modes in the Eastern Mediterranean and their synchronization with the North Atlantic Oscillation. J Clim 23:4060–4079
Feliks Y, Groth A, Robertson A et al (2013) Oscillatory Climate Modes in the Indian Monsoon, North Atlantic and Tropical Pacific. J Clim 26:9528–9544
Fiedler PC (2002) Environmental change in the eastern tropical Pacific Ocean: review of ENSO and decadal variability. Mar Ecol Prog Ser 244:265–283
García-Acosta V, Pérez-Zevallos JM, Molina del Villar A (2003) Desastres agrícolas en México. Catálogo Histórico. Tomo I. Épocas prehispánica y colonial (958–1822). Centro de Investigaciones y Estudios Superiores de Antropología Social, Fondo de Cultura Económica, México
García-Jiménez F, Fuentes-Mariles O, Matías-Ramírez LG (2002) Sequías. Serie Fascículos, vol 14. SEGOB–CENAPRED, México
Garcia ES, Swann ALS, Villegas JC et al (2016) Synergistic ecoclimate teleconnections from forest loss in different regions structure: global ecological responses. PLoS One 11:e0165042
Gastil G, Minch J, Phillips R (1983) The geology and ages of islands. In: Case TJ, Cody ML (eds) Island biogeography in the sea of Cortez. University of California Press, Berkeley, pp 13–25
Gaxiola-Castro G, García-Cordova J, Valdéz-Holguin JE et al (1995) Spatial distribution of chlorophyll a and primary productivity in relation to winter physical structure in the Gulf of California. Cont Shelf Res 15:1043–1059
Gervais F (2016) Anthropogenic CO_2 warming challenged by 60–year cycle. Earthsci Rev 155:129–135
Gilbert JY, Allen WE (1943) The phytoplankton of the Gulf of California obtained by the "E. W. Scripps" in 1939 and 1940. J Mar Res 5:89–110
González-Yajimovich O, Douglas RG, Gorsline DS (2005) The preserved carbonate record in Holocene sediments of the Alfonso and Pescadero basins, Gulf of California, Mexico. Proc Geol Assoc 116:315–330
González-Yajimovich OE, Gorsline DS, Douglas RG (2007) Frequency and sources of basin floor turbidites in Alfonso Basin, Gulf of California, Mexico: products of slope failures. Sediment Geol 199:91–105
González-Yajimovich OE (2004) Holocene sedimentation in the southern gulf of California and its climatic implications. Ph.D. Dissertation, University of Southern California, Los Angeles
Goñi MA, Hartz DM, Thunell RC et al (2001) Oceanographic considerations in the application of the alkenone–based paleotemperature $U^{K'}_{37}$ index in the Gulf of California. Geochimica Cosmochimica Acta 65:547–557
Goni MA, Thunell RC, Woodwort MP et al (2006) Changes in wind-driven upwelling during the last three centuries: interocean teleconnections. Geophys Res Lett 33:L15604
Haykin S (1998) Neural network: a comprehensive foundation. Macmillan College, New York
Hays CG, Richardson AJ, Robinson C (2005) Climate change and marine plankton. Trends Ecol Evol 20(6):337–344

Herrera-Cervantes H, Lluch-Cota SE, Lluch-Cota DB et al (2010) ENSO influence on satellite-derived chlorophyll trends in the Gulf of California. Atmosfera 23:253–262
Hollmen J (1996) Process modelling using the self-organizing map. Dissertation, Helsinki University of Technology
Holmgren-Urba D, Baumgartner TR (1993) A 250-year history of pelagic fish abundance from anaerobic sediments of the central Gulf of California. CalCOFI Rep 34:60–68
Honjo S, Manganini SJ, Krishfield RA et al (2008) Particulate organic carbon fluxes to the ocean interior and factors controlling the biological pump: a synthesis of global sediment trap programs since 1983. Prog Oceanogr 76:217–285
Honjo S, Eglinton TI, Taylor C et al (2014) Understanding the role of the biological pump in the global carbon cycle: an imperative for ocean science. Oceanography 27:10–16
INEGI (1995) Estudio hidrológico del Estado de Sinaloa. In: Instituto Nacional de Estadística. Geografía e Informática, México
IPCC (2007) Climate change 2007. In: Solomon SD, Qin D, Manning M et al (eds) The physical science basis. Contribution of working group I to the fourth assessment report of the intergovernmental panel on climate change. Cambridge University Press, Cambridge
IPCC (2014) Climate change 2014. Synthesis report. In: Pachauri RK, Meyer LA (eds) Contribution of Working Groups I, II and III to the Fifth Assessment Report of the Intergovernmental Panel on Climate Change. Switzerland, Geneva
Juillet-Leclerc A, Schrader H (1987) Variations of upwelling intensity recorded in varved sediment from the Gulf of California during the past 3000 years. Nature 329:146–149
Karhu M, Marinone SG, Lluch-Cota SE et al (2004) Ocean color variability in the Gulf of California: scales from the El Niño–La Niña cycle to tides. Deep Sea Res II Top Stud Oceanogr 51:139–146
King JR, Ivanov VV, Kurashov V et al (1998) General circulation of the atmosphere over the North Pacific and its relationship to the Aleutian low. North Pacific Anadromons Fish Commission Doc. 318, Canada
Klyashtorin LB, Lyubushin AA (2007) Cyclic climate changes and fish productivity, 1st edn. VNIRO publishing, Moscow
Knauer GA, Martin JH, Bruland KW (1979) Fluxes of particulate carbon, nitrogen and phosphorus in the upper water column of the northeast Pacific. Deep Sea Res A Oceanogr Res Pap 26:97–108
Kohonen T (2001) Self-organizing maps. 3rd extended edition. Springer, Germany
Kohonen T, Hynninen J, Kangas J et al (1996) SOM_PAK: The Self-Organizing Map Program Package. Technical Report A31, Helsinki University of Technology
Kowalewski M, Simões MG, Torello FF et al (2000) Drill holes in shells of Permian benthic invertebrates. J Paleontol 74:532–543
Lara-Lara JR, Valdez-Holguín JE, Jiménez-Pérez C. (1984). Plankton studies in the Gulf of California during the 1982-1983 El Niño. Trop. Ocean Atmos. Newslett. 28: 16–17
Lavín MF, Castro R, Beier E et al (2014) Surface circulation in the Gulf of California in summer from surface drifters and satellite images (2004–2006). J Geophys Res Oceans 119:4278–4290
Lavín MF, Marinone SG (2003) An overview of the physical oceanography of the Gulf of California. In: Lavín MF, Marinone SG, Velasco-Fuentes O et al (eds) Nonlinear processes in geophysical fluid dynamics. Kluwer Academic Publishers, Netherlands, pp 173–204
Lindegren M, Checkley DM, Rouyer T et al (2013) Climate, fishing, and fluctuations of sardine and anchovy in the California Current. Proc Natl Acad Sci 110:13672–13677
Liu Z, Peng C, Xiang W et al (2010) Application of artificial neural networks in global climate change and ecological research: an overview. Chin Sci Bull 55:3853–3863
Lluch-Belda D, Del-Monte-Luna P, Lluch-Cota SE (2009) 20th century variability in the Gulf of California SST. Cal Coop Ocean Fish 50:147–154
Lluch-Belda D, Lluch-Cota DB, Lluch-Cota SE (2003) Scales of interannual variability in the California Current System: associated physical mechanisms and likely ecological impacts. Cal Coop Ocean Fish Rep 44:76–85

Lluch-Cota SE, Salvadeo C, Lluch-Cota DB et al (2017) Impacts of climate change on Mexican Pacific fisheries. In: Phillips BF, Pérez-Ramírez M (eds) Climate change impacts on fisheries and aquaculture: a global analysis. Wiley, pp 219–238
Lluch-Belda D, Lluch-Cota DB, Lluch-Cota SE (2005) Changes in marine faunal distributions and ENSO events in the California Current. Fish Oceanogr 14:458–467
Lluch-Belda D, Schwartzlose RA, Serra R et al (1992a) Sardine and anchovy regime fluctuations of abundance in four regions of the world oceans: a workshop report. Fish Oceanogr 1:339–347
Lluch-Belda D, Lluch-Cota DB, Hernández-Vázquez et al (1992b) Sardine population expansion in eastern boundary systems of the Pacific Ocean as related to sea surface temperature. S Afr J Mar Sci 12:147–155
Lluch-Cota SE (2000) Coastal upwelling in the eastern Gulf of California. Oceanol Acta 23:731–740
Lohmann G, Rimbu N, Dima M (2004) Climate signature of solar irradiance variations: analysis of long-term instrumental, historical, and proxy data. Int J Climatol 24:1045–1056
Macdonald GM, Case RA (2005) Variations in the Pacific Decadal Oscillation over the past millennium. Geo Res Lett 32:L08703
Mantua NJ, Hare SR, Zhang Y et al (1997) A Pacific interdecadal climate oscillation with impacts on salmon production. Bull Am Meteorol Soc 78:1069–1080
Matherne A (1982) Paleoceanography of the Gulf of California: a 350-year diatom record. M.Sc. thesis, Corvallis, Oregon, Oregon State University, 111 p
Martínez JA, Allen JS (2004) A modeling study of coastal-trapped wave propagation in the Gulf of California, Part 1: response to remote forcing. J Phys Oceanogr 34:1313–1331
Martínez-López A, Álvarez-Gómez IG, Durazo R (2012) Climate variability and silicoflagellate fluxes in Alfonso Basin (Southern Gulf of California). Bot Mar 55:177–185
Martínez-López A, Álvarez-Gómez IG, Pérez-Cruz L et al (2016) Production, exportation and preservation of silicoflagellates in Alfonso Basin, Gulf of California. J Sea Res 109:52–62
Molina-Cruz A (1988) Late Quaternary oceanography of the mouth of the Gulf of California: the Polycystine connection. Paleoceanography 3:447–459
Molina-Cruz A, Welling L, Caudillo-Bohorquez A (1999) Radiolarian distribution in the water column, southern Gulf of California, and its implication in thanatocoenose constitution. Marine Micropaleontolog 37:149–171
Matías-Ramírez LG, Fuentes-Mariles OA, García-Jiménez F (2001) Heladas. Serie Fascículos. SEGOB–CENAPRED, México
Mazzarella A (2007) The 60–year solar modulation of global air temperature: the Earth's rotation and atmospheric circulation connection. Theor Appl Climatol 88:193–199
Mazzarella A (2008) Solar forcing of changes in atmospheric circulation, earth's rotation and climate. Open Atmos Sci J 2:181–191
McGowan JA (1990) Climate and change in oceanic ecosystems: the value of time–series data. Trends Ecol Evol 5:293–299
McGowan JA, Chelton DB, Conversi A (1996) Plankton patterns, climate, and change in the California current. Calif Coop Ocean Fish Investig Rep 37:45–68
Mercado-Santana JA, Santa María del Ángel E, González-Silvera S et al (2017) Productivity in the Gulf of California large marine ecosystem. Environ Dev 22:18–29
Minobe S (1999) Resonance in bidecadal and pentadecadal climate oscillations over the North Pacific: role in climatic regime shifts. Geophys Res Lett 26:855–858
Molina-Cruz A, Pérez-Cruz L, Monreal-Gómez MA (2002) Laminated sediments in the Bay of La Paz, Gulf of California: a depositional cycle regulated by pluvial flux. Sedimentology 49:1401–1410
Munk W, Bills B (2007) Tides and the climate: some speculations. J Phys Oceanogr 37:135–147
Murray DW (1982) Paleo-oceanography of the Gulf of California Based on Silicoflagellates from Marine Varved Sediments. M.Sc. Dissertation, Oregon State University, Corvallis
Murray D, Schrader H (1983) Distribution of Silicoflagellates in Plankton and Core Top Samples from the Gulf of California. Mar Micropaleontol 7:517–539

Nava-Sánchez E (1997) Modern Fan Deltas of the West Coast of the Gulf of California, Mexico. Ph.D. Dissertation, University of Southern California, LA
Onodera J, Takahashi K (2005) Silicoflagellates fluxes and environmental variations in the northwestern North Pacific during December 1997–May 2000. Deep Sea Res I 52:371–388
Ortega-Noriega S (1999) Breve historia de Sinaloa. México. Fondo de Cultura Económica, El Colegio de México
Páez-Osuna F, Sanchez-Cabeza JA, Ruiz-Fernández AC et al (2016) Environmental status of the Gulf of California: a review of responses to climate change and climate variability. Earth-Sci Rev 162:253–268
Pages 2k Consortium (2013) Continental-scale temperature variability during the past two millennia. Nat Geosci 6:339–346
Palafox-Ávila G, Herrera-Barrientos J, Ladrón de Guevara-Torres MA et al (2014) Riesgos potenciales de inundaciones en la ciudad de Guasave, Sinaloa. In: Flores-Campaña L, Morán-Angulo RE, Karam-Quiñones C (eds) Sinaloa ante el cambio climático global. México. Universidad Autónoma de Sinaloa. Instituto de Apoyo a la Investigación e Innovación, pp 145–160
Pasini A (2009) Neural network modeling in climate change studies. In: Haupt SE, Pasini A, Marzban C (eds) Artificial intelligence methods in the environmental sciences. Springer, Netherlands, pp 235–254
Pegau WS, Boss E, Martínez A (2002) Ocean color observations of eddies during the summer in the Gulf of California. Geophys Res Lett 29:1295
Pérez-Cruz L, Molina-Cruz A (1988) El Niño 1983: effect on the distribution of silicoflagellates in the Gulf of California. Cienc Mar 14:39–38
Pérez-Cruz L, Urrutia-Fucugauchi J (2009) Magnetic mineral study of Holocene marine sediments from the Alfonso Basin, Gulf of California –implications for depositional environment and sediment sources. Geofis Int 48:305–318
Pérez-Cruz L (2006) Climate and ocean variability during mid–late Holocene recorded in laminated sediments from Alfonso Basin, Gulf of California, Mexico. Quat Res 65:401–410
Pérez-Cruz L (2013) Hydrological changes and paleoproductivity in the Gulf of California during middle and late Holocene and their relationship with ITCZ and North American Monsoon variability. Quat Res 79:138–151
Pérez-Cruz L, Urrutia-Fucugauchi J (2010) Holocene laminated sediments from the southern Gulf of California: geochemical, mineral magnetic and microfossil study. J Quat Sci 25:989–1000
Philander SG (1990) El Niño, La Niña, and the Southern Oscillation. Academic Press, London
Pike J, Kemp AES (1997) Early Holocene decadal-scale ocean variability recorded in Gulf of California laminated sediments. Paleoceanography 12:227–238
Pineda J (1995) An internal tidal bore regime at nearshore stations along western USA: predictable upwelling within the lunar cycle. Cont Shelf Res 15:1023–1041
Pizarro O, Montecinos A (2004) Interdecadal variability of the thermocline along the west coast of South America. Geophys Res Lett 31:L20307
Poelchau HS (1976) Distribution of Holocene silicoflagellates in north Pacific sediments. Micropaleontology 22:164–193
Pride C, Thunell R, Sigman D et al (1999) Nitrogen isotopic variations in the Gulf of California since the last deglaciation: response to global climate change. Paleoceanography 14:397–409
Pugh DT (1987) Tides, surges and mean sea–level. John Wiley and Sons, Wiltshire
Rebert JP, Donguy JR, Eldin G et al (1985) Relations between sea level, thermocline depth, heat content, and dynamic height in the tropical Pacific Ocean. J Geophys Res Oceans 90:11719–11725
Revelle RR (1939) Sediments of the Gulf of California. Geol Soc Am Bull 50:1929
Ripa P, Velázquez G (1993) Modelo unidimensional de la marea en el Golfo de California. Geofis Int 32:41–56
Roden GI (1964) Oceanographic aspects of Gulf of California. In: van Andel TH, Shor Jr GG (eds) Marine Geology of the Gulf of California. American Association of Petroleum Geologists, Tulsa, Okla, p 30–58
Salas-Salinas MA, Jiménez-Espinosa M (2004) Inundaciones. Serie Fascículos. SEGOB–CENAPRED, México

Salvadeo C, Flores-Ramírez A, Gómez-Gallardo U, Lluch-Belda D, Jaume-Schinkel S, Urbán-Ramírez J (2011) El rorcual de Bryde (Balaenoptera edeni) en el suroeste del Golfo de California: su relación con la variabilidad del ENOS y disponibilidad de presas. Ciencias Marinas 37: 215–225
Saldívar-Lucio R, Di Lorenzo E, Nakamura M et al (2016) Macro-scale patterns in upwelling/downwelling activity at North American west coast. PLoS One 11:e0166962
Saldívar-Lucio R, Salvadeo C, Monte-Luna D et al (2015) Patrones históricos y escenarios térmicos futuros en mares mexicanos. Rev Biol Mar Oceanogr 50:331–345
Sancetta C (1995) Diatoms in the Gulf of California: seasonal flux patterns and the sediment record for the last 15,000 years. Paleoceanography 10:67–84
Santa María del Ángel E, Álvarez-Borrego S, Müller-Karger FE (1994b) The 1982–1984 El Niño in the Gulf of California as seen in coastal zone color scanner imagery. J Geophys Res 99:7423–7431
Santa María del Ángel E, Álvarez-Borrego S, Müller-Karger FE (1994a) Gulf of California biogeographic regions based on coastal zone color scanner imagery. J Geophys Res 99:7411–7421
Schrader H, Baumgartner T (1983) Decadal variation of upwelling in the central Gulf of California. In: Thiede J, Suess E (eds) Coastal upwelling: its sediment record. Part B. Plenum Publishing Corporation, New York, pp 247–276
Schrader H, Pisias N, Cheng G (1986) Seasonal variation of silicoflagellates in phytoplankton and varved sediments in the Gulf of California. Mar Micropaleontol 10:207–233
Segovia-Zavala JA, Delgadillo-Hinojosa F, Lares-Reyes ML et al (2009) Atmospheric input and concentration of dissolved iron in the surface layer of the Gulf of California. Cienc Mar 35:75–90
SGM (2016) Panorama Minero del Estado de Sinaloa. Secretaría de Economía, México
Silverberg N, Shumilin E, Aguirre-Bahena F et al (2007) The impact of hurricanes on sedimenting particulate matter in the semi–arid Bahía de La Paz, Gulf of California. Cont Shelf Res 27:2513–2522
Silverberg N, Aguirre-Bahena F, Mucci A (2014) Time-series mea-surements of settling particulate matter in Alfonso Basin, La Paz Bay, southwestern Gulf of California. Cont Shelf Res 84:169–187
Smoak JM, Moore WS, Thunell RC et al (1999) Comparison of 234Th, 228Th, and 210Pb fluxes with fluxes of major sediment components in the Guaymas Basin, Gulf of California. Mar Chem 65:177–194
Soutar A, Crill PA (1977) Sedimentation and climatic patterns in the Santa Bárbara Basic during the 19th and 20th centuries. Geol Soc Am Bull 88:1161–1172
Soutar A, Kling SA, Crill PA et al (1977) Monitoring the marine environment through sedimentation. Nature 266:136–139
Staines-Urías F, Douglas RG, Gorsline DS (2009) Oceanographic variability in the southern gulf of California over the past 400 years: evidence from faunal and isotopic records from planktic foraminifera. Palaeogeogr Palaeoclimatol Palaeoecol 284:337–354
Steemann N, Jensen EA (1957) Primary oceanic production. The autotrophic production of organic matter in the oceans. Galathea Rep 1:49–136
Takahashi K, Blackwelder PL (1992) The spatial distribution of silicoflagellates in the region of the Gulf Stream warm-core ring 82B: application to water mass tracer studies. Deep Sea Res A 39:327–346
Thunell R (1998) Seasonal and annual variability in particle fluxes in the Gulf of California: a response to climate forcing. Deep Sea Res I Oceanogr Res Pap 45:2059–2083
Thunell R, Pride C, Tappa E et al (1993) Varve formation in the Gulf of California: insights from time series sediment trap sampling and remote sensing. Quat Sci Rev 12:451–464
Thunell R, Pride C, Ziveri P et al (1996) Plankton response to physical forcing in the Gulf California. J Plankton Res 18:2017–2026
Thunell RC, Pride CJ, Tappa E et al (1994) Biogenic silica fluxes and accumulation rates in the Gulf of California. Geology 22:303–306

Torres-Torres F, Rodríguez-Velázquez D, Ibarra-Pellegrín F et al (1996) Desastres naturales. Aspectos sociales para su prevención y tratamiento en México. UNAM–IIE–CCS–UAS–CONACYT, Mexico

Tréguer P, Bowler C, Moriceau B et al (2018) Influence of diatom diversity on the ocean biological carbon pump. Nat Geosci 11:27–37

Trenary LL, Han W (2012) Intraseasonal to interannual variability of South Indian Ocean sea level and thermocline: remote versus local forcing. J Phys Oceanogr 42:602–627

Ultsch A, Siemon, HP (1990) Kohonen's self organizing feature maps for exploratory data analysis. Proceedings INNC'90 Int. Neural Network Conf. Kluwer, Dordrecht

Venrick EL, Reid FMH, Lange CB (2003) Siliceous phyto-plankton in the Santa Barbara Channel: a seven-year comparison of species in a near-bottom sediment trap and in water samples from the euphotic layer. Calif Coop Ocean Fish Invest Rep 44:107–122

Venrick EL, Lange CB, Reid FMH et al (2008) Temporal patterns of species composition of siliceous phytoplankton flux in the Santa Barbara Basin. J Plankton Res 30:283–297

Vesanto J, Himberg J, Alhoniemi E et al (2000) SOM Toolbox for Matlab 5. Technical Report A57, Helsinki University of Technology

Wahr JM (1988) The Earth's rotation. Annu Rev Earth Planet Sci 16:231–249

Wooster WS, Fluharty DL (eds) (1985) El Niño north: Niño effects in the eastern subarctic Pacific Ocean, 1st edn. University of Washington, Seattle

Zeitzschel B (1969) Primary productivity in the Gulf of California. Mar Biol 3:201–207

Ziveri P, Thunell RC (2000) Coccolithophore export production in Guaymas Basin, Gulf of California Response to climate forcing. Deep Sea Res II 47:2073–2100

Chapter 3
Holocene Hydroclimate of the Subtropical Mexico: A State of the Art

Priyadarsi D. Roy, Jesús David Quiroz-Jiménez, Claudia M. Chávez-Lara, and José Luis Sánchez-Zavala

Abstract The semiarid and arid subtropical Mexico has heterogeneous modern climate, and its different parts receive rainfall associated with the North American Monsoon (NAM) system, tropical storms, and westerlies. Increase in the number of paleoclimate registers with multidisciplinary studies has improved our understanding of this complex desert ecosystem during the intervals of millennial-scale global climate changes that occurred over the Holocene. This chapter describes and synthesizes the information of vegetation and dynamics of precipitation–desertification reconstructed from biological, physical, and chemical proxies in geological deposits (e.g., lacustrine deposit, paleosol, pack rat midden, sand dune, and speleothem) into four different climate regions (i.e., northeastern Mexico, NAM, desert, and south Baja) of the subtropical Mexico. The proxy records are compared with different climate forcings such as insolation strength, mean position of Intertropical Convergence Zone, El Niño-Southern Oscillation (ENSO) activity, and sea surface temperatures of the Atlantic and Pacific Oceans. Heterogeneity in hydroclimates of these climate regions is evaluated with respect to previously proposed hypotheses and a new hypothesis associated with size of the Atlantic warm pool. Tropical storms and their trajectories played important roles and contributed to the variable hydroclimates. ENSO was the dominant forcing during the late Holocene, and it generally caused reduction in total annual precipitation and enhancement in the aridity. Some parts, however, received more winter precipitation.

Keywords Paleoclimate · Paleovegetation · Register · Proxy · Hypothesis

---

P. D. Roy (✉) · J. L. Sánchez-Zavala
Instituto de Geología, Universidad Nacional Autónoma de México, Ciudad Universitaria, Mexico City, Mexico
e-mail: roy@geologia.unam.mx

J. D. Quiroz-Jiménez
Área Académica de Ciencias de la Tierra y Materiales, Universidad Autónoma del Estado de Hidalgo, Carr. Pachuca-Tulancingo, Mineral de la Reforma, Mexico

C. M. Chávez-Lara
Organic Geochemistry Unit, School of Chemistry, School of Earth Sciences, Cabot Institute for the Environment, University of Bristol, Bristol, UK

N. Torrescano-Valle et al. (eds.), *The Holocene and Anthropocene Environmental History of Mexico*, https://doi.org/10.1007/978-3-030-31719-5_3

Introduction

More than one-third of the national territory of Mexico, in the northern part, has subtropical climate. It extends from the Tropic of Cancer in south to political boundary with the USA in north (Fig. 3.1). This semiarid and arid region of the North America receives more than 60% of total annual precipitation during the warm season through the North American Monsoon (NAM) and tropical storm systems, and significant winter precipitation occurs only in its northwestern margin (Adams and Comrie 1997; Higgins et al. 1999; Mitchell et al. 2002). Topographies of the Sierra Madre Oriental and Sierra Madre Occidental Mountains limit transportation of moisture from the Atlantic and Pacific Oceans, and the intermediate rain shadow region hosts the Chihuahuan Desert. During the severe drought of AD 2003–2013, the water level in reservoirs declined to their historical minimum, and it caused water shortages in many rural communities (National Meteorological Service of Mexico, SMN). We observed desiccation of several ephemeral lacustrine basins, reduced groundwater resources, and subsidence in sedimentary basins during the multiple field expeditions over the last decade. Model projections suggest that the anthropogenic global warming might cause further enhancement in aridity in this

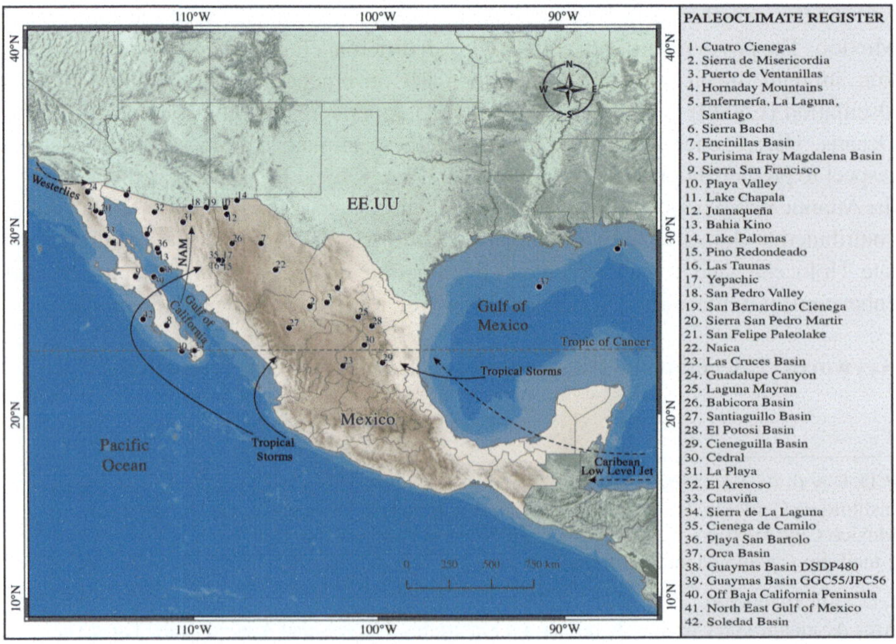

Fig. 3.1 Subtropical Mexico extends from the Tropic of Cancer in south to political boundary with the USA in north. It receives more precipitation during the warm season through the North American Monsoon (NAM) and tropical storm systems, and significant winter precipitation occurs only in its northwestern margin. Tree rings and other geological registers (marked as filled circle) have extended the climate information to the Holocene and beyond

drought-prone region by increasing the mean temperature and reducing the average annual rainfall in the near future (Wang et al. 2011; Cavazos et al. 2013; Cook et al. 2013; Seager et al. 2013; Seager et al. 2014). The annually resolved tree-ring records and the Palmer Drought Severity Index provide information about winter to early-summer soil moisture balance since the fifteenth century (Villanueva-Diaz et al. 2007; Stahle et al. 2011, 2016). Both indicate that the droughts have been more frequent and severe in this part compared to the tropical central-southern Mexico (Villanueva-Diaz et al. 2007).

Geological registers have extended the climate information to the Holocene and beyond. Fossil pollen preserved in sediments of the Cuatro Cienegas Basin provided the first record of vegetation from this region, although it did not have a robust chronological control (Meyer 1973). Subsequent studies were focused on fossil pollen in lacustrine sediments as well as in middens (Betancourt et al. 1990; Sirkin et al. 1994; Anderson and Van Devender 1995; Metcalfe et al. 1997). Poor preservation of this biological proxy, hiatus, or discontinuity in geological deposits and limited dating control led to lack of understanding about the response of this desert ecosystem to the Pleistocene–Holocene transition. The number of geological registers and the type of proxies applied to these registers have increased over the last two decades, and the new studies involved several multidisciplinary methods (Murillo De Nava et al. 1999; Caballero et al. 2005; Castiglia and Fawcett 2006; Ortega-Rosas et al. 2008a; Pigati et al. 2009; Brunelle et al. 2010; Chávez-Lara et al. 2012, 2015, 2018; Cruz-y-Cruz et al. 2014; Gázquez et al. 2013; Roy et al. 2010, 2013a, 2015, 2016). Some of the previously studied sites (Ortega-Guerrero et al. 1999; Metcalfe et al. 2002) were reanalyzed using multidisciplinary approaches in order to reconstruct millennial-scale variations in near-surface processes like erosion, water column salinity, and total organic productivity (Roy et al. 2010, 2013b). However, the lack of higher-resolution records (e.g., speleothems) and analytical methods to discriminate the seasonality (e.g., δD in leaf wax) are some of the major challenges for paleoclimate research in this region (Metcalfe et al. 2015). In this chapter, we present a state of the art of the vegetation composition and hydrological conditions in the subtropical Mexico over the Holocene considering the divisions of Comrie and Glenn (1998). Temporal variations in precipitation are evaluated with respect to previously proposed hypotheses and a new hypothesis presented in this study, and all of them are compared with several climate forcings.

Modern Climate

Summer and autumn rainfalls control the modern climate of this region. The winter precipitation is associated with westerlies, and it is restricted to the Baja California Peninsula in the northwestern margin (Friedman et al. 1992; Douglas et al. 1993; Stensrud et al. 1995; Higgins et al. 1997; Mitchell et al. 2002; Amador et al. 2006; Farfán and Fogel 2007; Cayan et al. 2007; Caldwell 2010; Ritchie et al. 2011; Neelin et al. 2013). The dominant warm season moisture is sourced from the Atlantic and Pacific Oceans by easterly waves and tropical storms. Both of them are driven by northward migration of the Intertropical Convergence Zone (ITCZ), convectional uplift generated by thermal low over the Mexican Plateau, and warmer sea surface temperature (SST) of both the oceans (Higgins et al. 1998; Mitchell et al. 2002; Li and Wang 2005). The amounts of summer rainfall decrease during warmer phase (El Niño year) of the El Niño-Southern Oscillation (ENSO), and it is associated with enhanced subsidence in northern Mexico and southward shift in the ITCZ (Magaña et al. 2003). A northward shift in the ITCZ and weaker subsidence over the northern Mexico during the cooler phase of ENSO (La Niña year), however, increase the total annual precipitation (Magaña and Quintanar 1997; Magaña et al. 2003; Pavia et al. 2006). The Sierra Madre Oriental Mountain acts as barrier for moisture transported from the Atlantic Ocean into the western part (Adams and Comrie 1997; Comrie and Glenn 1998). The Sierra Madre Occidental Mountain channels and confines the low-level moisture from the tropical and subtropical Pacific into the western part, considered as the core region of NAM (Comrie and Glenn 1998). Based on the meteorological data, Comrie and Glenn (1998) divided the subtropical Mexico into four different climate regions, that is, northeastern Mexico, NAM, desert, and south Baja. Figure 3.2 shows distribution of precipitation at sites present within each of these regions.

The Gulf of Mexico (GoM) and Caribbean Sea are the main moisture sources for the "northeastern Mexico" (Higgins et al. 1998). The precipitation here is bimodal with an intermediate midsummer dry spell, known as *caniculas* (Comrie and Glenn 1998). The early-summer precipitation is associated with moisture transported from the Caribbean Sea and Gulf of Mexico by northward branch of the Caribbean low-level jet (Mestas-Nuñez et al. 2005, 2007). Tropical storms, originating in the Atlantic Ocean, provide the autumn moisture (Jones et al. 2003; Wang et al. 2006, 2011). The interannual variability of warm pool (SST $\geq$ 28.5 °C) that forms during the autumn in the tropical Atlantic Ocean controls the activity and trajectory of these storms (Wang et al. 2006, 2011). The high-pressure cell migrates toward northeast over the Atlantic Ocean during the years of large warm pool, and it reduces the possibility of these storms to landfall in northeastern Mexico (Wang et al. 2011).

Comrie and Glenn (1998) have also divided the "NAM" region into four different subregions, and the northwestern Mexico hosts three of them (Fig. 3.2). The subregion 1 (monsoon south) is characterized by unimodal summer rains (June–August), mainly monsoonal. It also receives late-summer and early-fall tropical storms. The subregion 2 (monsoon west) shows longer dry period, with monsoon rainfall during the summer and minor amounts of winter precipitation. The distribution in subregion 4 (monsoon east) is similar to subregion 1, but it receives significantly less annual rainfall. SSTs of the Gulf of California (GoC) and the Pacific Ocean off the Baja California Peninsula provide constraints on ocean moisture and geographic expressions in the "NAM" region (Higgins and Shi 2001; Mitchell et al. 2002; Corbosiero et al. 2009). The "desert" region represents mostly northern part of the Baja California Peninsula. It remains dry with marginally more winter precipitation compared to the "south Baja" and lacks the monsoon precipitation peak. Extreme precipitation events in the recent past are associated with the El Niño

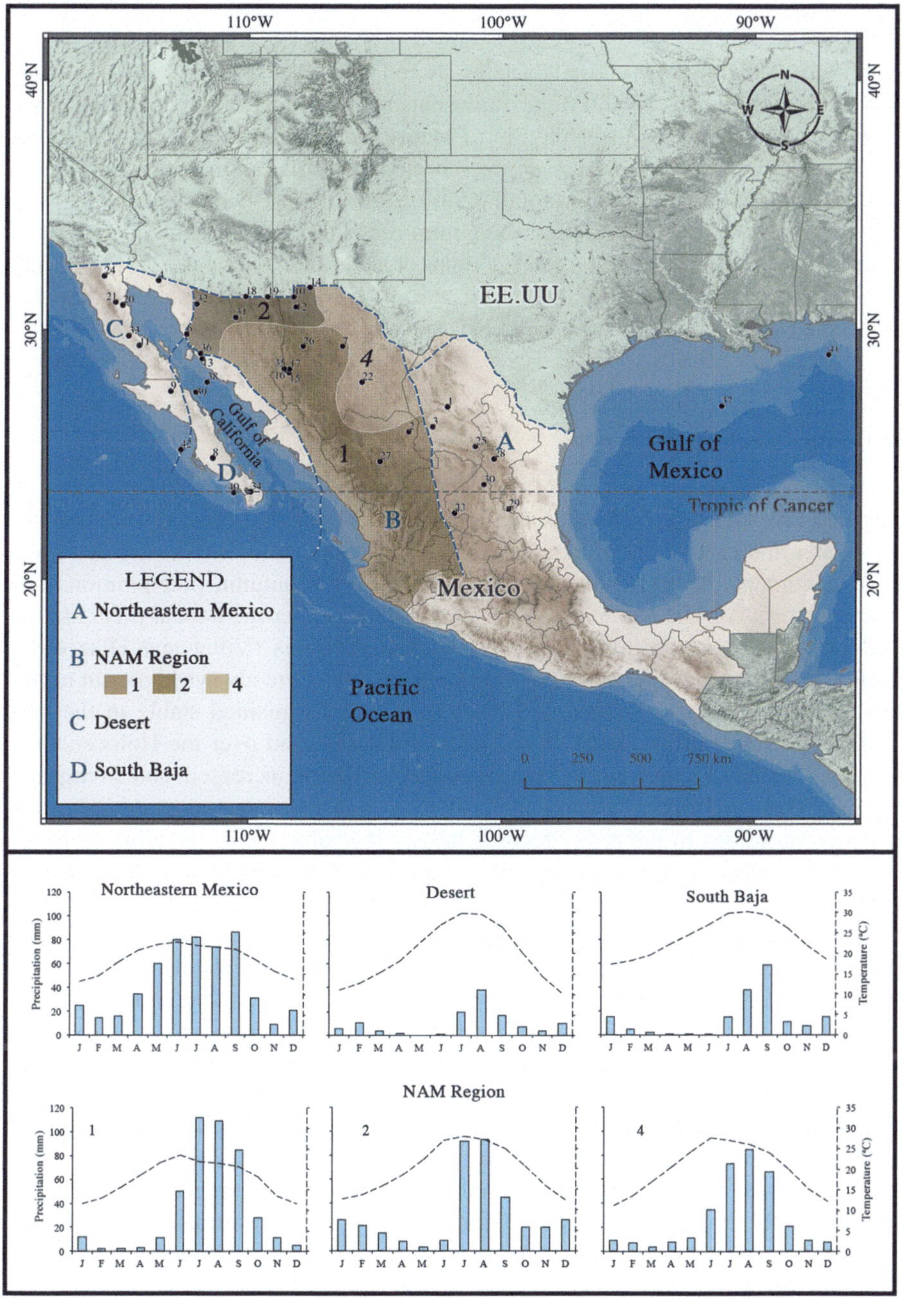

Fig. 3.2 Distributions of precipitation at sites present within four different climate regions of the subtropical Mexico (northeastern Mexico, NAM, desert, and south Baja). The "NAM" region has four different subregions, and the northwestern Mexico hosts three (1, 2, and 4) of them. The subregion 1 (monsoon south) receives unimodal summer rains, and the subregion 2 (monsoon west) shows longer dry period, with monsoon rainfall during the summer and minor amounts of winter precipitation. The subregion 4 (monsoon east) is similar to subregion 1, but it receives significantly less annual rainfall

conditions, and they have caused flash floods and landslides (e.g., AD 1993; Cerezo-Mota et al. 2006). The "south Baja" represents southern part of the Baja California Peninsula and western part of the Sierra Madre Occidental Mountains. It receives summer–autumn as well as winter precipitations, and the autumn precipitation is more compared to the summer rainfall (Comrie and Glenn 1998). The summer rainfall comes through the NAM, and the autumn moisture is contributed by tropical storms in the eastern Pacific Ocean (Douglas et al. 1993; Xu et al. 2004; Farfán and Fogel 2007). The winter rainfall is associated with southward displacement of the subtropical westerly jet (Cavazos and Rivas 2004).

Register and Hypothesis

The tree-ring-based drought atlas of Stahle et al. (2011, 2016) provides registers of the winter and early-summer precipitations of last six centuries. It could not, however, document the soil moisture associated with the autumn precipitation. The tropical storms of late summer and autumn generally alter the annual precipitation budget and reverse the long-term hydrological droughts (Villanueva-Diaz et al. 2007). The frequency and severity of decadal drought were always higher in northern Mexico, and the influence of ENSO on drought remained stable in the past 150 years (Stahle et al. 2016). In the millennial scale and over the Holocene, the forcings of ENSO and Atlantic Multidecadal Oscillation increased, and strength of the insolation declined (Metcalfe et al. 2015). Comparison and geographic expressions of variations in the proxy records helped in evaluation of the climate-driven surface processes as well as proposal of new hypotheses explaining the hydroclimate variations. Information of the Holocene are mostly sourced from lacustrine deposits and middens. Limited studies are also focused on eolian deposit, beach ridge, paleosol, speleothem, bog sediment, and groundwater deposit (Murillo De Nava et al. 1999; Caballero et al. 2005; Castiglia and Fawcett 2006; Ortega-Rosas et al. 2008a, b, c; Pigati et al. 2009; Brunelle et al. 2010; Cruz-y-Cruz et al. 2014; Gázquez et al. 2013; Roy et al. 2013a, 2016; Chavez-Lara et al. 2018) (Table 3.1). Most of the sediment sequences lacked preservation of fossil pollen (e.g., Lozano-García et al. 2002; Metcalfe et al. 2002). The new studies involved multidisciplinary tools such as stratigraphy, geomorphology, charcoal content, chemical composition, stable isotopes, magnetic property, associations of fossil diatom and ostracods, and more recently the lipid biomarkers. Some of the previously studied sites (Ortega-Guerrero et al. 1999; Metcalfe et al. 2002) were investigated again by using multidisciplinary approaches in order to reconstruct the millennial-scale variations of near-surface processes like erosion via eolian and fluvial activities, water column salinity through proxy registers of evaporation, and allochthonous and autochthonous organic productivities (Roy et al. 2010, 2013b).

The longer records of paleo-precipitation and desertification were generated from sediments deposited in the Babicora, Santiaguillo, and San Felipe Basins (Roy et al. 2010, 2013b; Quiroz-Jiménez et al. 2017). Paleo-precipitation was inferred

Table 3.1 Location of paleoclimate registers used for reconstructing the environmental conditions of the subtropical Mexico over the Holocene

Map key	Site name	Latitude (N)	Longitude (W)	Altitude (m)	State	Register	Proxy	Age	References
1	Cuatro Cienegas	26° 55′	102° 09′	740	Coahuila	Lacustrine sediment	Pollen	Holocene	Meyer (1973)
2	Sierra de Misericordia	25° 55′	103° 40′	1310	Durango	Midden	Pollen	Holocene	Betancourt et al. (1990)
3	Puerto de Ventanillas	26° 07′	102° 44′	1370	Coahuila	Midden	Pollen	Holocene	Betancourt et al. (1990)
4	Hornaday Mountains	31° 59′	113° 36′	240	Sonora	Midden	Pollen	Holocene	Betancourt et al. (1990)
5	Enfermeria, La Laguna, Santiago	23° 24′	110°	–	Baja California Sur	Lacustrine sediment	Pollen	Holocene	Sirkin et al. (1994)
6	Sierra Bacha	29° 50′	112° 28′	400	Sonora	Midden	Pollen	Holocene	Van Devender et al. (1994), Anderson and Van Devender (1995)
7	Encinillas Basin	29° 20′	106° 17′	1530	Chihuahua	Lacustrine sediment	Pollen, diatom, geochemistry	Holocene	Metcalfe et al. (1997)
8	Purisima–Iray–Magdalena Basin	24° 52′	111° 25′	100	Baja California Sur	Eolian deposit	Stratigraphy, geomorphology	Holocene	Murillo De Nava et al. (1999)
9	Sierra San Francisco	27° 32′	113° 06′	780	Baja California Sur	Midden	Pollen	Pleistocene-Holocene transition	Rhode (2002)
10	Playa Valley	31° 19′	108° 12′	1500	Chihuahua	Midden	Pollen	Holocene	Holmgren et al. (2003)
11	Lake Chapala	29° 22′	114° 21′	660	Baja California	Lacustrine sediment	Stratigraphy	Holocene	Davis (2003)

(continued)

Table 3.1 (continued)

Map key	Site name	Latitude (N)	Longitude (W)	Altitude (m)	State	Register	Proxy	Age	References
12	Juanaqueña	30° 55′	108° 07′	1350	Chihuahua	Paleosol	Stable isotope, geomorphology	Holocene	Nordt (2003)
13	Bahia Kino	28° 50′	111° 51′	–	Sonora	Lacustrine sediment	Pollen, diatom, foraminifera, ostracode, magnetic property, geochemistry	Holocene	Caballero et al. (2005)
14	Lake Palomas	31° 42′	107° 34′	1210	Chihuahua	Beach ridge	Geomorphology	Holocene	Castiglia and Fawcett (2006)
15	Pino Redondeado	28° 22′	108° 23′	1945	Chihuahua	Bog sediment	Pollen	Holocene	Ortega-Rosas et al. (2008a)
16	Las Taunas	28° 23′	108° 33′	1700	Sonora	Bog sediment	Pollen	Holocene	Ortega-Rosas et al. (2008a, 2017)
17	Yepachic	28° 25′	108° 22′	1810	Chihuahua	Bog sediment	Pollen	Holocene	Ortega-Rosas et al. (2008a)
18	San Pedro Valley	31° 20′	110° 08′	1300	Sonora	Groundwater discharge deposit	Geochemistry, stable isotopes, ostracodes, gastropods, sediments	Holocene	Pigati et al. (2009)
19	San Bernardino Cienega	31° 19′	109° 15′	1150	Sonora	Bog sediment	Pollen, grain size, charcoal	Holocene	Minckley and Brunelle (2007), Minckley et al. (2011), Brunelle et al. (2010)
20	Sierra San Pedro Martir	31° 00′	115° 00′	647–899 m	Baja California	Middens	Pollen	Holocene	Holmgren et al. (2011)
21	San Felipe Paleolake	31° 07′	115° 17′	400	Baja California	Lacustrine sediment	Magnetic property, grain size, geochemistry, pollen	Holocene	Ortega-Guerrero et al. (1999), Lozano-García et al. (2002), Roy et al. (2010, 2012b)
22	Naica	27° 54′	105° 31′	–	Chihuahua	Speleothem	Stable isotope	Holocene	Gázquez et al. (2013)
23	Las Cruces Basin	22° 39′	101° 52′	2100	San Luis Potosi	Lacustrine sediment	Geochemistry, magnetic property	Holocene	Roy et al. (2013a)

24	Guadalupe Canyon	32° 10′	115° 44′	150–290	Baja California	Midden	Pollen	Holocene	Holmgren et al. (2014)
25	Laguna Mayran	25° 20′	101° 03′	1774	Coahuila	Lacustrine sediment	Pollen, stable isotope	Holocene	Albert (2015)
26	Babicora Basin	29° 20′	107° 50′	2130	Chihuahua	Lacustrine sediment	Geochemistry, sedimentology, magnetic property, pollen, ostracodes, diatom	Holocene	Metcalfe et al. (1997, 2002), Urrutia-Fucugauchi et al. (1997), Ortega-Ramírez et al. (1998, 2000), Palacios-Fest et al. (2002), Chávez-Lara et al. (2012), Roy et al. (2012a, 2013b), Romero-Mayén and Carreño (2016)
27	Santiaguillo Basin	24° 44′	104° 48′	1960	Durango	Lacustrine sediment	Geochemistry, magnetic property, ostracode	Holocene	Roy et al. (2014b, 2015), Cruz-y-Cruz et al. (2015, 2018), Quiroz-Jiménez et al. (2017)
28	El Potosi Basin	24° 50′	100° 19′	1880	Nuevo Leon	Lacustrine sediment	Geochemistry, magnetic properties	Holocene	Roy et al. (2016)
29	Cieneguilla Basin	22° 50′	99° 45′	1045	Tamaulipas	Lacustrine sediment	Geochemistry, stable isotopes	Holocene	Roy et al. (2019)
30	Cedral	23° 48′	100° 43′	1700	San Luis Potosi	Paleosol	Stable isotope, fossil bone	Holocene	Cruz-y-Cruz et al. (2016)
31	La Playa	30° 30′	110° 32′	1226	Sonora	Paleosol	Stable isotope, fossil bone	Holocene	Cruz-y-Cruz et al. (2016)
32	El Arenoso	31° 02′	112° 04′	545	Sonora	Paleosol	Stable isotope, fossil bone	Holocene	Cruz-y-Cruz et al. (2016)
33	Cataviña	29° 46′	114° 46′	550	Baja California	Midden	Pollen	Antropocene	Sankey et al. (2001)

(continued)

Table 3.1 (continued)

Map key	Site name	Latitude (N)	Longitude (W)	Altitude (m)	State	Register	Proxy	Age	References
34	Sierra de La Laguna	23° 30′	110° 20′	–	Baja California Sur	Tree ring	Dendroclimatology	Antropocene	Díaz et al. (2001)
35	Cienega de Camilo	28° 26′	108° 35′	1550	Sonora	Bog sediment	Pollen	Antropocene	Van Devender et al. (2003), Ortega-Rosas et al. (2008b, 2008c)
36	Playa San Bartolo	29° 03′	111° 55′	28	Sonora	Eolian deposit	Geochemistry, magnetic property, stratigraphy	Antropocene	Ortega-Guerrero et al. (2013)
37	Orca Basin	25° 57′	91° 20′	−2280	Gulf of Mexico	Marine sediment	Foraminifera, Mg/Ca, stable isotopes	Deglacial	Flower et al. (2004)
38	Guaymas Basin-DSDP480	27° 54′	111° 39′	−655	Gulf of California	Marine sediment	Calcium carbonate, biogenic opal, diatoms, silicoflagellates	Holocene	Barron et al. (2005)
39	Guaymas Basin-GGC55/JPC56	27° 30′	112° 06′	−818	Gulf of California	Marine sediment	Calcium carbonate, biogenic opal, diatoms, silicoflagellates	Holocene	Barron et al. (2005)
40	Off Baja California Peninsula	23° 28′	110° 36′	−606	Baja California Sur	Marine sediment	Geochemistry, magnetic property	Holocene	Blanchet et al. (2007)
41	North East Gulf of Mexico	29° 01′	87° 08′	−847	Gulf of Mexico	Marine sediment	Foraminifera, Mg/Ca	Holocene	Ziegler et al. (2008)
42	Soledad Basin	25° 12′	112° 42′	−290	Gulf of California	Marine sediment	Foraminifera, Mg/Ca	Holocene	Marchitto et al. (2010)

from reconstructions of erosion by estimating concentration of Ti and other insoluble elements and, in some cases, comparison of the erosion proxy with diatom and ostracod-based paleo-salinity records (Metcalfe et al. 2002; Chávez-Lara et al. 2012, 2015) and estimation of chemical alteration of the silicate fractions (e.g., feldspars) present in the bulk sediment (Roy et al. 2015). Records of the desertification were reconstructed directly from geochronology of eolian deposits (Murillo De Nava et al. 1999) and indirectly from abundances of Zr-bearing as well as Si-bearing clastic minerals transported into the lacustrine basins from the arid watersheds (Roy et al. 2012a, 2016) and precipitation of authigenic minerals like carbonates (i.e., calcite, proto-dolomite, and gaylussite; Roy et al. 2014a) and the relative abundance of sulfate compared to carbonate (i.e., gypsum/calcite; Roy et al. 2013a). Abundance of magnetic minerals estimated from magnetic susceptibility of the sediments did not differentiate between pluvial/fluvial and eolian activities in the surroundings.

Figure 3.3 shows the tendencies of proxies indicating erosion in the watershed and variations in water column salinity of the Las Cruces Basin (precipitation, 380 mm/a; geology, volcanic and clastic sedimentary rocks; Roy et al. 2013a) located in the "northeastern Mexico." The abundance of clastic minerals is estimated from Ti/Ca ratio in bulk sediments, and it is used here as a proxy to estimate erosion in the watershed. Abundance of CO_3 (carbonates) is the proxy to estimate salinity of the water column. The inverse trends of both indicate gradually decreasing erosion in the watershed over the Holocene as the water column became increasingly saline. This suggests that erosion was mainly controlled by the runoff.

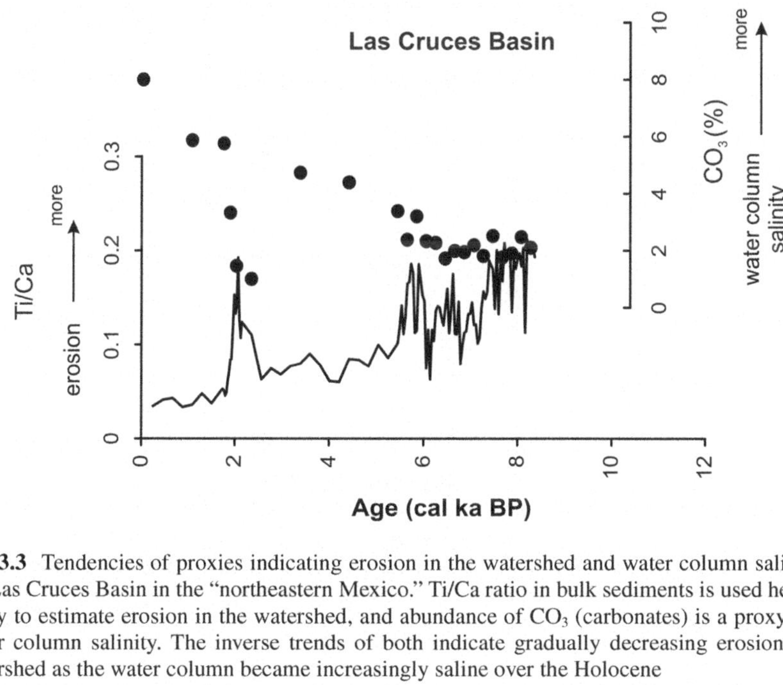

Fig. 3.3 Tendencies of proxies indicating erosion in the watershed and water column salinity of the Las Cruces Basin in the "northeastern Mexico." Ti/Ca ratio in bulk sediments is used here as a proxy to estimate erosion in the watershed, and abundance of CO_3 (carbonates) is a proxy of the water column salinity. The inverse trends of both indicate gradually decreasing erosion in the watershed as the water column became increasingly saline over the Holocene

Increasing water column salinity and reduced erosion indicate that the amounts of rainfall in surroundings of the Las Cruces Basin decreased, and the overall warmer conditions caused more evaporation since the middle Holocene with the exception of an abrupt wet event at ca. 2 cal ka BP.

Proxies have also reconstructed distinct hydrological conditions of other lacustrine basins during the generally arid Holocene, and many of them might be local, rather than regional. Even in sediments with poor organic carbon preservation (total organic carbon (TOC) ≤1.0 wt %), the C/N ratio provided key information about the relative contributions of lacustrine algae and terrestrial plants. Organic carbon present in sediments of the Santiaguillo Basin in "NAM" region (precipitation, 430 mm/a; geology, volcanic and clastic sedimentary rocks; Roy et al. 2014b) was dominantly sourced from terrestrial vegetation at ca. 9–8 cal ka BP (C/N > 20; Talbot and Johannessen 1992), and the combinations of lacustrine and terrestrial plants contributed to total organic productivity during rest of the Holocene (Fig. 3.4, Roy et al. 2015). Zr/Ti ratio in sediments of the San Felipe Basin in relatively more arid "desert" region (precipitation, 220 mm/a; geology, igneous–metamorphic rocks; Roy et al. 2012b) indicated significant amplification in the eolian activity at ca. 8–7.5 cal ka BP, and it transported more clastic sediments (Fig. 3.5). Chemical index of alteration (CIA; Nesbitt and Young 1999) could also detect the minor reduction in interaction between the siliciclastic fraction of the sediments and inflowing water even within a regime of low weathering over the Holocene (CIA, 50–60). The minimal values of total organic carbon (TOC) in sediments suggest that most of the preserved organic matter was oxidized during this interval of more arid condition and periodic events of desiccation (Roy et al. 2010, 2012b).

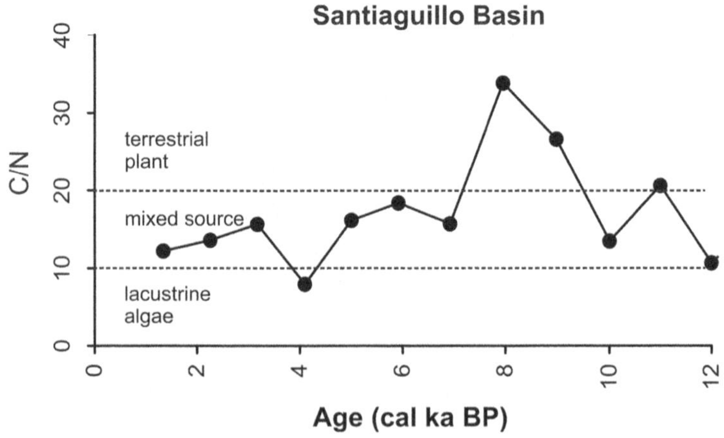

Fig. 3.4 The C/N ratio in sediments with poor organic carbon preservation (TOC ≤ 1.0 wt %) in sediments of the Santiaguillo Basin in "NAM" region suggests relative contributions of lacustrine algae and terrestrial plants. Organic matter of ca. 9–8 cal ka BP was dominantly sourced from the terrestrial vegetation

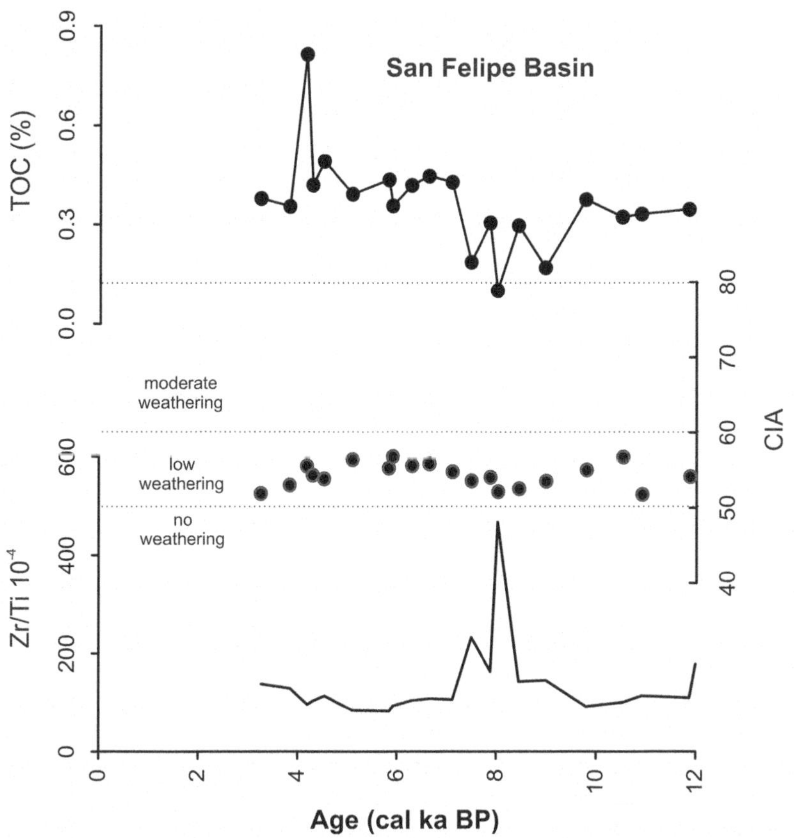

Fig. 3.5 Zr/Ti ratio, total organic carbon (TOC), and chemical index of alteration (CIA) in sediments of the San Felipe Basin in relatively more arid "desert" region. Significant amplification of eolian activity at ca. 8–7.5 cal ka BP transported more Zr-bearing clastic sediments into the basin, and CIA detected the minor reduction in interaction between the siliciclastic fractions and inflowing water even within a regime of low weathering. Minimal values of TOC suggest oxidation of most of the preserved organic matter during this arid interval

During the earliest Holocene, the expansion of C_4 grasses and summer-flowering plants in northern part of the Baja California Peninsula ("desert") suggests that the warm season moisture and the stronger winds came mainly through an active NAM (Holmgren et al. 2011; Barron et al. 2012; Roy et al. 2014b). Barron et al. (2012) associated this interval to increased number of tropical cyclones and proposed that it was forced by differences in SSTs between the California Current system off the Baja California Peninsula and the GoC. Warmer SST of the subtropical Pacific Ocean favored more summer precipitation in the region west of 114° W longitude during >8 cal ka BP. SST of the GoC became warmer at 7.5 cal ka BP, and it fueled the northward surge of tropical moisture establishing the modern NAM climatology to the east of 114° W longitude. The fossil pollen preserved in peat bogs from the Sierra Madre

Occidental, however, provided information contrary to this observation of Barron et al. (2012). Ortega-Rosas et al. (2008a) suggested that winter precipitation persisted until 9.2 cal ka BP, and Antinao and McDonald (2013) suggested that enhanced frontal storm activity in the subtropical Pacific Ocean was responsible for alluvial fan aggradation during ca. 11.5–9 cal ka BP. Both these sedimentary registers are located to the east of 114° W longitude. Rhode (2002) also suggested that the Pleistocene–Holocene transition in the "desert" had cooler and dryer summer and wetter winter than today.

Vegetation Composition

Fossil pollen preserved in lacustrine sediments and pack rat middens provide information on the paleovegetation composition, and some important information also came from the lipid biomarkers. Studies are scarce in the "northeastern Mexico," and most of the information came from sites located within the "NAM," "desert," and "south Baja" (Fig. 3.2 and Table 3.1).

Northeastern Mexico

Pollen record from the Cuatro Cienegas Basin revealed vegetation without significant variability (Meyer 1973). Vegetation composition reflected gradual transition from a phase of cooler summer and wetter winter to another phase of expansion in aridity during the late Holocene. The succulents and desert scrubs were more, and the vegetation assemblage did not contain woodland plants (Puerto de Ventanillas; Betancourt et al. 1990).

NAM

Presence of conifer forest at elevations of >1700 m asl in the Sierra Madre Occidental Mountains reflected cooler conditions during the earliest Holocene (i.e., Cienega Las Taunas, Cienega Yepachic, and Cienega Pino Redondeado; Ortega-Rosas et al. 2008a, b). Warm mixed forest appeared post 9.2 cal ka BP in these higher elevations (Ortega-Rosas et al. 2008a, b). The desert species occurred as small and rare populations within dry microsites of the chaparral and pinyon–juniper–oak woodlands, and they extended rapidly (Holmgren et al. 2014). Chavez-Lara et al. (2018) successfully studied lipid biomarkers in organic-poor sediments of the Santiaguillo Basin and provided evidences of disappearance of woodland plants and establishment of desert scrub. Input of leaf decreased, and contribution from the woody plant tissues and bacterial biomass increased over early and middle Holocene (ca. 11.5–4 cal ka BP). Expansion of modern grassland during ca. 11.5–9 cal ka BP reflected the aridity of early Holocene, and the terrestrial organic matter was predominantly sourced from woody material during the extended drought interval of ca. 9–7.9 cal ka BP. Over the dryer and warmer middle Holocene, the Sierra de la Misericordia had mesic spring marshes (Betancourt et al. 1990), and the Santiaguillo Basin hosted less C_4 grasses (Chávez-Lara et al. 2018).

The subtropical dry forest was established over the late Holocene, and it included species of the *Burseraceae* family and grasses (Chávez-Lara et al. 2018). Increased supply of plant litters from resinous trees and grasses reflected establishment of the modern vegetation at ca. 4 cal ka BP. Pines included *Pinus strobiformis* (cool and humid habitats), and the herbaceous vegetation was composed of *Poaceae*, *Cyperaceae*, and *Pteridophyta* (Ortega-Rosas et al. 2008c). Warm mixed forest and conifer forest extended at higher elevations of the Sierra Madre Occidental Mountains (Ortega-Rosas et al. 2008a, b). Pollen data from the upper Nazca Basin and Parras Basin in eastern margin of the Sierra Madre Occidental Mountains indicated that shortgrasses and park woodlands were replaced by more flora in higher-energy depositional environment during the late Holocene, and both reached their maximum values at ca. 2 cal ka BP (Albert 2015). In the Cienega de Camilo, the pine–oak forest with vegetation reflecting more xeric conditions has been persisting at least since the last millennium (Van Devender et al. 2003). The Cienega San Marcial had tropical tree and shrub with very low abundance of aquatic taxa during this interval (Ortega-Rosas et al. 2016).

Desert

Vegetation was represented by both chaparral and woodland plants during the early Holocene (i.e., Sierra San Francisco; Rhode 2002). In the Sierra Bacha, *Fouquieria columnaris* continued, and other species from the genus *Prosopis*, *Eriogonum*, and *Fouquieria* began to appear (Van Devender et al. 1994; Anderson and Van Devender 1995). Pack rat midden records from the Hornaday Mountains, however, were an exception to this cold environment vegetation. They had brittlebush and Mormon tea without woodland plants or other succulent plants (Betancourt et al. 1990). The vegetation of more arid and warmer middle Holocene was represented by *Bursera microphylla*, *Fouquieria splendens*, *Hyptis emoryi*, *Jatropha cuneata*, *Solanum hindsianum*, and *Trixis californica* (i.e., Sierra Bacha; Van Devender et al. 1994). Expansion of *Opuntia arbuscula* and *Erioneuron pulchellum* suggested greater warm season rainfall (Van Devender et al. 1994). Presence of *Cercidium* and *Fouquieria splendens* indicate the establishment of persistent mesic conditions since ca. 5.4 cal ka BP (Anderson and Van Devender 1995).

Desert scrub appeared during the late Holocene (Holmgren et al. 2014). Vegetation similar to the modern conditions (i.e., mesquite, creosote bush, desert lavender, goldeneye, jojoba, joint fir, wishbone bush, and *Encelia californica*) has been continuing at least since ca. 1.8 ka BP in Cataviña (Sankey et al. 2001).

South Baja

Mangrove dominated the coastal areas and pine–oak dominated the uplands during the early Holocene (Sirkin 1994). Vegetation shifted to thorn forest with pine and thorn at ca. 7 cal ka BP. Abundance of red mangrove increased at ca. 5.6 cal ka BP, suggesting rise in the sea level (Sirkin 1994). Main vegetation of the Bahia Kino was characterized by wetlands surrounded by desert trees and tall shrub taxa (Caballero et al. 2005). The abundance of pine forest and non-arboreal pollen increased over the last 3 cal ka (Sirkin 1994).

Hydrological Variation and Climate Forcing

The $\delta^{18}O$ record of ice cores from the Greenland, in general, showed uniform values over the Holocene, except for the negative excursion during the Pleistocene–Holocene transition (NGRIP project members 2004). We evaluated hydrological conditions of the subtropical Mexico by considering proxy records from the climate divisions of Comrie and Glenn (1998). The proxy records of runoff from the northeastern Mexico are compared with the climate forcings. The paleohydrological conditions of NAM, desert, and south Baja regions are presented separately, and we evaluated them as a group with respect to possible climate forcings as all of them are located in the northwestern part of Mexico.

Northeastern Mexico

Sediment archives from the Las Cruces, Cieneguilla, and El Potosi Basins provided important information, and all of them show evidences of desiccation, depositional hiatus, and poor conditions of preservation. Archive from the Cieneguilla Basin had depositional hiatus between ca. 6.8 and 0.5 cal ka BP, and the record from Las Cruces Basin reached only up to ca. 8.4 cal ka BP. Ti/Ca ratios in these sedimentary sequences suggest different abundances of siliciclastic minerals (Fig. 3.6). Different values in sediments of these basins were due to variable geology of their watersheds (Roy et al. 2013a, 2015, 2016). Limestone is the dominant lithology in watersheds of the Cieneguilla and El Potosi Basins, and the presence of volcanic and clastic sedimentary rocks is minor. Abundance of limestone is also more in watershed of the El Potosi Basin compared to watershed of the Cieneguilla Basin. In watershed of the Las Cruces Basin, the volcanic rocks (rhyolite, andesite, and basalt) as well as clastic sedimentary rocks (sandstone) are more abundant, and the exposure of limestone is minimal. The Las Cruces Basin received more siliciclastic minerals from erosion of these lithologies, and its sediment shows seven- to tenfold higher Ti/Ca values compared to sediments of the Cieneguilla and El Potosi Basins.

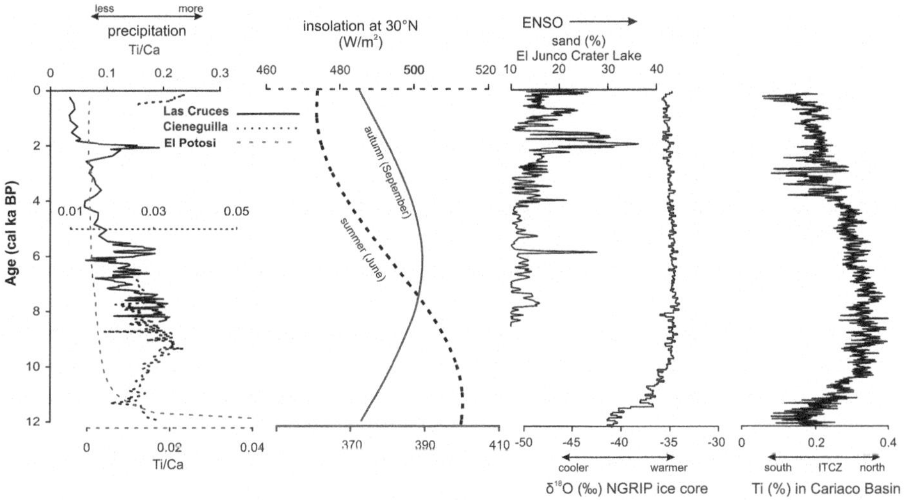

Fig. 3.6 Comparison of the proxy records of precipitation (Ti/Ca) in sediment archives from the Las Cruces, Cieneguilla, and El Potosi Basins with summer (June) and autumn (September) insolations, ENSO activity, North Hemisphere temperature, and mean position of the ITCZ. Summer insolation, mean position of ITCZ, and ENSO activity forced the millennial-scale hydroclimate of the northeastern Mexico. The wet event at ca. 2 cal ka BP was an exception to this as precipitation increased during this event of frequent El Niño conditions

Even with differences related to geology of these sedimentary basins and independently constructed age models for each sedimentary archive, this proxy in general shows millennial-scale tendencies of gradually decreasing runoff over the Holocene, and it is similar to the tendency of summer (June) insolation. More runoff of the early Holocene occurred during an interval of higher summer insolation, and the runoff decreased as the summer insolation gradually decreased over the middle and late Holocene. Offset between the tendencies of precipitation and summer insolation during the earliest Holocene was contemporary to the events of melt water pulses that occurred from the Laurentide Ice Sheet into the Atlantic Ocean (e.g., Aharon 2003; Rasmussen and Thomsen 2012). Cooler SST during these melt water pulses possibly reduced the annual precipitation in this region, even though the insolation remained stronger (Roy et al. 2016).

Variations in runoff also show positive relationship with tendency of the ITCZ position (e.g., Peterson and Haug 2006) and negative relationship with the ENSO activity (e.g., Conroy et al. 2008). Concentration of Ti in sediments of the Cariaco Basin is universally considered as a proxy of latitudinal shift in the ITCZ after the pioneer works of Haug et al. (2001) and Peterson and Haug (2006). Similarly, Conroy et al. (2008) presented sand content in sediments of the El Junco Crater Lake of the Galápagos Islands as a proxy of ENSO activity. Precipitation in the northeastern Mexico was generally more as the ITCZ was located in the northerly latitudes. The paleo-ENSO record indicates weak or nonexistent ENSO during the middle Holocene. Shifting of the ITCZ to southerly latitudes over the middle and late Holocene caused reduction in the amounts of precipitation, and this interval was also characterized by increase in frequency and magnitude of the ENSO (last ca. 4.2 cal ka). Higher precipitation occurred during the interval of weak or nonexistent ENSO, and precipitation in the region decreased as ENSO became more frequent. This suggests that summer insolation, mean position of ITCZ, and ENSO activity forced the hydroclimate of the northeastern Mexico over the Holocene. The wet event at ca. 2 cal ka BP was an exception to this as precipitation increased during this event of more frequent El Niño-like conditions.

Meteorological observations of Wang et al. (2006, 2011) suggest that formation of warm pool (SST ≥ 28.5 °C) in the Atlantic Ocean has been modifying trajectories of the tropical storms and affecting the hydroclimate of this region. It might also be an additional force that possibly modulated the amount of total annual precipitation during the Holocene. We evaluated the proposal of Roy et al. (2016) and compared the variations in hydroclimate conditions with the sizes of this warm pool. Proxy SST records from the Orca Basin (GoM; Flower et al. 2004), Colombia Basin (Caribbean Sea; Schmidt et al. 2004), and Tobago Basin (tropical Atlantic Ocean; Rühlemann et al. 1999) helped to reconstruct the warm pool and its size in the millennial-scale resolution (Fig. 3.7). The North Atlantic switched to warmer phase in AD 1995 (Goldenberg et al. 2001), and this resulted in formation of consistent large warm pools since then (Wang et al. 2006). As the modern-era warm pool is considered as large, we classified its reconstructed sizes over the Holocene as intermittent, small, and intermediate. It formed intermittently during the early Holocene (up to ca. 8 cal ka BP) with SST in the GoM reaching the threshold value of 28.5 °C only during brief events. SST in the Caribbean Sea became warmer post ca. 8 cal ka BP,

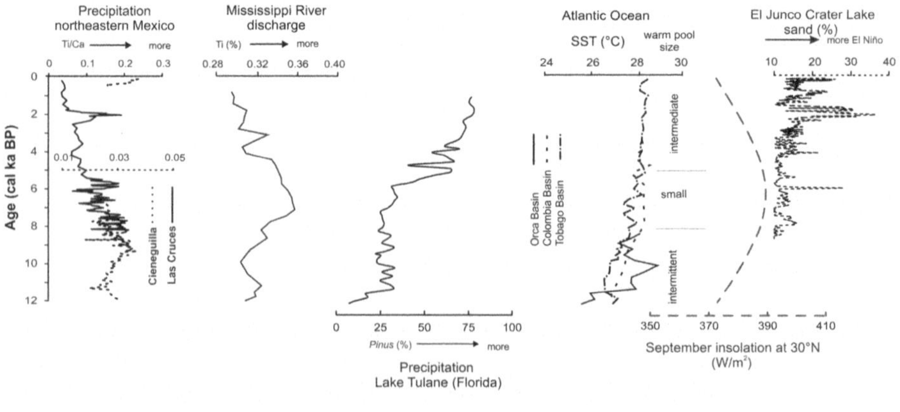

Fig. 3.7 Evaluation of proxy precipitation records from the northeastern Mexico and central-southern USA with respect to warm pool sizes in the Atlantic Ocean. Proxy SST records from the Orca Basin (GoM; Flower et al. 2004), Colombia Basin (Caribbean Sea; Schmidt et al. 2004), and Tobago Basin (tropical Atlantic Ocean; Rühlemann et al. 1999) helped to reconstruct the warm pool and its size in the millennial-scale resolution. It became permanent and small as the autumn insolation increased. The intermediate size over the late Holocene was contemporary to declined autumn insolation and intense ENSO activity

and a small warm pool formed during the middle Holocene (i.e., ca. 8–5 cal ka BP). The size of this warm pool increased further, and it became intermediate over the late Holocene as the tropical Atlantic Ocean in the Tobago Basin also reached the threshold value post ca. 5 cal ka BP. The size of our reconstructed warm pool shows positive relationship with orbital-scale variations of autumn (September) insolation. It formed intermittently during the interval of lowest insolation and became permanent and small as the autumn insolation increased. However, the warm pool still increased in size and became intermediate as the autumn insolation declined over the late Holocene. Intermediate size over the late Holocene was also contemporary to the interval of higher ENSO activity. Modern instrumental data of AD 1950–2003 indicate that about two-thirds of the overall warm pools were unrelated to ENSO (Bell and Chelliah 2006; Wang et al. 2006). Intermediate size of this warm pool during the late Holocene, however, was contemporary to the interval of enhanced ENSO activity. It reflects the existence of positive relationship between both the climate forcings, possibly in the millennial scale.

Considering the modern meteorological observations and trajectories of some of the hurricanes in the last decade (e.g., Hurricane Harvey and Hurricane Irma), we worked with a new hypothesis that increase in the size of warm pool over the Holocene would shift position of the subtropical high-pressure cell across the North Atlantic Ocean. This would steer away the tropical storms from the northeastern Mexico toward the central-southern USA. Concentration of Ti in sediments of the GoM reconstructed the Mississippi River discharge into the Atlantic Ocean (Tripsanas et al. 2013), and abundance of *Pinus* pollen in sediments of the Lake Tulane reconstructed the paleo-precipitation in coastal Florida (Grimm et al. 1993; Donders et al. 2011). Figure 3.7 compares proxy records from the northeastern Mexico and central-southern USA. The northeastern Mexico received more precipitation during the early Holocene in an interval of intermittently formed warm pool. During this interval, Ti/Ca values of both the Las Cruces (8.4–8 cal ka BP) and Cieneguilla (ca. 9.5–8.2 cal ka BP) Basins were the highest, suggesting more runoff and more precipitation compared to the middle and late Holocene. During the interval of small and permanent warm pool over the middle Holocene, precipitation in the northeastern Mexico decreased, and the riverine discharge into the GoM increased. This might be indication of higher precipitation in the Mississippi River Basin between ca. 7.6 and 5.2 cal ka BP (e.g., Tripsanas et al. 2013). This warm pool further increased in size over the late Holocene, and discharge of the Mississippi River decreased. Coastal Florida received its highest annual precipitation of the entire Holocene during later part of the Holocene. It is indicated by less Ti concentrations in sediments of the GoM and higher *Pinus* pollen in sediments of the Lake Tulane over the last ca. 5 cal ka (Grimm et al. 1993, 2006; Donders et al. 2011; Tripsanas et al. 2013). The northeastern Mexico received less precipitation, and the regimes of higher precipitation reached different parts across the central-southern USA as the warm pool formed and its size gradually increased. Drier conditions in the northeastern Mexico during intervals of small and intermediate warm pools were possibly due to declined influence of the tropical storms. Most of the tropical storms were steered to landfall in the Mississippi River Basin during the middle

Holocene. It further shifted to coastal Florida during the late Holocene, and rainfall in watersheds of the Mississippi River decreased. This might have occurred due to migration of storm tracks toward the northeast as the warm pool gradually increased in size. This observation, as well as our interpretation of possible millennial-scale relationship between ENSO and the warm pool, needs further attention from paleoclimate community as well as modeling studies.

NAM

Studies of speleothem and paleosols are limited. A complex gypsum–carbonate speleothem from the Cueva de las Espadas at Naica revealed dynamics of the groundwater circulation in the aquifer (Gázquez et al. 2013). The Holocene history was constrained by only one Th/U date of 7.9 ± 0.1 ka. Generally, drier climate and less aquifer recharge led to a fall in the groundwater table. Composition of deuterium was interpreted as influence of more monsoonal summer moisture from the Gulf of Mexico. The paleosol units in general indicate more desertification over the Holocene compared to the late Pleistocene (Cruz-y-Cruz et al. 2015). Carbon isotope in carbonate and fossil bones in some of these paleosols inferred relatively wetter interval during ca. 10.2–7.7 cal ka BP (increase in C_3 vegetation) and marked seasonality and occurrence of mixed C_3/C_4 vegetation at ca. 4.6 cal ka BP (Cruz-y-Cruz et al. 2016). The gray soil (Gleysols) at El Arenoso represents an interval of pedogenic processes during the middle Holocene (ca. 7–4.5 cal ka BP). In the Laguna project, Butzer et al. (2008) studied soil profiles located between both the Sierra Madre Mountains and suggested that these soil geo-archives documented the environmental histories of alternating wet cycles, possibly due to switching between the winter and summer storms. They also proposed that the hypothesis of human activity triggering the erosion in these dry settings should be reevaluated as the alternate intervals of severe drought and heavy rainfall between ca. AD 1050–1200 destroyed the ground cover and led to ecological aridification and enhanced erosion. This happened well before the Spanish miners and settlers came to this region.

Ti/Ca in sedimentary archives from the Santiaguillo and Babicora Basins reconstructed the temporal variations in precipitation of southern and central parts of the NAM region over the Holocene (Roy et al. 2012a, 2014b). Charcoal particles and pollen in sediments of the San Bernardino Ciénega reconstructed the fire activity and intervals of variable precipitation in the northern part of NAM region (Brunelle et al. 2010). Precipitation record of the Babicora Basin was almost contrary to the record of fire activity in surroundings of the San Bernardino Ciénega, and both these records are different from the tendency of proxy precipitation in the Santiaguillo Basin (Fig. 3.8). The Santiaguillo Basin located in southern part of the NAM region experienced some events of higher precipitation during ca. 11.5–9 cal ka BP, and they were associated with higher monsoonal precipitation by Roy et al. (2014b). Fire events were less frequent in the Sierra Madre Occidental Mountain and *Abies*, *Quercus*, and ferns represented the vegetation composition (Ortega-Rosas et al.

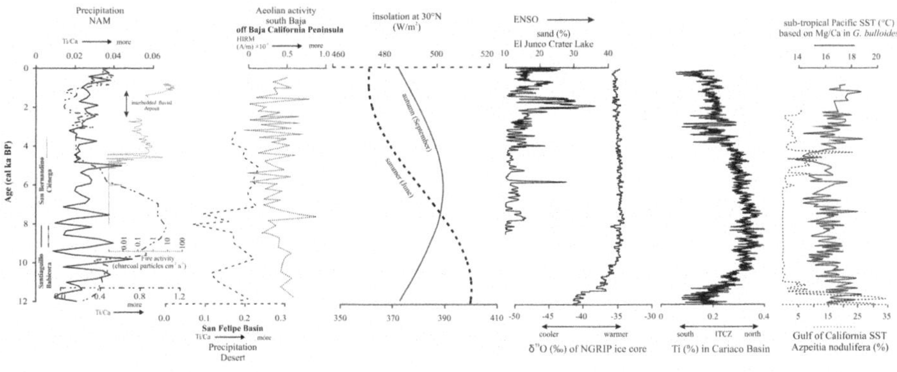

Fig. 3.8 Comparison of the proxy records from the NAM (Santiaguillo Basin, Babicora Basin, and San Bernardino Ciénega), desert (San Felipe Basin), and south Baja (southern Baja California Peninsula) regions with summer (June) and autumn (September) insolations, ENSO activity, North Hemisphere temperature, and mean position of the ITCZ and SSTs of the subtropical Pacific Ocean and Gulf of California

2008a, b, c). Modern pollen rain of this dominant summer-precipitation-fed region contains 10–20% of *Quercus* within the pine forest (Douglas et al. 1993; Ortega-Rosas et al. 2008a, b, c). Presence of significant *Quercus* ($\leq$10–20%) in the fossil pollen record suggests that this region also received more summer precipitation during the earliest Holocene, and minor contribution possibly came from winter rainfall. Based on recent satellite data, Mitchell et al. (2002) suggested that SST in the GoC should reach the threshold values of 25–26 °C for the summer precipitation to commence in the northern Mexico. Summer (June) insolation was higher in this interval. The abundance of tropical diatom *Azpeitia nodulifera* (Barron et al. 2005) and U^{k}_{37} alkenones (McClymont et al. 2012) suggested tropical-like warmer water in the Gulf of California, and the SST possibly went beyond 25 °C. Roy et al. (2015) proposed that strengthening of the NAM system caused more rainfall in a broader region extending from the central-northern Mexico up to the continental interiors of southwestern USA. It possibly reached one of its greatest spatial extents during the Pleistocene–Holocene transition and the earliest Holocene.

Hydroclimates of the central and northern parts of the NAM region, however, were comparable. The interval of more precipitation in the Babicora Basin was contemporary to rare fire activities between ca. 8.1 and 5.3 cal ka BP in the San Bernardino Ciénega (Fig. 3.8). This interval had minimal winter-precipitation taxa suggesting more warm season precipitation. We evaluated it with respect to the hypothesis of Barron et al. (2012). Warmer SST in the subtropical Pacific Ocean compared to GoC would favor the transportation of more precipitation through higher number of tropical cyclones in the subtropical Pacific Ocean. Mg/Ca in *G. bulloides* of the Soledad Basin provides SST of the subtropical Pacific Ocean (Marchitto et al. 2010). The abundance of *Azpeitia nodulifera* is a proxy SST record of the GoC (Barron et al. 2005). SST of GoC did not have the threshold values for the summer rainfall to commence, and the subtropical Pacific Ocean was warmer compared to GoC. More precipitation during the middle Holocene was also contemporary to the interval of ITCZ located at northern latitudes as well as higher values of autumn (September) insolation. The interaction between midlatitude upper-level troughs and northward-moving tropical cyclones in the autumn drops large quantities of precipitation (Jones et al. 2003). Compilation of meteorological data of AD 1992–2005 indicates that the highest numbers of tropical cyclones remnants were formed during the autumn (Ritchie et al. 2011). More frequent tropical cyclones possibly provided more precipitation in the central and northern parts of the NAM region during the middle Holocene. Precipitation reduced and the frequency of fire activity increased post ca. 5.3 cal ka BP. The late Holocene was contemporaneous to an interval of gradually lower autumn insolation and shifting of the ITCZ to southerly latitudes. It was also associated with higher winter-precipitation taxa and more ENSO activity, suggesting less warm season precipitation and more winter rainfall. High-magnitude flooding events during ca. 2.5–1.1 cal ka BP in the San Bernardino Ciénega might be due to more winter storms. The Babicora Basin either did not receive these winter storms or its sediment archive did not preserve these wet events. The GoC became warmer, and differences in SST between the California Current system off the Baja California Peninsula and the GoC decreased. Conditions were favorable for the GoC to contribute more moisture during the warm season, even though it was not sufficient for improving hydrological budgets of the lacustrine basins.

Desert

Stratigraphic exposures in natural profiles, water wells, and archaeological excavations in surroundings of the Lake Chapala suggested high lake stand during the earliest Holocene (>ca. 10.2 cal ka BP) compared to ca. 10.2–7.8 cal ka BP (Davis 2003). Desiccation became more frequent, and rapid growth of the sand dunes occurred by ca. 8.4 cal ka BP, and both continued till TL 2.5 ka (Davis 2003), and the modern sand dunes have been forming at least since the last millennium (ca. 1.1 cal ka BP). Sedimentation in the nearby San Felipe Basin during the Holocene was 4–12 times faster compared to the sediments deposited during late Pleistocene, and the mafic and REE-bearing heavy minerals were transported by more summer precipitation into the basin (Roy et al. 2010). Higher pluvial discharge also caused more erosion of catchment rocks. The reduced vegetation cover (e.g., Lozano-García et al. 2002) might also have affected the hill slope stability, hence erosion in the drainage basin. Holmgren et al. (2006) and Minckley et al. (2011) reported several events of more winter precipitation related to higher ENSO activities. Low rate of sedimentation in this register, however, did not allow identification of these events. Additionally, the summer precipitation and tropical storms controlled the hydrology of this basin. The high-altitude mountain ranges present to the west of this basin obstruct the westerlies, and hence, the winter precipitation was minimal.

Ti/Ca in sedimentary archive of the same basin provided temporal variation in precipitation till 4 cal ka BP (Roy et al. 2010). It identified an interval at ca. 8.5–7.5 cal ka BP that received less precipitation compared to rest of the records (Fig. 3.8). It suggests that the dynamics of precipitation in the desert region was different from the proxy precipitation records from the NAM region. The ENSO activity was weak, and ITCZ was located at northerly latitudes in this interval. Proxy SST records suggested that the subtropical Pacific Ocean was warmer compared to the GoC, and SST in the GoC also did not reach the threshold values required for the summer precipitation to commence. This interval had conditions favorable for transportation of more moisture from the subtropical Pacific Ocean through higher number of tropical storms. Most of these storms possibly did not have the rainfall swaths to bring precipitation to the desert region at ca. 8.5–7.5 cal ka BP, and their expanded influence brought more rainfall to the region after ca. 7.5 cal ka BP.

South Baja

Thermoluminescence (TL), amino-acid epimerization, and radiocarbon chronologies of dune fields in the Purísima–Iray–Magdalena sedimentary basin and the adjacent lagoon reconstructed the arid intervals, wind intensities, and wind directions (Murillo De Nava et al. 1999). Younger dates obtained in older dunes indicate two major intervals of dune emplacements as a result of reactivation and redeposition of the late Pleistocene dune fields. The first one continued during the early Holocene with TL dates of 10.4 ± 1.9 ka and 8.7 ± 1.5 ka. Formation of linear dunes suggests that the northwesterly winds (possibly westerlies) were stronger. The second interval caused formation of mega-barchan dunes during the middle and late Holocene with TL dates of 7.1 ± 0.8 ka, 5.5 ± 0.6 ka, and 3.7 ± 0.8 ka. Different wind directions and intensities characterized this arid interval. Over the middle Holocene, the northwesterly winds decreased in intensity, and dune field orientations were associated with enhanced monsoonal circulation and greater insolation seasonality. Crescentic dunes of the late Holocene (TL 1.5 ± 0.5 ka) were also deposited by northwesterly and westerly winds, and this arid interval was interrupted by major storms (possible hurricanes). Changes in coercivity (HIRM) of sediments deposited in the subtropical Pacific Ocean reflected the strength of eolian activity in the south Baja (Blanchet et al. 2007). Abrupt increase in the eolian activity at ca. 7.5 cal ka BP was contemporary to the event of reduced precipitation in the desert region (Fig. 3.8). Arid conditions in the south Baja and desert regions and more precipitation in the central and northern parts of the NAM region might be useful to reconstruct the geographic expansion of rainfall swaths of the tropical storms. Increase in the frequency and amplitude of eolian activity over the late Holocene was contemporary to the interval of higher ENSO activity.

Conclusions

The drought-prone subtropical Mexico might experience more aridity in the near future due to the anthropogenic global warming. The modern climate of this region is heterogenous with different parts receiving different amounts of summer, autumn, and winter precipitations. The tree-ring records provide climate information of last six centuries, and the geological registers extended this information to the Holocene. The number of geological registers and type of proxies applied to these registers have increased over the last two decades, and the new studies involved multidisciplinary methodologies in order to reconstruct the millennial-scale dynamics of near-surface processes like erosion via eolian and fluvial activities, water column salinity through evaporation, and allochthonous and autochthonous organic productivities during the intervals of global climate change over the Holocene. This chapter presents a state of the art of the available information on vegetation composition and hydrological conditions considering the modern-era meteorological divisions and grouped them into northeastern Mexico, NAM, desert, and south Baja. Heterogeneity in temporal variations in precipitation in each of them was evaluated in terms of previously proposed hypotheses as well as a new hypothesis. All of them were also compared with several climate forcings.

This new hypothesis evaluated hydroclimate of the northeastern Mexico with respect to the size of warm pool in the Atlantic Ocean. Shifts in position of the subtropical high-pressure cell over the Holocene possibly steered the tropical storms to landfall in central-southern USA, and the northeastern Mexico remained drier as the warm pool gradually increased in size. ENSO was the dominant forcing on hydroclimate of the NAM, desert, and south Baja regions. It caused reduction in the total annual precipitation and enhancement in the aridity, even though the amounts of winter precipitation increased in some parts. The subtropical dry forest was established over the late Holocene, and it included species of the *Burseraceae* family, and the northern margin of NAM region received more winter precipitation. Abundance of red mangrove suggested rise in the sea level in south Baja. Our evaluation of hydroclimate of northeast Mexico, as well as the interpretation of possible relationship between ENSO and the warm pool in millennial scale, needs further attention from paleoclimate community as well as modeling studies.

Acknowledgments Data presented in this chapter were generated with financial support from DGAPA-Papiit-UNAM project IN102217, and some data were obtained in previous projects of Conacyt (CB-237579 and CB-83800). The undergraduate and postgraduate students of the Laboratory of Paleoenvironment and Paleoclimate have helped in sample preparation and geochemical analysis. Valuable technical assistances came from Irma Vargas, Alejandra Chavez, Guillermo Vera, and Fernando Ibañez (Facultad de Ingeniería, UNAM). We are thankful to suggestions and comments from the reviewers.

References

Adams DK, Comrie AC (1997) The North American Monsoon. Am Meteorol Soc 23:1453–1464

Aharon P (2003) Meltwater flooding events in the Gulf of Mexico revisited: implications for rapid climate changes during the last deglaciation. Paleoceanography 18:1079

Albert BM (2015) Prehistoric and colonial land-use in the Chihuahuan Desert as inferred from pollen and sedimentary data from Coahuila, Mexico. Quat Int 377:38–51

Amador JA, Alfaro EJ, Lizano OG et al (2006) Atmospheric forcing of the eastern tropical Pacific: a review. Prog Oceanogr 69:101–142

Anderson R, Van Devender TR (1995) Vegetation history and paleoclimates of the coastal lowlands of Sonora, Mexico, pollen records from packrat middens. J Arid Environ 30:295–306

Antinao JL, McDonald E (2013) An enhanced role for the tropical Pacific on the humid Pleistocene-Holocene transition in southwestern North America. Quat Sci Rev 78:319–341

Barron JA, Bukry D, Dean WE (2005) Paleoceanographic history of the Guaymas Basin, Gulf of California, during the past 15,000 years based on diatoms, silicoflagellates, and biogenic sediments. Mar Micropaleontol 56:81–102

Barron JA, Metcalfe SE, Addison JA (2012) Response of the North American monsoon to regional changes in ocean surface temperature. Paleoceanography 27:PA3206

Bell GD, Chelliah M (2006) Leading tropical modes associated with interannual and multidecadal fluctuations in North Atlantic hurricane activity. J Clim 19:590–612

Betancourt J, Devender T Van, Martin P (1990) Packrat middens: the last 40,000 years of biotic change. University of Arizona Press, Tucson

Blanchet CL, Thouveny N, Vidal L et al (2007) Terrigenous input response to glacial/interglacial climatic variations over southern Baja California: a rock magnetic approach. Quat Sci Rev 26:3118–3133

Brunelle A, Minckley T, Blissett S et al (2010) A∼ 8000 year fire history from an Arizona/Sonora borderland ciénega. J Arid Environ 74:475–481

Butzer KW, Abbott JT, Frederick CD et al (2008) Soil-geomorphology and "wet" cycles in the Holocene record of North-Central Mexico. Geomorphology 101:237–277

Caballero M, Peñalba MC, Martínez M et al (2005) A Holocene record from a former coastal lagoon in Bahía Kino, Gulf of California, Mexico. The Holocene 15:1236–1244

Caldwell P (2010) California wintertime precipitation bias in regional and global climate models. J Appl Meteorol Climatol 49:2147–2158

Castiglia PJ, Fawcett PJ (2006) Large Holocene lakes and climate change in the Chihuahuan Desert. Geology 34:113–116

Cavazos T, Rivas D (2004) Variability of extreme precipitation events in Tijuana, Mexico. Clim Res 25:229–243

Cavazos T, Salinas JA, Martínez B et al (2013) Actualización de escenarios de cambio climático para México como parte de los productos de la Quinta Comunicación Nacional. Informe, Instituto Nacional de Ecología y Cambio Climático, México

Cayan DR, Maurer EP, Dettinger MD et al (2007) Climate change scenarios for the California region. Clim Chang 87:21–42

Cerezo-Mota R, Cavazos T, Farfán LM et al (2006) Numerical simulation of heavy precipitation in northern Baja California and Southern California. J Hydrometeorol 7:137–148

Chávez-Lara CM, Roy PD, Caballero M et al (2012) Lacustrine ostracodes from the Chihuahuan Desert of Mexico and inferred Late Quaternary paleoecological conditions. Rev Mex Cienc Geol 29:422–431

Chávez-Lara CM, Roy PD, Pérez L et al (2015) Ostracode and C/N based paleoecological record from Santiaguillo basin of subtropical Mexico over last 27 cal kyr BP. Rev Mex Cienc Geol 32:1–10

Chavez-Lara CM, Holtvoeth J, Roy PD et al (2018) A 27cal ka biomarker-based record of ecosystem changes from lacustrine sediments of the Chihuahua Desert of Mexico. Quaternary Science Review 191: 132–143.

Comrie A, Glenn E (1998) Principal components-based regionalization of precipitation regimes across the Southwest United States and northern Mexico, with an application to monsoon precipitation variability. Clim Res 10:201–215

Conroy JL, Overpeck JT, Cole JE et al (2008) Holocene changes in eastern tropical Pacific climate inferred from a Galápagos lake sediment record. Quat Sci Rev 27:1166–1180

Cook J, Nuccitell D, Green SA et al (2013) Quantifying the consensus on anthropogenic global warming in the scientific literature. Environ Res Lett 8:024024

Corbosiero KL, Dickinson MJ, Bosart LF (2009) The contribution of eastern North Pacific tropical cyclones to the rainfall climatology of the Southwest United States. Mon Weather Rev 137:2415–2435

Cruz-y-Cruz T, Sedov S, Sánchez G et al (2014) Late Pleistocene-Holocene palaeosols in the north of Sonora, Mexico: chronostratigraphy, pedogenesis and implications for environmental history. Eur J Soil Sci 65:455–469

Cruz-y-Cruz T, Sánchez G, Sedov S et al (2015) Spatial variability of Late Pleistocene-Early Holocene soil formation and its relation to early human paleoecology in Northwest Mexico. Quat Int 365:135–149

Cruz-y-Cruz T, Pérez-Crespo VA, Pustovoytov K et al (2016) Paleosol (organic matter and pedogenic carbonates) and paleontological d13C records applied to the paleoecology of late Pleistocene–Holocene in Mexico. Quat Int 418:147–164

Davis LG (2003) Geoarchaeology and geochronology of Pluvial Lake Chapala, Baja California, Mexico. Geoarchaeology 18:205–223

Díaz SC, Touchan R, Swetnam TW (2001) A Tree-Ring reconstruction of past precipitation for Baja California Sur, Mexico. Int J Climatol 21:1007–1019

Donders TH, De Boer HJ, Finsinger W et al (2011) Impact of the Atlantic warm pool on precipitation and temperature in Florida during North Atlantic cold spells. Clim Dyn 36:109–118

Douglas MW, Maddox RA, Howard K et al (1993) The Mexican monsoon. J Clim 6:1665–1677

Farfán LM, Fogel I (2007) Influence of tropical cyclones on humidity patterns over southern Baja California, Mexico. Mon Weather Rev 135:1208–1224

Flower BP, Hastings DW, Hill HW et al (2004) Phasing deglacial warming and Laurentide Ice Sheet meltwater in the Gulf of Mexico. Geology 32:597–600

Friedman I, Smith GI, Gleason JD et al (1992) Stable isotope composition of waters in southeastern California 1. Modern precipitation. J Geophys Res 97:5795–5812

Gázquez F, Calaforra JM, Stoll H et al (2013) Isotope and trace element evolution of the Naica aquifer (Chihuahua, Mexico) over the past 60,000 yr revealed by speleothems. Quat Res 80:510–521

Goldenberg SB, Landsea CW, Mestas-Nuñez AM et al (2001) The recent increase in Atlantic hurricane activity: causes and implications. Science 293:474–479

Grimm EC, Jacobson GL, Watts WA et al (1993) A 50,000-year record of climate oscillations from Florida and its temporal correlation with the Heinrich events. Science 261:198–200

Grimm EC, Watts WA, Jacobson GL Jr et al (2006) Evidence for warm wet Heinrich events in Florida. Quat Sci Rev 25:2197–2211

Haug GH, Hughen KA, Sigman DM et al (2001) Southward migration of the intertropical convergence zone through the Holocene. Science 93:1304–1308

Higgins RW, Shi W (2001) Intercomparison of the principal modes of interannual and intraseasonal variability of the North American Monsoon System. J Clim 14:403–417

Higgins RW, Yao Y, Wang XL (1997) Influence of the North American Monsoon System on the U.S. summer precipitation regime. J Clim 10:2600–2622

Higgins RW, Mo KC, Yao Y et al (1998) Interannual variability of the U.S. summer precipitation regime with emphasis on the southwestern monsoon. J Clim 11:2582–2606

Higgins RW, Chen Y, Douglas AV et al (1999) Interannual variability of the North American warm season precipitation regime. J Clim 12:653–680

Holmgren C, Peñalba M, Rylander KA et al (2003) A 16,000 14C yr BP packrat midden series from the USA–Mexico Borderlands. Quat Res 60:319–329

Holmgren CA, Betancourt JL, Rylander KA (2006) A 36,000-yr vegetation history from the Peloncillo Mountains, southeastern Arizona, USA. Palaeogeogr Palaeoclimatol Palaeoecol 240:405–422

Holmgren CA, Betancourt JL, Rylander KA (2011) Vegetation history along the eastern, desert escarpment of the Sierra San Pedro Mártir, Baja California, Mexico. Quat Res 75:647–657

Holmgren CA, Betancourt JL, Peñalba MC et al (2014) Evidence against a Pleistocene desert refugium in the Lower Colorado River Basin. Riddle B, editor. J Biogeogr 41:1769–1780

Jones SC, Harr PA, Abraham J et al (2003) The extratropical transition of tropical cyclones: forecast challenges, current understanding, and future directions. Weather Forecast 18:1052–1092

Li T, Wang B (2005) A review on the western North Pacific monsoon: synoptic-to-interannual variabilities. Terr Atmos Ocean Sci 16:285

Lozano-García MS, Ortega-Guerrero B, Sosa-Nájera S (2002) Mid-to late-Wisconsin pollen record of San Felipe basin, Baja California. Quat Res 58:84–92

Magaña VO, Quintanar A (1997) On the use of a general circulation model to study regional climate, 2nd UNAM-CRAY Supercomputing on Earth Sciences. Cambridge University Press, Cambridge

Magaña VO, Vázquez JL, Párez JL et al (2003) Impact of El Niño on precipitation in Mexico. Geofis Int 42:313–330

Marchitto TM, Muscheler R, Ortiz JD et al (2010) Dynamical response of the tropical Pacific Ocean to solar forcing during the early Holocene. Science 330:1378–1381

McClymont EL, Ganeshram RS, Pichevin LE et al (2012) Sea-surface temperature records of Termination 1 in the Gulf of California: challenges for seasonal and interannual analogues of tropical Pacific climate change. Paleoceanography 27:PA2202

Mestas-Nuñez AM, Zhang C, Enfield DB (2005) Uncertainties in estimating moisture fluxes over the intra-Americas sea. J Hydrometeorol 6:696–709

Mestas-Nuñez AM, Enfield DB, Zhang C (2007) Water vapor fluxes over the intra-Americas sea: seasonal and interannual variability and associations with rainfall. J Clim 20:1910–1922

Metcalfe SE, Bimpson A, Courtice AJ et al (1997) Climate change at the monsoon / westerly boundary in northern Mexico. J Paleolimnol 17:155–171

Metcalfe S, Say A, Black S (2002) Wet conditions during the last glaciation in the Chihuahuan. Quat Res 101:91–101

Metcalfe SE, Barron JA, Davies SJ (2015) The Holocene history of the North American Monsoon: "known knowns" and "known unknowns" in understanding its spatial and temporal complexity. Quat Sci Rev 120:1–27

Meyer EF (1973) Late Quaternary paleoecology of the Cuatro Cienegas Basin. Ecology 54:982–995

Minckley TA, Brunelle A (2007) Paleohydrology and growth of a desert ciénega. J Arid Environ 69:420–431

Minckley TA, Brunelle A, Blissett S (2011) Holocene sedimentary and environmental history of an in-channel wetland along the ecotone of the Sonoran and Chihuahuan Desert grasslands. Quat Int 235:40–47

Mitchell DL, Ivanova D, Severe N (2002) Gulf of California sea surface temperatures and the North American monsoon: mechanistic implications from observations. J Clim 15:2261–2281

Murillo De Nava JM, Gorsline DS, Goodfriend GA et al (1999) Evidence of Holocene climatic changes from aeolian deposits in Baja California Sur, Mexico. Quat Int 56:141–154

Neelin JD, Langenbrunner B, Meyerson JE et al (2013) California winter precipitation change under global warming in the coupled model intercomparison project phase 5 ensemble. J Clim 26:6238–6256

Nesbitt HW, Young GM (1999) Formation and diagenesis of weathering profiles. J Geol 97:129–147

Nordt L (2003) Late Quaternary fluvial landscape evolution in desert grasslands of northern Chihuahua, Mexico. Bull Geol Soc Am 115:596–606

North Greenland Ice Core Project members (2004) High resolution record of Northern Hemisphere climate extending into the last interglacial period. Nature 431:147–151

Ortega-Guerrero B, Caballero-Miranda M, Lozano-García S et al (1999) Palaeoenvironmental record of the last 70 000 yr in San Felipe Basin, Sonora desert, Mexico. Geofis Int 38:153–163

Ortega-Guerrero B, Schaaf P, Murray A et al (2013) Eolian deposition cycles since AD 500 in Playa San Bartolo lunette dune, Sonora, Mexico: Paleoclimatic implications. Aeolian Res 11:1–13
Ortega-Ramírez JR, Valiente-banuet A, Urrutia-fucugauchi J et al (1998) Paleoclimatic changes during the late Pleistocene – Holocene in Laguna Babícora, near the Chihuahuan Desert, México. Can J Earth Sci 35:1168–1179
Ortega-Ramírez J, Urrutia-Fucugauchi J, Valiente-Banuet A (2000) The Laguna de Babícora basin: a late Quaternary paleolake in northwestern Mexico. AAPG Stud Geol 46:569–580
Ortega-Rosas CI, Peñalba MC, Guiot J (2008a) Holocene altitudinal shifts in vegetation belts and environmental changes in the Sierra Madre Occidental, Northwestern Mexico, based on modern and fossil pollen data. Rev Palaeobot Palynol 151:1–20
Ortega-Rosas CI, Guiot J, Peñalba MC et al (2008b) Biomization and quantitative climate reconstruction techniques in northwestern Mexico-With an application to four Holocene pollen sequences. Glob Planet Chang 61:242–266
Ortega-Rosas CI, Peñalba MC, López-Sáez JA et al (2008c) Retrospectiva del bosque de pino y encino de la Sierra Madre Occidental, Sonora, noroeste de México, hace 1000 años. Acta Bot Mex 83:69–92
Ortega-Rosas CI, Peñalba MC, Guiot J (2016) The Lateglacial interstadial at the southeastern limit of the Sonoran Desert, Mexico: vegetation and climate reconstruction based on pollen sequences from Ciénega San Marcial and comparison with the subrecent record. Boreas 45:773–789
Ortega-Rosas C, Vidal-Solano J, Williamson D et al (2017) Geochemical and magnetic evidence of change from winter to summer rainfall regimes at 9.2calkaBP in northwestern Mexico. Palaeogeogr Palaeoclimatol Palaeoecol 465:64–78
Palacios-Fest M, Carreño AL, Ortega-Ramirez J et al (2002) A paleoenvironmental reconstruction of Laguna Babícora, Chihuahua, Mexico based on ostracode paleoecology and trace element shell chemistry. J Paleolimnol 27:185–206
Pavia EG, Graef F, Reyes J (2006) PDO – ENSO effects in the climate of Mexico. J Clim 19:6433–6438
Peterson LC, Haug GH (2006) Variability in the mean latitude of the Atlantic Intertropical Convergence Zone as recorded by riverine input of sediments to the Cariaco Basin (Venezuela). Palaeogeogr Palaeoclimatol Palaeoecol 234:97–113
Pigati JS, Bright JE, Shanahan TM et al (2009) Late Pleistocene paleohydrology near the boundary of the Sonoran and Chihuahuan Deserts, southeastern Arizona, USA. Quat Sci Rev 28:286–300
Quiroz-Jiménez JD, Roy PD, Beramendi-Orosco LE et al (2017) Orbital-scale droughts in central-northern Mexico during the late Quaternary and comparison with other subtropical and tropical records. Geol J 53:230–242
Rasmussen TL, Thomsen E (2012) Changes in planktic foraminiferal faunas, temperature and salinity in the Gulf stream during the last 30,000 years: influence of meltwater via the Mississippi River. Quat Sci Rev 33:42–54
Rhode D (2002) Early Holocene juniper woodland and chaparral taxa in the Central Baja California Peninsula, Mexico. Quat Res 57:102–108
Ritchie EA, Wood KM, Gutzler DS et al (2011) The influence of eastern pacific tropical cyclone remnants on the southwestern United States. Mon Weather Rev 139:192–210
Romero-Mayén VA, Carreño AL (2016) Holocene paleoenvironmental reconstruction of Laguna Babícora, Chihuahua, Northern Mexico, based on Ostracode Ecology and Shell Chemistry. Paleontología Mexicana 5:111–122
Roy PD, Caballero M, Lozano R et al (2010) Geochemical record of Late Quaternary paleoclimate from lacustrine sediments of paleo-lake San Felipe, western Sonora Desert, Mexico. J S Am Earth Sci 29:586–596
Roy PD, Jonathan MP, Pérez-Cruz LL et al (2012a) A millennial-scale late Pleistocene-Holocene palaeoclimatic record from the western Chihuahua desert, Mexico. Boreas 41:707–718
Roy PD, Caballero M, Lozano S et al (2012b) Provenance of sediments deposited at paleolake San Felipe, western Sonora Desert: implications to regimes of summer and winter precipitation during last 50 cal kyr BP. J Arid Environ 81:47–58

Roy PD, Rivero-Navarette A, Lopez-Balbiaux N et al (2013a) A record of Holocene summer-season palaeohydrological changes from the southern margin of Chihuahua Desert (Mexico) and possible forcings. The Holocene 23:1105–1114

Roy PD, Quiroz-Jiménez JD, Pérez-Cruz LL et al (2013b) Late Quaternary paleohydrological conditions in the drylands of northern Mexico: a summer precipitation proxy record of the last 80 cal ka BP. Quat Sci Rev 78:342–354

Roy PD, Charles-Polo MP, López-Balbiaux N et al (2014a) Last glacial hydrological variations at the southern margin of sub-tropical North America and a regional comparison. J Quat Sci 29:495–505

Roy PD, Quiroz-Jiménez JD, Chávez-Lara CM et al (2014b) Humid Pleistocene-Holocene transition and early Holocene in sub-tropical northern Mexico and possible Gulf of California forcing. Boreas 43:577–587

Roy PD, Chávez-Lara CM, Beramendi-Orosco LE et al (2015) Paleohydrology of the Santiaguillo Basin (Mexico) since late last glacial and climate variation in southern part of western subtropical North America. Quat Res 84:335–347

Roy PD, Rivero-Navarrete A, Sánchez-Zavala JL et al (2016) Atlantic Ocean modulated hydroclimate of the subtropical northeastern Mexico since the last glacial maximum and comparison with the southern US. Earth Planet Sci Lett 434:141–150

Roy PD, Vera-Vera G, Curtis JH et al (2019) Response of surface processes in drought-prone northeast Mexico to climate forcings between the latest Pleistocene and middle Holocene. Earth Surf Process Landf 44:2211. https://doi.org/10.1002/esp.4645

Rühlemann C, Mulitza S, Müller PJ et al (1999) Warming of the tropical Atlantic Ocean and slowdown of thermohaline circulation during the last deglaciation. Nature 402:511–514

Sankey JT, Van Devender TR, Clark WH (2001) Late Holocene plants, Catavina, Baja California. Southwest Nat 46:1–7

Schmidt MW, Spero HJ, Lea DW (2004) Links between salinity variation in the Caribbean and North Atlantic thermohaline circulation. Nature 428:160–163

Seager R, Ting M, Li C et al (2013) Projections of declining surface-water availability for the southwestern United States. Nat Clim Chang 3:482–486

Seager R, Goddard L, Nakamura J et al (2014) Dynamical causes of the 2010/11 Texas–northern Mexico drought. J Hydrometeorol 15:39–68

Sirkin L, Pedrín-Avíles S, Padilla-Arredondo G et al (1994) Holocene vegetation and climate of Baja California Sur, México. Revista Mexicana de Ciencias Geológicas 11:79–86

Stahle DW, Diaz JV, Burnette DJ et al (2011) Major Mesoamerican droughts of the past millennium. Geophys Res Lett 38:L05703

Stahle DW, Cook ER, Burnette DJ et al (2016) The Mexican drought atlas: tree-ring reconstructions of the soil moisture balance during the late pre-hispanic, colonial, and modern eras. Quat Sci Rev 149:34–60

Stensrud DJ, Gall RL, Mullen SL et al (1995) Model climatology of the Mexican monsoon. J Clim 8:1775–1794

Talbot MR, Johannessen T (1992) A high resolution palaeoclimatic record for the last 27,500 years in tropical West Africa from the carbon and nitrogen isotopic composition of lacustrine organic matter. Earth Planet Sci Lett 110:23–37

Tripsanas EK, Karageorgis AP, Panagiotopoulos IP et al (2013) Paleoenvironmental and paleoclimatic implications of enhanced Holocene discharge from the Mississippi River based on the sedimentology and geochemistry of a deep core (JPC-26) from the Gulf of Mexico. PALAIOS 28:623–636

Urrutia-Fucugauchi J, Ortega-Ramirez J, Cruz-Gatica R (1997) Rock-magnetic study of late Pleistocene-Holocene sediments from the Babícora lacustrine basin, Chihuahua, northern Mexico. Geofis Int 36:77–86

Van Devender TR, Burgess TL, Piper JC et al (1994) Paleoclimatic implications of holocene plant remains from the sierra bacha, Sonora, Mexico. Quat Res 41:99–108

Van Devender TR, Reina GAL, Peñalba G et al (2003) The ciénega de Camilo: a threatened habitat in the Sierra Madre Occidental of Eastern Sonora, Mexico. Madrono 50:187–195

Villanueva-Diaz J, Stahle DW, Luckman BH et al (2007) Winter-spring precipitation reconstructions from tree rings for Northeast Mexico. Clim Chang 83:117–131
Wang C, Enfield DB, Lee SK et al (2006) Influences of the Atlantic warm pool on western hemisphere summer rainfall and Atlantic hurricanes. J Clim 19:3011–3028
Wang C, Liu H, Lee SK et al (2011) Impact of the Atlantic warm pool on United States landfalling hurricanes. Geophys Res Lett 38:L19702
Xu J, Shuttleworth WJ, Gao X et al (2004) Soil moisture-precipitation feedback on the North American monsoon system in the MM5-OSU model. Q J R Meteorol Soc 130:2873–2890
Ziegler M, Nürnberg D, Karas C et al (2008) Persistent summer expansion of the Atlantic warm pool during glacial abrupt cold events. Nat Geosci 1:601–605

Chapter 4
The Environment of Ancient Cloud Forests in the Mexican Pacific

Blanca L. Figueroa-Rangel, Miguel Olvera-Vargas, and Ana P. Del Castillo-Batista

Abstract Two ancient cloud forests in the Mexican Pacific were reconstructed to reveal signals of past ecosystem responses to climate anomalies during the Late Holocene, Little Ice Age (~AD 1350–1850) and Medieval Climate Anomaly (~AD 800–1350). Applying palaeoecological techniques, two organic sediment cores were extracted from forest hollows: one from the cloud forest of the *Acer Forest State Park* (AFSP) and one from the *Sierra de Manantlan Biosphere Reserve* (SMBR), both in the state of Jalisco, Mexico.

Fossil pollen was used as proxy for tree, herb and fern abundances, whilst carbonate and organic matter content, magnetic susceptibility and geochemical elemental composition were used as proxies for soil environment. Generalized additive models (GAMs) and tree regression models were used to discern the best soil–environmental variable that could explain tree, herb and fern abundances. Results showed a great diversity in taxa composition (110 taxa) with important spatial and temporal heterogeneity between the two forests. Carbonates and Mg were the most important variables explaining these plant assemblages. The driest period for the Little Ice Age was detected from ~AD 1650 to 1730 in AFSP and from AD 1528 to 1831 in SMBR; trees experienced a remarkable decreased, whilst herbs increased. For the warmest period of the MCA (AD 800–1200), herbs and ferns expanded, whilst trees contracted. The late MCA (AD 1200–1350) was acknowledged as a wet event with the highest expansion in trees.

Keywords Ancient cloud forest · Fossil pollen · Late Holocene · Mexican Pacific · Geochemical elemental

B. L. Figueroa-Rangel (✉) · M. Olvera-Vargas · A. P. Del Castillo-Batista
Departamento de Ecología y Recursos Naturales, Centro Universitario de la Costa Sur, Universidad de Guadalajara, Autlán de Navarro, Jalisco, Mexico
e-mail: bfrangel@cucsur.udg.mx; molvera@cucsur.udg.mx; ana.delcastillo@academicos.udg.mx

N. Torrescano-Valle et al. (eds.), *The Holocene and Anthropocene Environmental History of Mexico*, https://doi.org/10.1007/978-3-030-31719-5_4

The Mexican Pacific

The Mexican Pacific (MP) is a territory characterized by high tectonic and volcanic activity, frequent tropical cyclone disturbances and significant sea-level changes. It is also a region of frequent earthquakes and tsunamis where altitudinal gradient is sharp from the coast towards the Sierra Madre Occidental. These are primordial drivers of the vast spatial and temporal environmental heterogeneity where distinct vegetation types evolved, mainly shaped by a series of climate change events and anthropogenic activities during the Holocene.

In terms of biogeography, the MP belongs to the Mexican Pacific Coast province (Morrone 2001) encompassing the states of Sinaloa, Nayarit, Jalisco, Colima, Michoacán, Guerrero, Oaxaca and Chiapas. The most characteristic vegetation types are humid and dry forests, savannas and palm forests (Dinerstein et al. 1995). According to parsimony analysis of endemicity based on plants, insects and birds, this province is closely related to the Sierra Madre del Sur, Balsas Depression and the Trans-Mexican Volcanic Belt (Morrone et al. 1999).

The Past Environments in the Mexican Pacific

Palaeoclimatic and palaeoecological records extracted from sediments in lakes, lagoons and forest hollows in the MP provide signals of past ecosystem responses to environmental changes (Lozano-Garcia et al. 2015). In particular, for the Holocene, different proxies have revealed striking information with key baselines to foresee future actions, either to enhance ecosystems conservation or to prevent human risks triggered by extreme climate change events. Taken the Mexican Pacific Coast province as a geographic framework, a number of palaeoecological and palaeoclimatic studies have been undertaken using forest hollows (Figueroa-Rangel et al. 2010; Del Castillo-Batista et al. 2016), in coastal lagoons (Bianchette 2014; Figueroa-Rangel et al. 2016) and in several lakes (Alcántara et al. 2010; Ortega et al. 2010; Correa-Metrio et al. 2012; Torres-Rodriguez et al. 2012; Caballero-Rodríguez et al. 2018). In some of the previous localities for the mid-Holocene (~5000 years BP), there are evidences of climatic oscillations associated to the Little Ice Age (LIA) (AD 1350–1850) and to the Medieval Climate Anomaly (MCA) (AD 800–1200). Depending on the ecosystem under reconstruction, as well as on the sample resolution, the MCA presented a dry interval at the beginning and a small humid interval at the end. However, in other records, the MCA was humid along the whole period, yet there was a coincidence for the dry environment reported for the LIA (Lozano-Garcia et al. 2015).

The Present Cloud Forest

Cloud forest is a community included in the humid forest ecosystems of Mexico. It is found over steep slopes around 1000–3000 m asl where moist winds come from the sea (Villaseñor 2010), but it is better represented between 1000 and 1500 m asl (Challenger, A., y Soberón 2008), in places with high humidity due to cloud condensation. The structure and floristic composition in cloud forests is an intricate combination of plant strata composed by trees, shrubs, herbs and ferns forming heterogeneous spatial patterns (Meave et al. 1992). Due to this complexity, their classification is ambiguous (Gual-Díaz and González-Medrano 2014). However, one particular feature is their location in altitudes above tropical lowlands and below temperate ecosystems such as *Pinus* forest, *Quercus* forest and *Pinus-Quercus* forest (Villaseñor 2010). In terms of floristic richness, it contains 6790 species distributed in 1625 genera and 238 families; this figure represents 82% of the families, 52% of the genera and 10% of the total number of the flora in Mexico (Villaseñor 2003, 2004). Regarding biogeography, it contains species from Holarctic and Neotropical origin, with few exceptions, most of them are evergreen (Rzedowski 2006). The understory includes trees, shrubs and ferns fundamentally of Tropical affinity (Challenger and Soberón 2008).

The cloud forest has a restricted and fragmented distribution in Mexico; it grows in deep and shallow soils with high organic matter content, humid along the whole year, pH from 4 to 6 with sandy and clayey texture (Rzedowski 2006). Geographical distribution pertains 20 of the 32 states embracing 180,000 km^2. Jalisco is the fifth state with the highest number of species (2802) (Villaseñor 2010). Most recurrent tree genera according to recent studies are *Ardisia*, *Carpinus*, *Cordia*, *Inga*, *Matudea*, *Picramnia*, *Piper*, *Quercus*, *Styrax*, *Symplococarpon*, *Fraxinus*, *Hedyosmum*, *Ilex*, *Juglans*, *Magnolia*, *Oreopanax*, *Persea*, *Prunus*, *Rondeletia*, *Symplocos*, *Triumfetta*, *Trophis*, *Xylosma*, *Dendropanax*, *Cinnamomum*, *Clethra* and *Zinowiewia* (Morales-Arias et al. 2018; Sánchez-Rodriguez et al. 2018).

The Ancient Cloud Forests

The present-day spatial distribution of cloud forests in Mexico and particularly in the Mexican Pacific is rather well documented (Catalán-Heverástico et al. 2003; Mejía-Domínguez 2006; Santiago-Pérez et al. 2009; Felson and Lane 2010; Sotelo et al. 2010; Morales-Arias et al. 2018; Sánchez-Rodriguez et al. 2018). However, temporal floristic changes have been barely reported, except for three palaeoecological studies (Figueroa-Rangel et al. 2010, 2012; Del Castillo-Batista et al. 2016) on centurial scales and one study (Sánchez-Rodriguez et al. 2018) using permanent plots in 12-year interval. Efforts to compare different sites are largely scarce, as well as the response of cloud forest vegetation to their past environment, except the study carried out by Ponce-Reyes et al. (2012), who portrayed the distribution of cloud forests in Mexico according to their vulnerability to climate change.

In this chapter, in order to evaluate temporal changes in floristic composition using fossil pollen as a proxy, as well as vegetation response to environmental changes using sediment geochemical composition as a proxy, two ancient forests, spatially close to each other (~50 km), were studied in the state of Jalisco (Fig. 4.1). Records of both forests have been previously published (Figueroa-Rangel et al. 2010; Del Castillo-Batista et al. 2016).

The first, *Acer Forest State Park* (hereafter referred as AFSP), is a remnant cloud forest located at 104° 49′W, 20°23′N in Talpa de Allende, Jalisco, at 1800 m asl. It belongs to a protected area under the denomination of a State Park since 2016 (Decreto 001/2016 2016) (Fig. 4.1). Predominant soils are Regosol, Cambisol and Leptosol. The climate is temperate subhumid with mean annual temperature of 21 °C and mean annual precipitation of 1003 mm (Del Castillo-Batista et al. 2016). Some of the tree species include *Abies jaliscana, Acer bizanyedii, Carpinus tropicalis, Clethra flagrans, Cleyera integrifolia, Clusia salvinii, Conostegia volcanalis, Cornus disciflora, Dendropanax arboreus, Ilex brandegeana, Ilex dugesii, Ostrya virginiana, Persea hintonii, Pinus* sp., *Podocarpus matudae, Quercus salicifolia, Symplocos citrea, Ternstroemia lineata, Cyathea costaricensis* and *Zinowiewia concinna*. Other plant components include ferns, mosses, lichens, orchids and bromeliads (Del Castillo-Batista et al. 2016). Main human activities carried out in the study area are agriculture, livestock, mining and logging (Vargas-Rodriguez et al. 2010). Pre-Hispanic populations of Nahuatl origin have been settled in the study area since 1000 years BC (Mountjoy 1987).

The second cloud forest, *Sierra de Manantlan Biosphere Reserve* (which will be referred as SMBR), is located at 104°16′W, 19°35′N in the Sierra de Manantlan Biosphere Reserve in Jalisco (Fig. 4.1). Climate is a temperate subhumid with temperatures from 8 to 25° and precipitation from 1000 to 1700 mm (Vázquez-García et al. 1995). This is a highly diverse forest with around 483 species (Cuevas-Guzmán et al. 2004). Tree species include *Carpinus tropicalis, Clethra vicentina, Cornus disciflora, Ilex brandegeana, Magnolia iltisiana, Ostrya virginiana, Persea hintonii, Pinus douglasiana, Quercus xalapensis, Quercus candicans, Quercus castanea, Quercus xalapensis, Tilia americana* var. *mexicana* and *Zinowiewia concinna*. Non-arboreal taxa encompass herbs, epiphytic ferns, climbers, mosses, lichens, orchids and bromeliads (Figueroa-Rangel et al. 2010). From 1900 to 1987, the area was extensively logged, and since 1987 to present, it is a protected area under the scheme of a Biosphere Reserve (SEMARNAP 2000). Pre-Hispanic tools from the Autlan, Mylpa and Cofradia ceramic complexes were found in archaeological sites near to the cloud forest dating from AD 1200 to 1500 (Kelly 1945).

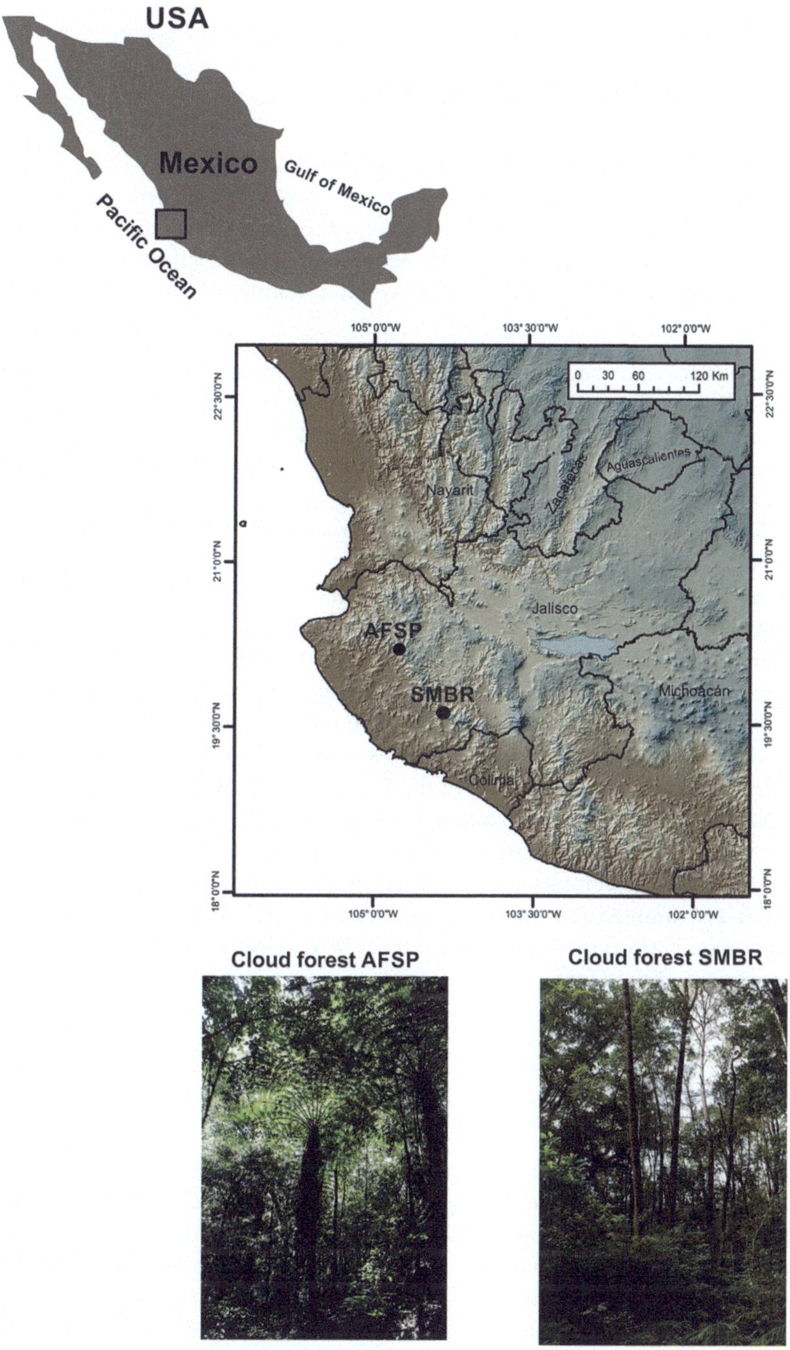

Fig. 4.1 Map of the study area with the location of the two cloud forests of the Mexican Pacific. AFSP, *Acer* Forest State Park; SMBR, Sierra de Manantlan Biosphere Reserve

Methods

Field Sampling and Laboratory Analyses

Two sedimentary cores were extracted from forest hollows in each of the two cloud forests previously described; a 37 cm long core was extracted from the AFSP site and one 96 cm long core from the SMBR site. Both cores were qualitatively assessed in the field and then transported to the laboratory to be sectioned every 1 cm in AFSP and every 2 cm in SMBR, in order to obtain samples to carry out the following analyses:

1. For pollen fossil extraction, a standard acetolysis method was used (Bennett and Willis 2001); to reach a statistically significant sample, at least 400 pollen grains and spores were counted per depth (Maher 1972). The sequence was analysed every 1 cm with a total of 37 samples for AFSP and every 2 cm with a total of 48 samples for SMBR. Fossil pollen identification was undertaken using the Mexican Reference Collection of the Global Pollen Project (Figueroa-Rangel 2018).
2. Magnetic susceptibility was measured using a Bartington MS2 device (Bartington Instruments, Witney, Oxford, UK) to determine the presence of iron-bearing minerals within the sediments (Thompson and Oldfield 1986).
3. Loss-on-ignition (LOI) analysis was undertaken in each core to determine the percentage of organic matter and carbonate content (Dean 1974).
4. Geochemical elemental composition was estimated by X-ray fluorescence with a *Thermo Scientific Niton XL3t* in the Environmental Chemistry Laboratory of the National Autonomous University of Mexico for AFSP. For SMBR, a modified technique of the acid digestion method (Bengtsson and Enell 1986) was carried out. Soluble concentrations of different elements were determined using inductively coupled plasma mass spectrometer (ICP-MS) with a *Perkin Elmer spectrometer* (Perkin Elmer, Waltham, MA, USA) in the Earth Sciences Department of the University of Oxford.

Chronology

Organic samples (2 g) determined the chronology using accelerator mass spectrometry (AMS) radiocarbon dating; for AFSP, three samples were dating in Beta Analytic Laboratory in Florida, USA; for SMBR, six samples were dating in two laboratories, Poznan, Poland, and East Kilbride, UK. Radiocarbon dates were calibrated with the curve IntCal13 for the northern hemisphere (Reimer et al. 2004). The age–depth relationship was modelled using linear interpolation (Bennett 1994).

Statistical Analyses

Pollen counts were converted into percentage values and then plotted in a pollen percentage diagram. Pollen sum included trees, herbs and ferns. Geochemical elemental composition, magnetic susceptibility, carbonates and organic matter values were also plotted in a similar diagram. All plots were designed using psimpoll 4.25 and pscomb 1.03 software (Bennett 2005).

Multiple regression analyses were applied using soil-related variables (geochemical elemental composition, organic matter, carbonates and magnetic susceptibility) as explanatory variables and the sum of trees, sum of herbs and sum of ferns as single response variables. All explanatory and response variables were log transformed prior to the analysis to correct for non-normal distributions and to minimize the potential effect of extreme values.

We first look at all the correlations amongst the whole set of environmental variables to avoid collinearity; subsequently, we applied tree regression models to see whether complex interactions between the explanatory variables were indicated (Crawley 2005). Then we applied generalized additive models (GAMs). GAMs allow the use of non-normally distributed response variables to be fitted with parametric and nonparametric smoothing terms. Therefore, both linear and highly non-linear relationships between the response and the explanatory variables were modelled (Birks 2012). All analyses were processed using R software v. 3.4.3 with the mgcv library (R Development Core Team 2017).

Results

Stratigraphy and Chronology

Both cores were almost entirely composed of rich organic material with small amounts of sand grains and dark-brown colour. Radiocarbon dating of the sedimentary sequences revealed that, for AFSP, 1 cm of sediment accumulated every 19 years, starting approximately from AD 1230. For SMBR, the sequence started at around AD 674; it accumulated 1 cm of sediment every 16 years, resulting in a sampling interval of 32 years (Table 4.1) (results previously published elsewhere; (Figueroa-Rangel et al. 2010; Del Castillo-Batista et al. 2016)).

Plant Assemblages of the Ancient Cloud Forest

In total, 110 fossil pollen taxa were identified in the two cloud forests of the MP; 49 taxa corresponded to trees, 38 to herbs and 23 to ferns. Thirty-two, out of the 49 taxa, were shared between the two forests; 18 corresponded to trees, 10 to herbs and

Table 4.1 Radiocarbon dates (by accelerator mass spectrometry, AMS) for the two cloud forests

Cloud forest	Depth (cm)	Laboratory code	^{14}C age (BP)	Calibrated age (cal yr BP) (2σ)
Acer Forest State Park	3	Beta-367699	102.4 ± 0.3pMC	65 ± 8
	15	Beta-343632	230 ± 30	290 ± 20
	37	Beta-339534	720 ± 30	675 ± 15
Sierra de Manantlan	16	SUERC-5872	207 ± 24	162 ± 10
	40	SUERC-5873	586 ± 21	615 ± 20
	60	SUERC-5874	743 ± 24	678 ± 10
	75	SUERC-5877	925 ± 21	879 ± 30
	82	Poz-10750	1100 ± 30	1040 ± 10
	96	Poz-4747	1635 ± 35	1540 ± 30

Source: Figueroa-Rangel et al. (2010), Del Castillo-Batista et al. (2016)

only 4 to ferns. Fifty taxa were exclusive to AFSP and 28 to SMBR (Table 4.2). Forty-six, out of 49 trees, were identified to genus level; for herbs and ferns, around 50% were identified to genus level as well (Table 4.2). Life-form distribution varied along time as follows:

Acer Forest State Park

Tree pollen percentages in AFSP varied from 22% to 70% with a mean of 52%; herbs presented a mean of 23.6%, ranging from 8% to 50%; ferns' mean was 14%, fluctuating from 2.6% to 28% (Fig. 4.2a). The highest decrease in tree pollen taxa (22%) occurred along the years enclosing the LIA (AD 1350–1850). However, at AD 1250, tree pollen also declined up to 37%, and ferns declined at the end of the LIA. This lowering was also evident for herbs, but their highest drop was concomitant with the biggest increase in pollen trees around AD 1610. For ferns, lower values were present at ~AD 1761 and two high values at AD 1252 and AD 1319, respectively (Fig. 4.2a).

Sierra de Manantlan Biosphere Reserve

In SMBR, pollen of trees oscillated from 35% to 82% with a mean of 60%; herbs from 0% to 16%, with a mean of 4%; and ferns from 12% to 52% with a mean of 27% (Fig. 4.3a).

Compared to AFSP, tree pollen percentages in SMBR fluctuated more along the last ~1300 years of the sequence; however, the highest drop in tree pollen occurred around AD 1121, a time period within the MCA, associated to the highest rise in ferns and following the second highest peak in herb pollen percentages (Fig. 4.3a).

Table 4.2 Fossil pollen taxa found in the two cloud forests of the Mexican Pacific

Trees	Herbs	Ferns
Abies[a, b]	Acanthaceae[b]	Pteridaceae[b]
Acer[b]	Amaranthaceae[b]	*Adiantum*[a]
Alnus[a, b]	Araceae[b]	*Anogramma*[a]
Arbutus[a, b]	*Arisaema*[b]	Aspleniaceae[b]
Celastraceae[a]	Asteraceae[b]	*Blechnum*[b]
Celastrus[b]	*Begonia*[a, b]	*Botrychium*[a]
Cestrum[a]	Boraginaceae[b]	Bromeliaceae[b]
Clethra[a, b]	Caryophyllaceae[a, b]	*Cheilanthes*[a]
Cleyera[a, b]	Chenopodiaceae[a, b]	Dennstaedtiaceae[b]
Clusia[b]	Commelinaceae[b]	Dryopteridaceae[b]
Conostegia[b]	Convolvulaceae[b]	*Dryopteris*[a]
Cornus[b]	*Crotalaria*[a]	*Equisetum*[a, b]
Cyathea[b]	Cucurbitaceae[b]	Grammitidaceae[b]
Dendropanax[a, b]	*Cuphea*[b]	*Huperzia*[b]
Fabaceae[b]	Cyperaceae[a, b]	Hymenophyllaceae[a, b]
Salicaceae[a]	*Dalea*[a]	Ophioglossaceae[b]
Fraxinus[a, b]	Euphorbiaceae[b]	*Osmunda*[b]
Fuchsia[a]	Juncaceae[a]	Polypodiaceae[a, b]
Hedyosmum[a]	Lamiaceae[b]	*Pteridium*[a]
Ilex[a, b]	Lentibulariaceae[b]	*Pteris*[a]
Juglans[a, b]	Liliaceae[a, b]	*Selaginella*[b]
Juniperus[b]	Melastomataceae[b]	*Sphagnum*[b]
Lauraceae[a, b]	*Passiflora*[b]	Thelypteridaceae[a, b]
Magnolia[a, b]	*Peperomia*[b]	
Melastomataceae[a]	*Phaseolus*[b]	
Meliosma[a, b]	*Piper*[b]	
Mimosa[a]	*Plantago*[a, b]	
Moraceae[b]	Poaceae[a, b]	
Ostrya-Carpinus[a, b]	Polygalaceae[a]	
Parathesis[a, b]	*Polygonum*[a, b]	
Pinus[a, b]	*Reseda*[b]	
Podocarpus[b]	*Salvia*[a]	
Prunus[a]	Solanaceae[a, b]	
Quercus[a, b]	Umbelliferae[a]	
Rhamnus[a]	Urticaceae[b]	
Rondeletia[a]	*Utricularia*[b]	
Rubus[a, b]	*Zea*[a, b]	
Salix[a]	Zingiberaceae[b]	
Saurauia[a]		
Sebastiana[a]		
Styrax[b]		
Symplococarpon[b]		

(continued)

Table 4.2 (continued)

Trees	Herbs	Ferns
Ternstroemia[b]		
Tilia[a, b]		
Tournefortia[a]		
Trophis[a]		
Turpinia[b]		
Xylosma[b]		
Zinowiewia[b]		

[a]Sierra de Manantlan Biosphere Reserve
[b]*Acer* Forest State

Remarkably, tree pollen showed a consistent decreased at the beginning and at the end of the LIA, with an important descend following the Spaniards' arrival (after AD 1524). Tree pollen thrived better in the boundary between the MCA and the LIA (AD 1200–1350), reaching consistent high percentages with small fluctuations (Fig. 4.3a).

Herb pollen followed a similar intermittent pattern than trees in terms of drops and peaks. There was also an apparent drop at the beginning and at the end of both the LIA and the MCA; high values ascended in the MCA, specifically at AD 928 and AD 1011. For ferns, MCA enclosed the highest percentages as well (Fig. 4.3a).

The Soil Environment of the Ancient Cloud Forest

Soil environment in both forests presented dissimilar patterns according to the proxy used. Organic matter (OM) varied from 3% to 19.18% in AFSP (Fig. 4.2b) and from 0.63% to 37.5% in SMBR (Fig. 4.3b). Lower values in AFSP were present before the LIA, from AD 1230 to around AD 1300; the highest peak occurred in present times (~AD 2012) (Fig. 4.2b). For the SMBR forest, low values involved the MCA, but the major decline was evident from AD 1145 to 1270; the highest peak was present at AD 1802, almost at the end of the LIA (Fig. 4.3b).

Carbonates (Carb) were very low in SMBR (0.012–0.07%) and slightly higher in AFSP (0.36–3.42%). In AFSP, minimum (0.36%) and maximum (3.42%) values were present during the LIA, the highest at AD 1775 and a second-high peak (2.62%) at AD 1319. A rather similar arrangement was apparent in SMBR, with the highest peak at AD 1774 and a second increased from AD 1227 to 1257 (Figs. 4.2b and 4.3b).

Magnetic susceptibility (MS) shifted from 103 to 980 A/m in AFSP, whilst it was persistently constant in SMBR with values from 686.6 to 741.7 A/m. For AFSP, two of the lowest values occurred around the beginning (AD 1386) and at the end (AD 1816) of the LIA; the highest (AD 1761) peak was also inside this climate anomaly.

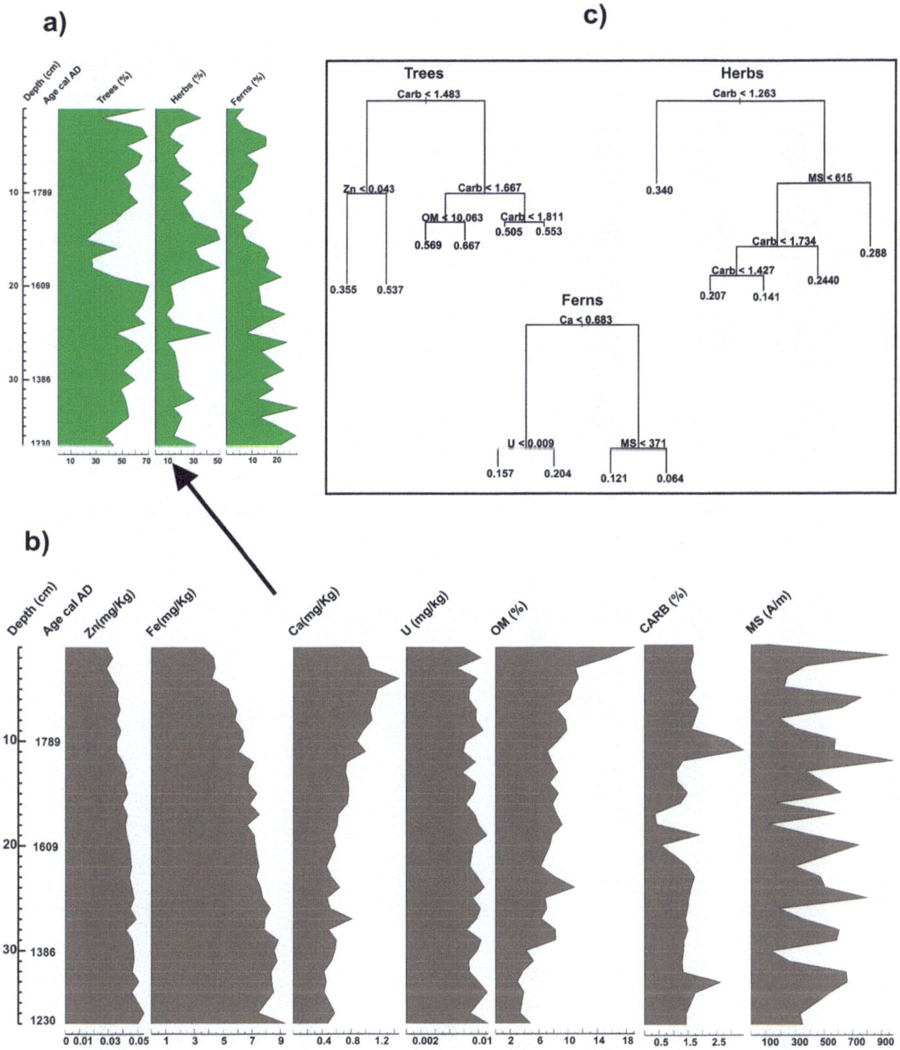

Fig. 4.2 Plant assemblages and soil–environmental variables of the cloud forest in the Acer Forest State Park. (**a**) Diagrams of pollen percentages for trees, herbs and ferns along time; (**b**) diagrams of soil–environmental variables including geochemical elements, Zn, zinc; Fe, iron; Ca, calcium; U, uranium; OM, organic matter; Carb, carbonates and MS, magnetic susceptibility in amperes per meter (Am/m). (**c**) Tree regression model diagrams for trees, herbs and ferns as response variables

For SMBR, the highest value was present at the end of the LIA around AD 1831, and a similar value arose at AD 537 (Figs. 4.2b and 4.3b).

Correlation amongst the whole set of soil variables showed a number of highly correlated geochemical elements; therefore, those highly correlated variables were

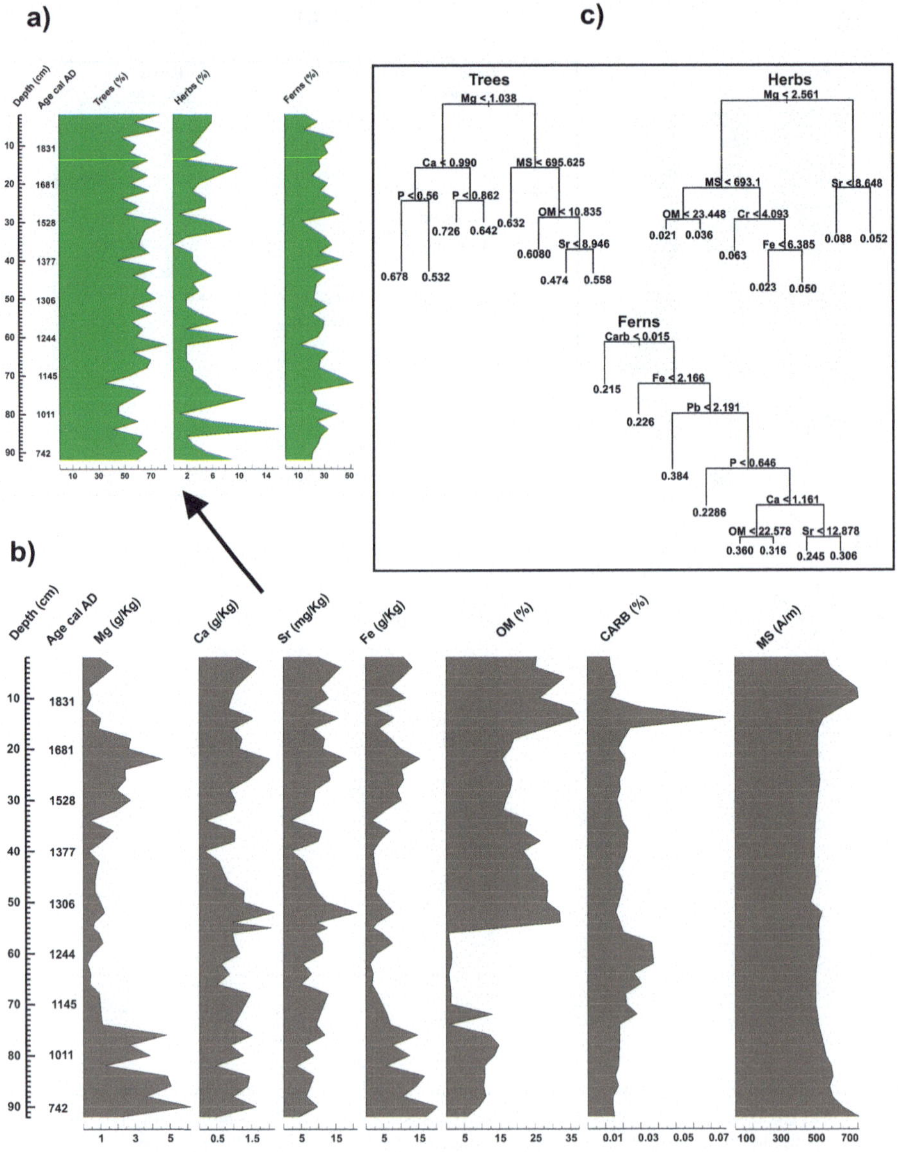

Fig. 4.3 Plant assemblages and soil–environmental variables of the cloud forest in the Sierra de Manantlan Biosphere Reserve. (**a**) Diagrams of pollen percentages for trees, herbs and ferns along time; (**b**) diagrams of soil–environmental variables including geochemical elements, Mg, magnesium; Ca, calcium; Sr, strontium; Fe, Iron; OM, organic matter; Carb, carbonates and MS, magnetic susceptibility in amperes per meter (Am/m). (**c**) Tree regression model diagrams for trees, herbs and ferns as response variables

removed. Hence, for AFSP, only Zn, Fe, Ca and U were considered for further analyses; for SMBR, they were Mg, Fe, Ca and Sr.

In AFSP, Zn values were constant along the LIA, but minimum and maximum concentrations of this element were evident after and before the LIA. Fe increased before the LIA and continued high along this anomaly; nonetheless, it descended slightly in present times. Ca behaved opposite to Fe, whilst U varied in a constant fashion along the last 720 years, with a highest peak at AD 1230 (Fig. 4.2b).

In SMBR, Fe and Mg highest values were present during years of both the LIA and the MCA; however, both elements were somewhat higher in the MCA. Opposed to this previous distribution, Ca and Sr were marginally higher in the interval between the MCA and the LIA. Conversely, some peaks arose in both anomalies; minimum values for both elements were constant at the beginning and at the end of the MCA and the LIA (Fig. 4.3b).

Plant Assemblages and Soil–Environment Relationship

Tree regression models indicated that, regarding AFSP, Carb was the main factor explaining sum of trees. However, when carbonate was <1.48%, Zn interacted with Carb (Fig. 4.2c). This factor was also important explaining sum of herbs; but at values >1.26%, magnetic susceptibility (MS) interacts as well (Fig. 4.2c). For sum of ferns, Ca was the most important factor; when Ca <0.683 mg/Kg, U interacted with Ca, but at values >0.683, MS was also important (Fig. 4.2c).

For SMBR, Mg was the most important variable explaining sum of trees, interacting with Ca at low values of Mg (<1.03 g/kg) and with MS at high values (Fig. 4.3c). For sum of herbs, Mg was also the most significant variable, but the interaction arose with MS when Mg <2.561 (Fig. 4.3c). In terms of sum of ferns, carbonates were more important, but at values higher than 0.015%, it interacted with Fe (Fig. 4.3c).

Results of the GAM models are summarized in Table 4.3 for each of the three response variables (trees, herbs and ferns) in the two cloud forests. Table 4.3 includes information on the deviance explained by the model prediction, the Akaike information criterion (AIC) and the adjusted R^2. Only those models with at least one significant soil–environmental variable of the smoothed term <0.05 are shown, except the models for AFSP predicting sum of herbs, which resulted non-significant.

For AFSP, Carb contributed to explain (15.7% of the deviance; AIC, 44.94) sum of trees, but none of the models were significant to explain sum of herbs; Ca, together with MS, explained sum of ferns (66.9%). Nevertheless, a higher deviance (68.7%) with a small AIC (113.618) was accomplished when adding U, but this element was not significant at $p < 0.05$ (Table 4.3).

For SMBR, to explained sum of trees, the highest deviance (13.6%) was reached with Mg, Ca and MS, although only Mg was significant at $p < 0.05$. To explain sum of herbs, Mg was significant at $p < 0.05$ in several model combinations with other

Table 4.3 Generalized additive models for the sum of tree, herb and fern pollen as a response variable and soil variables as explanatory variables (Carb, carbonate; MS, magnetic susceptibility; Ca, calcium; U, uranium; Mg, magnesium; Sr, strontium; Fe, iron; Pb, lead; P, phosphorous) for the *Acer* Forest State Park and the Sierra de Manantlan Biosphere Reserve

Model	Deviance explained	Adjusted R^2	AIC
Acer Forest State Park (AFSP)			
Models predicting sum of tree pollen			
Sum of trees~log(Carb[a])	15.7	0.132	44.94
Models predicting sum of herb pollen			
Sum of herbs~log(Carb) + log(MS)	40.1	0.209	51.2574
Sum of herbs~(Carb) + (MS)	12.5	0.058	49.7135
Models predicting sum of fern pollen			
Sum of ferns~log(Ca[a]) + log(MS[a]) + log(U)	68.7	0.603	113.618
Sum of ferns~log(Ca[a]) + log(MS[a])	66.9	0.597	113.778
Sierra de Manantlán Biosphere Reserve (SMBR)			
Models predicting sum of tree pollen			
Sum of trees~log(Mg[a]) + log(Ca) + log(MS)	13.6	0.065	74.8626
Sum of trees~log(Mg[a]) + log(Ca)	11.3	0.061	75.5499
Models predicting sum of herb pollen			
Sum of herbs~log(Mg[a]) + log(MS) + log(Sr)	28.4	0.151	188.255
Sum of herbs~log(Mg[a]) + log(MS)	15.7	0.104	189.426
Sum of herbs~log(Mg[a]) + log(Sr)	17.1	0.118	190.196
Sum of herbs~log(Mg[a])	15.9	0.121	191.092
Models predicting sum of fern pollen			
Sum of ferns~(Carb[a]) + (Fe) + (Pb) + (P) + (Ca)	53.3	0.315	101.745
Sum of ferns~(Carb[a]) + (Pb) + (Ca)	49.5	0.316	103.218

Model selection was obtained using the function *mgcv* package in R
[a]Represents those soil–environmental variables significant at $p < 0.05$

soil–environmental variables; however, the highest deviance (28.4%), together with the smallest AIC (188.255), was reached in the combination with Mg, MS and Sr. Finally, to explain sum of ferns, the model with Carb, Fe, Pb, P and Ca obtained the highest deviance (53.3%) with a low AIC (101.745), but only Carb was significant at $p < 0.05$ (Table 4.3).

Discussion

Spatio-temporal Floristic Heterogeneity in Cloud Forests of the Mexican Pacific

A conspicuous spatial heterogeneity in fossil pollen taxa composition was observed in both ancient cloud forests of Jalisco, with 29% of taxa shared between them. This is in agreement with the high structural and floristic heterogeneity widely recognized for the cloud forests in Mexico (Villaseñor 2010); its fragmentary geographic distribution, similar to archipelagos, allows a high spatial species turnover (Vázquez-García 1995; Vega et al. 1999).

Results also showed a high floristic divergence between SMBR and AFSP, mainly in canopy components such as trees: as previously reported in this chapter, only 18 out of 49 tree taxa were common between the two forests. In addition, some of the taxa correspond to Holarctic affinity (e.g. *Abies*, *Alnus*, *Arbutus*, *Fraxinus*, *Juglans*, *Ostrya-Carpinus*, *Pinus*, *Quercus* and *Tilia*), whilst some correspond to Tropical affinity (e.g. *Clethra*, *Cleyera*, *Dendropanax*, *Ilex*, *Meliosma* and *Rubus*), a fact that reflected the mixture of Nearctic and Neotropical origin floras in the MP humid mountains where this forest developed (González-Espinosa et al. 2012; Morales-Arias et al. 2018; Sánchez-Rodriguez et al. 2018). Concerning life forms such as herbs and ferns involved in this study, their identification to genus level represented a difficult task. Therefore, a more precise comparison in floristic composition similarity between the two forests was rather challenging.

In terms of number of fossil pollen taxa, trees were more diverse followed by herbs and finally ferns. AFSP was more diverse than SMBR in herbs and ferns but similar in tree taxa. At present, ferns are the most diverse life form in the cloud forests of Mexico with 800 species; followed by herbs and shrubs, each with 600 species; and finally trees with 450 species (Rzedowski 2006). The discrepancy of the two ancient forests in terms of life-form diversity, compared to present-day cloud forest assemblages, could be related to the level of fossil pollen identification, particularly with very diverse herb families, e.g. Asteraceae family contains 417 genera and 3113 species (Villaseñor 2018). However, fossil pollen of Asteraceae species is very similar; therefore, they were grouped as a family, and in few cases, they were identified to genus level. Other examples of equally highly diverse herb families in Mexico include Poaceae with 1047 species, Euphorbiaceae with 787 and Solanaceae with 428 (Villaseñor et al. 2007). A comparable case concerns ferns, with families such as Pteridaceae with 218 species, Dryopteridaceae with 166, Polypodiaceae with 92, Aspleniaceae with 89 and Hymenophyllaceae with 48 (Martínez-Salas and Ramos 2014).

Changes in the abundances of trees, herbs and ferns in SMBR and AFSP forests also revealed a heterogeneous dynamics with floristic fluctuation patterns related to the LIA and the MCA.

For AFSP, it was problematic to determine a possible contraction or expansion in vegetation for the whole period of the LIA, as the three life forms analysed displayed irregular percentages. However, our results showed that trees experienced a remarkable decreased, whilst herbs followed an important increase. The LIA has been globally recognized as a cold period (Mann et al. 2009); in particular, for Mexico, it has been coupled to temperature decreases and recurrent droughts (Lozano-Garcia et al. 2007; Vázquez-Selem 2011). Therefore, we were expecting a more persistent decrease in trees and ferns and an increase in herbs along the whole LIA; however, this only occurred at a specific interval (~AD 1653–1734). Very close to these ages, Carb (AD 1775) and magnetic susceptibility (AD 1761) also presented the highest value. Therefore, we could interpret that this was the driest period of the LIA spotted in AFSP forest. This finding is supported by evidence of microfossil charcoal in the site used as a proxy for fire occurrences (Del Castillo-Batista et al. 2016). In La Luna Lake from Central Mexico, the coolest and driest episode of the LIA was also reported inside the same interval (~AD 1660–1760), and it was associated to the Maunder Minimum in solar activity (Cuna et al. 2014).

For years previous to the LIA (around AD 1252–1319), ferns exhibited clear higher percentages, indicating that wetter conditions were prevalent for these ages. Likewise, a diminution in fire and the highest Ti value occurred in this period (Del Castillo-Batista et al. 2016). This last element was used as a proxy to denote summer monsoonal rainfall in the Juanacatlan Lake from AD 1200 to 1350 (Metcalfe et al. 2010). For the same period, Curtis et al. (1996), using $\delta^{18}O$ records, also reported wet conditions in the Yucatan Peninsula.

Trees in the cloud forest of SMBR showed a distinct signature with a clear contraction related to the warm conditions of the beginning and to the middle of the MCA (AD 800–1200); herbs and ferns, on the contrary, expanded during this interval. The highest expansion for the arboreal component, together with high OM values, occurred in late MCA (AD 1200–1350).

This was an expected result due to the fact that cloud forest, at present, developed in the humid mountains of Mexico, particularly in places where physiography allows cloud condensation (Challenger and Soberón 2008; Villaseñor 2010; Gual-Díaz and González-Medrano 2014). The second critical contraction of trees occurred during the LIA, around AD 1528–1831, concomitant with a significant peak in herbs as well as in Mg, Ca, Sr and Fe but a decreased in OM. This array was analogous to the one observed in AFSP, also during an episode enclosing the Maunder Minimum in solar activity (~AD 1645–1715), associated to dry and cold conditions (Lean 2018). This event was also observed in Lake Santa Maria del Oro, Nayarit (Sosa-Najera et al. 2010). The dry environment constrained tree growth permitting, at the same time, herb increases due to canopy opening.

Only Carb were significant in explaining trees in AFSP, whereas Ca and MS were better for ferns. In SMBR, Carb explained ferns, whilst Mg was significant for trees and herbs. As our results showed in both forests, peaks in Carb were present during the LIA, around AD 1770, also very closed to the Maunder Minimum associated to droughts in Santa Maria del Oro (Sosa-Najera et al. 2010), as well as in Lake Chalco where high values in carbonates were associated to intervals of droughts from 85.000 to 10.800 cal yr. BP (Torres-Rodríguez et al. 2015). Cloud forest species are very sensitive to humid conditions; therefore, according to our results, carbonates seem to determine tree abundance in these two assemblages over the past climate anomalies (LIA and MCA) due to periods of aridity that were in detriment of tree growth.

The highest Mg in SMBR, during both the MCA and the LIA, coincided with decrements in trees and increases in herbs, particularly at the Maunder Minimum incidence (AD 1645–1715) associated to dry environments. Mg was one of the most important explanatory variables of tree species distribution in a humid forest of Jalisco, in particular for species *Styrax radians, Q. martinezi, I. vera* and *Symplococarpon purpusi* (Morales-Arias et al. 2018). In temperate ecosystems such as the transition oak-northern hardwood forest in north-west Connecticut, USA, high Mg concentrations were related to slow growth in *Fraxinus* (Bigelow and Canham 2002), while some studies have noted that drought conditions promote the release of inorganic nutrients such as Mg and Ca (Freeman et al. 1997).

Conclusion

The environmental history of two cloud forest assemblages located in the Mexican Pacific revealed a dynamic response of plant–soil relationship to climate change along Late Holocene. The high time resolution of the two sequences, every 19 years for *Acer Forest State Park* and every 32 years for *Sierra de Manantlan Biosphere Reserve*, disclosed the high sensitivity of cloud forests to climate change anomalies such as the LIA (~AD 1350–1850) and the MCA (~AD 800–1350). Spatial heterogeneity in past taxa composition and soil environment was observed between the two cloud forests. Carbonates and Mg were the most recurrent soil variables explaining trees, herbs and ferns, mostly associated to dry periods of the last millennia. Along dry periods, trees experienced a remarkable decreased, while herbs followed an important increase, e.g. during the LIA, at the Maunder Minimum interval (~AD 1645–1715) and at the first part of the MCA (AD 800–1200). In wet events such as late MCA (AD 1200–1350), trees experienced their highest expansion in both forests. The present evidence provides key baselines to foresee future actions to prevent these ecosystems from risks triggered by climate change events but also to enhance the importance of cloud forest conservation in the Mexican Pacific.

Acknowledgements The present research was supported by Conacyt (Mexican National Council for Science and Technology, project 106435) and CoecytJal (State Council for Science and Technology Jalisco, project 5-2010-875). The comments of Professor Hermann Behling substantially improved this chapter.

References

Alcántara II, Velázquez-Durán R, García MSL et al (2010) Evolución Paleolimnológica del Lago Cuitzeo, Michoacán durante el Pleistoceno-Holoceno. Bol la Soc Geol Mex 62:345–357

Bengtsson L, Enell M (1986) Chemical analysis. In: Berglund BE (ed) Handbook of Holocene palaeoecology and paleohydrology. Wiley, New York, pp 423–455

Bennett KD (1994) Confidence intervals for age estimates and deposition times in late-Quaternary sediment sequences. The Holocene 4:337–348. https://doi.org/10.1177/095968369400400401

Bennett K (2005) Psimpoll 4.25 and pscomb 1.03

Bennett KD, Willis K (2001) Pollen. In: Smol JP, Birks HJB, Last WM (eds) Tracking environmental change using lake sediments. Kluwer Academic Publishers, Dordrecht, pp 5–32

Bianchette T (2014) Holocene paleoenvironmental reconstruction from Mexico's Pacific Coast – a paleotempestological investigation. Louisiana State University, Baton Rouge

Bigelow SW, Canham CD (2002) Community organization of tree species along soil gradients in a north-eastern USA forest. J Ecol 90:188–200. https://doi.org/10.1046/j.0022-0477.2001.00655.x

Birks HJB (2012) Overview of numerical methods in Paleolimnology. In: Smol JP, Birks HJB, Last WM (eds) Tracking environmental change using lake sediments: data handling and numerical techniques. Springer, Dordrecht/New York, pp 19–92

Caballero-Rodríguez D, Correa-Metrio A, Lozano-García S et al (2018) Late-Quaternary spatio-temporal dynamics of vegetation in Central Mexico. Rev Palaeobot Palynol 250:44. https://doi.org/10.1016/j.revpalbo.2017.12.004

Catalán-Heverástico C, López-Mata L, Terrazas T (2003) Estructura, composicion floristica y diversidad de especies leñosas de un bosque mesófilo de montaña de Guerrero, México. An del Inst Biol Ser Botánica 74:209–230

Challenger A, Soberón J (2008) *Los ecosistemas terrestres, en Capital natural de México. vol. I: Conocimiento actual de la biodiversidad.* pp 87–108

Correa-Metrio A, Lozano-García S, Xelhuantzi-López S et al (2012) Vegetation in western Central Mexico during the last 50 000 years: modern analogs and climate in the Zacapu Basin. J Quat Sci 27:509–518. https://doi.org/10.1002/jqs.2540

Crawley MJ (2005) Statistics an introduction using R. Wiley, Sussex

Cuevas-Guzmán R, Koch S, García-Moya E et al (2004) Flora vascular de la Estación Científica Las Joyas. Univ Guadalajara:117–176

Cuna E, Zawisa E, Caballero M et al (2014) Environmental impacts of Little Ice Age cooling in central Mexico recorded in the sediments of a tropical alpine lake. J Paleolimnol 51:1–14. https://doi.org/10.1007/s10933-013-9748-0

Curtis JH, Hodell DA, Brenner M (1996) Climate variability on the Yucatan Peninsula (Mexico) during the past 3500 years, and implications for Maya cultural evolution. Quat Res 46:37–47. https://doi.org/10.1006/qres.1996.0042

Dean WE (1974) Determination of carbonate and organic matter in calcareous sediments and sedimentary rocks by loss on ignition. J Sediment Petrol 44:242–248. https://doi.org/10.1306/74D729D2-2B21-11D7-8648000102C1865D

Decreto 001/2016 (2016) Decreto por el que se establece área natural protegida bajo la categoría de “Parque Estatal Bosque de Arce”, con una superficie de 150.04 hectáreas, ubicada en el municipio de Talpa de Allende, Jalisco

Del Castillo-Batista AP, Figueroa-Rangel BL, Lozano-García S et al (2016) Historia florística y ambiental del bosque mesófilo de montaña en el centro-occidente de México durante la pequeña edad de hielo. Rev Mex Biodivers 87:216–229. https://doi.org/10.1016/j.rmb.2016.01.021

Dinerstein EDM, Olson DJ, Graham AL, Webster SA, Primm MPB & GL (1995) Una evaluación del estado de conservación de las ecorregiones terrestres de América Latina y el Caribe

Felson RB, Lane KJ (2010) Does violence involving women and intimate partners have a special etiology? Criminology 48:321–338. https://doi.org/10.1111/j.1745-9125.2010.00186.x

Figueroa-Rangel BL (2018) Mexican Reference Collection (Version 2). Digitised Palynological Reference Collection. In: Glob. Pollen Proj. https://globalpollenproject.org/Reference/006fa89d-ee86-4b59-a249-a30572b67358/2

Figueroa-Rangel BL, WIllis KJ, Olvera-Vargas M (2010) Cloud forest dynamics in the Mexican neotropics during the last 1300 years. Glob Chang Biol 16:1689–1704. https://doi.org/10.1111/j.1365-2486.2009.02024.x

Figueroa-Rangel BL, Willis KJ, Olvera-Vargas M (2012) Late-Holocene successional dynamics in a transitional forest of west-central Mexico. The Holocene 22:143–153. https://doi.org/10.1177/0959683611414929

Figueroa-Rangel BL, Valle-Martínez A, Olvera-Vargas M, Liu K (2016) Environmental history of mangrove vegetation in Pacific West-Central Mexico during the Last 1300 years. Front Ecol Evol 4. https://doi.org/10.3389/fevo.2016.00101

Freeman C, Liska G, Ostle NJ et al (1997) Enzymes and biogeochemical cycling in wetlands during a simulated drought. Biogeochemistry 39:177–187. https://doi.org/10.1023/A:1005872015085

González-Espinosa M, Meave JA, Ramírez-Marcial N et al (2012) Los bosques de niebla de México: conservación y restauración de su componente arbóreo. Ecosistemas 21:36–54

Gual-Díaz M, González-Medrano F (2014) Los bosques mesófilos de montaña en México, diversidad, ecología y manejo. Comisión Nacional para el Conocimiento y Uso de la Biodiversidad, México, D.F.

Kelly I (1945) The archaeology of the Autlan-Tuxcacuesco area of Jalisco. I: The Autlan Zone. Ibero-Americana 26:1–93

Lean JL (2018) Estimating solar irradiance since 850 CE. Earth Space Sci 5:133–149. https://doi.org/10.1002/2017EA000357

Lozano-Garcia MS, Caballero M, Ortega B et al (2007) Tracing the effects of the Little Ice Age in the tropical lowlands of eastern Mesoamerica. Proc Natl Acad Sci 104:16200–16203. https://doi.org/10.1073/pnas.0707896104

Lozano-Garcia S, Roy P, Correa-Metrio A et al (2015) Registros paleoclimáticos. In: Gay, García CRA (eds) Reporte Mexicano de Cambio Climático. Universidad Nacional Autonoma de México y Programa de Investigación en Cambio Climático, México, p 294

Maher LJJ (1972) Absolute pollen diagram of Redrock Lake, Boulder County, Colorado. Quat Res 2:531–553. https://doi.org/10.1016/0033-5894(72)90090-7

Mann ME, Zhang Z, Rutherford S et al (2009) Global signatures and dynamical origins of the little ice age and medieval climate anomaly. Science. (80-) 326:1256–1260. https://doi.org/10.1126/science.1177303

Martínez-Salas E, Ramos CH (2014) Biodiversidad de Pteridophyta en México. Rev Mex Biodivers 85:110. https://doi.org/10.7550/rmb.31827

Meave JA, Soto MA, Calvo IL et al (1992) Análisis ecológico del bosque mesófilo de montaña de Omiltemi, Guerrero. Bol la Soc Botánica México 1 52:31–77

Mejía-Domínguez N (2006) Dinámica de la comunidad de árboles de un bosque mesófilo de montaña en la Sierra Madre del Sur (Oaxaca), México. Rev Biol Neotrop 3:177–178

Metcalfe SE, Jones M, Davies SJ et al (2010) Climate variability over the last two millennia in the North American Monsoon region, recorded in laminated lake sediments from Laguna de Juanacatlán, Mexico. The Holocene 20:1195–1206. https://doi.org/10.1177/0959683610371994

Morales-Arias JG, Olvera-Vargas M, Cuevas-Guzman R et al (2018) Variación ambiental y composición florística de especies arbóreas en un bosque húmedo de montaña del occidente de México. Rev Mex Biodivers 89:769–783

Morrone JJ (2001) Biogeografía de América Latina y el Caribe. M&T - Manuales Tesis SEA 3:148

Morrone JJ, Organista DE, Zuniga CA, Bousquets JL (1999) Preliminary classification of the Mexican biogeographic provinces: a parsimony analysis of endemicity based on plant, insect, and bird taxa. Southwest Nat 44:507. https://doi.org/10.2307/3672351

Mountjoy JB (1987) Antiquity, interpretation and stylistic evolution of petro-glyphs in western México. Am Antiq 52:161–174

Ortega B, Vázquez G, Caballero M et al (2010) Late Pleistocene: Holocene record of environmental changes in Lake Zirahuen, Central Mexico. J Paleolimnol 44:745–760. https://doi.org/10.1007/s10933-010-9449-x

Ponce-Reyes R, Reynoso-Rosales VH, Watson JEM et al (2012) Vulnerability of cloud forest reserves in Mexico to climate change. Nat Clim Chang 2:448–452. https://doi.org/10.1038/nclimate1453

R Development Core Team R (2017) R: A language and environment for statistical computing

Reimer PJ, Baillie MGL, Bard E et al (2004) INTCAL04 terrertrial radiocarbon age calibration, 0-26 cal kyr BP. Radiocarbon 46:1029–1058

Rzedowski J (2006) Vegetación de México, Edición di. Comisión Nacional para el Conocimiento y Uso de la Biodiversidad

Sánchez-Rodriguez E, Figueroa-Rangel BL, Cuevas-Guzmán R et al (2018) Dinámica del bosque mesófilo del entro-sur del estado de Jalisco, México. Polibotánica 46

Santiago-Pérez AL, Jardel-Peláez EJ, Cuevas-Guzmán R, Huerta-Martínez FM (2009) Vegetación de bordes en un bosque mesófilo de montaña del occidente de México. Boletín la Soc Botánica México 85:31–49

SEMARNAP (2000) Programa de Manejo de la Reserva de la Biosfera Sierra de Manatlán. INE-SENARNAP

Sosa-Najera S, Lozano-Garcia S, Roy PD, Caballero M (2010) Registro de sequías históricas en el occidente de México con base en el análisis elemental de sedimentos lacustres: El caso del lago de Santa María del Oro. Boletín la Soc Geológica Mex 62:437–451

Sotelo DI, Casas FN, Camelo MG (2010) Borojó (Borojoa patinoi): Fuente de polifenoles con actividad antimicrobiana. Vitae 17:329–336. https://doi.org/10.5944/educxx1.17.1.10708
Thompson R, Oldfield F (1986) Environmental magnetism. Allen & Unwin, London
Torres-Rodriguez E, Lozano-Garcia S, Figueroa-Rangel BL et al (2012) Environmental change and vegetation responses during the last 17,000 years in central Mexico: The lake Zirahuen record. Rev Mex CIENCIAS Geol 29:764–778
Torres-Rodríguez E, Lozano-Garcia S, Roy PD et al (2015) Last Glacial droughts and fire regimes in the central Mexican highlands. J Quat Sci 30:88–99. https://doi.org/10.1002/jqs.2761
Vargas-Rodriguez YL, Platt WJ, Vázquez-García JA, Boquin G (2010) Selecting relict montane cloud forests for conservation priorities: the case of Western Mexico. Nat Areas J 30:156–173. https://doi.org/10.3375/043.030.0204
Vázquez-García JA (1995) Cloud forest archipelagos: preservation of fragmented montane ecosystems in tropical America. In: Hamilton LS, Juvik JO, Scatena FN (eds) Tropical montane cloud forest. Springer, New York, pp 315–332
Vázquez-García JA, Cuevas-Guzmán R, Cochrane TS, et al (1995) *Flora de Manantlan.* SIDA-Botanical Miscellany N.13
Vázquez-Selem L (2011) Las glaciaciones en las montan˜as del centro de México. In: Caballero-Miranda M, Ortega-Guerrero B (eds) Registros del Cuaternario en America Latina. Universidad Nacional Autonoma de México, pp 215–238
Vega IL, Ayala OA, Organista DE, Morrone JJ (1999) Historical relationships of the Mexican cloud forests: a preliminary vicariance model applying Parsimony Analysis of Endemicity to vascular plant taxa. J Biogeogr 26:1299. https://doi.org/10.1046/j.1365-2699.1999.00361.x
Villaseñor JL (2003) Diversidad y distribución de las Magnoliophyta de Mexico. Interciencia 28:160–167+184. https://doi.org/10.4414/smw.2017.14531
Villaseñor JL (2004) Los géneros de plantas vasculares de la flora de México. Boletín la Soc Botánica México 75:105–135. https://doi.org/10.17129/botsci.1694
Villaseñor JL (2010) El bosque húmedo de montaña en México y sus plantas vasculares: catálogo florístico-taxonómico. Universidad Nacional Autonoma de México, México, D.F
Villaseñor JL (2018) Diversidad y distribución de la familia Asteraceae en México. Bot Sci 96:332–358
Villaseñor JL, Maeda P, Rosell JA, Ortiz E (2007) Plant families as predictors of plant biodiversity in Mexico. Divers Distrib 13:871–876. https://doi.org/10.1111/j.1472-4642.2007.00385.x

Chapter 5
Sea Level Change and Its Influence on the Coastal Landscape in the Gulf of Mexico During the Holocene

Gabriela Domínguez-Vázquez and Dulce M. Bocanegra-Ramírez

Abstract The ecological services of mangrove ecosystems are globally recognized. Mangroves reduce hurricane damages and provide habitat for many organisms that inhabit coastal environments.

Mangroves are excellent sea-level indicators. They depend on the balance between land and sea, making them sensible indicators of sea-level variations. Mean sea-level rise has an immediate and direct effect on ecosystems of the intertidal zone, which showed an increase in the influence of marine processes on land ecological process.

During the Holocene, fluctuations of sea level and human impact on the vegetation have modified the extension of the mangrove ecosystem, changing the dynamics of the vegetation and the coastal landscape. We review the consequences of transgressions and regressions on the terrestrial and aquatic vegetation during the Holocene in the Gulf of México.

Keywords Sea level · Gulf of Mexico · Mangroves · Holocene · Transgressions · Regressions

Introduction

The Gulf of Mexico is the ninth largest body of water in the world. It covers an area of approximately 1,300,000 m^2 that contains a large number of riparian and estuarine systems (Giattina and Altsman 1999). The Grijalva-Usumacinta River, at the central part of the Gulf of Mexico, contributes with 30% of Mexico's total flow of freshwater and is Mexico's largest discharge of freshwater into the ocean (SEMARNAP 1997) (Fig. 5.1).

G. Domínguez-Vázquez (✉) · D. M. Bocanegra-Ramírez
Faculty of Biology, Universidad Michoacana de San Nicolás de Hidalgo.,
Edificio R. Gral. Francisco J. Múgica. CU. Felicitas del Río., Morelia, Mexico

N. Torrescano-Valle et al. (eds.), *The Holocene and Anthropocene Environmental History of Mexico*, https://doi.org/10.1007/978-3-030-31719-5_5

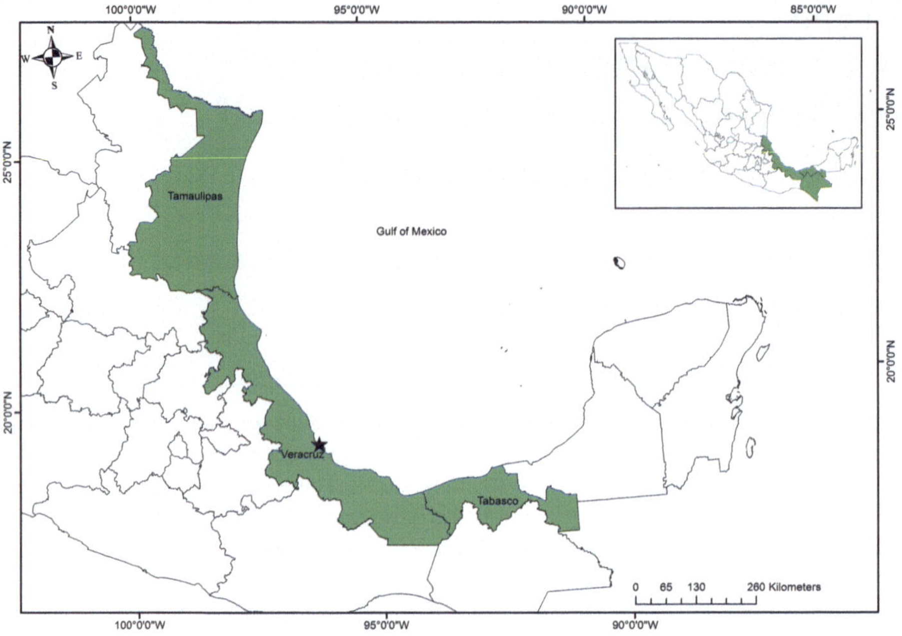

Fig. 5.1 Study area and coastal areas of Mexico bordering the Gulf of Mexico

Sea-level change during the Holocene has been the main driver of vegetation distribution in the coastal zone of the Gulf of Mexico. The lack of sedimentary records spanning the complete Holocene in the central Gulf of Mexico (Tamaulipas, Veracruz, Tabasco, Chiapas) hardens the reconstruction of sea-level fluctuations. There is a general agreement that in the northern Gulf of Mexico, the sea level remained low during the Holocene. Blum et al. (2003) suggested that the sea level along the shoreline of the Gulf of Mexico was very close to present elevations throughout the middle to late Holocene. On the contrary, a sea-level high stand has been described from Veracruz during the middle Holocene (Sluyter 1997; Sluyter and Dominguez 2006), which coincides with records from the Mexican Pacific Ocean (Bocanegra-Ramírez et al. 2019; Ramírez-Herrera et al. 2012).

The vegetation of the Gulf of Mexico comprises different vegetation types that range from tropical forest in the piedmont lowlands to mangroves and wetlands to sand dunes in the littoral area. The sea-level changes, and the impacts of early human activities on the environment during the Holocene had a strong influence in the distribution of the different vegetation types in the Gulf of Mexico.

Vegetation Types in the Gulf of Mexico

The distribution of different vegetation types in the Gulf of Mexico is the result of different processes of geomorphology, climate, ecology, and hydrology (Lot 2004). Miranda and Hernández (1963) classified the different vegetation types in the

Gulf of Mexico as (1) savanna, (2) palm forest, (3) tropical forest, (4) mangrove, (5) coastal dune vegetation; and (6) wetland. The first five vegetation units are equivalent to classifications by Lot (2004), defining them as *woody formation*. Meanwhile, Lot (2004) defined the wetland as a *herbaceous formation*.

Despite the fact that the Gulf of Mexico is one of Mexico's most diverse areas, human activity put most of its flora in danger. These industrial and agricultural activities have caused a strong change in land use. Large areas formerly covered by tropical forests and mangroves were replaced by industries, grassland, and secondary forest or fallows (Castillo-Campos et al. 2011).

Coastal dunes Stabilized coastal dunes occurred in areas where organic matter produced by plants has been integrated into the sand, allowing the establishment of more structured vegetation (Martínez et al. 2014). Some species that can be found in these environments are *Chrysobalanus icaco L.*, *Coccoloba uvifera* L., *C. barbadensis* Jacq., *Manilkara zapota* (L.) P. Royen, and *Quercus oleoides* Schltdl. & Cham. Different species like *Ficus*, *Pouteria hypoglauca* (Standl.) Baehni, *Bursera simaruba* (L.) Sarg., *Nectandra salicifolia* (Kunth) Nees, *Sabal mexicana* Mart, *Randia aculeata* L., *Diphysa americana* (Mill), *Karwinskia humboldtiana* (Schult.) Zucc., *Enterolobium cyclocarpum* (Jacq.)Griseb., and *Brosimum alicastrum* Sw are present (Moreno-Casasola 1991; Castillo and Moreno-Casasola 1998; Moreno-Casasola et al. 1998; Martínez et al. 2014; Espejel et al. 2013, 2017).

Savanna This vegetation included different species of grasses and scattered trees growing in flooded soils with a deficient drainage (Castillo-Campos et al. 2011). The main tree species are *Acacia angustissima (Mill) Kuntze* (acacia), *Byrsonima crassifolia* (L.) Kunth (nanche), *Curatella americana L.* (raspa viejo), *Crescentia alata Kunth* and *C. cujete L.* (jícaro), *Lonchocarpus cruentus* Lundell, *Pithecellobium pachypus Pittier* and *Senna spectabilis (Lam) Irwin & Barneby*.

Palm swamp Several palm species tolerant to the floods can be found in wetlands along the coastal floodplain in the Gulf of Mexico, mainly *Attalea liebmannii* (Becc.), *Roystonea dunlapiana* Allen, and *Sabal mexicana* Martius, which form monospecific patches or species mixed with other palm or tree species.

Tropical forest The tropical rainforest covered most of the coastal plain. Currently, most of the tropical forests have been deforested and fragmented due to agricultural and cattle activities (Castillo-Campos et al. 2011). The tropical forest presents a complex structure formed by different vegetation layers. Canopy species are dominated by *Brosimum alicastrum* Sw, *Cynometra retusa* Britton & Rose, *Dialium guianense* (Aubl.) Sandwith, *Lonchocarpus cruentus* Lundell, *L. guatemalensis* var. *mexicanus* (Pittier) F.J. Herm, *Ormosia panamensis* Benth ex Seem, *Pterocarpus rohrii Vahl*, *Poulsenia armata* (Miq.) Standl, *Vatairea lundellii* (Standl.) Killip ex Record, *Nectandra ambigens* (S.F. Blake) C.K. Allen, *N. lundellii* C.K. Allen, *N. cissiflora* Nees, and *Ocotea uxpanapana* Wendt and van der Werff y *Manilkara zapota* (L.) Royen.

Tropical dry forest grows frequently over the stabilized coastal dunes (Castillo-Campos et al. 2011). This vegetation loses their leaves during the dry season

(Miranda and Hernández 1963). The main species of this vegetation are *Antirhea aromatic* Castillo-Campos and Lorence, *Aphananthe monoica* (Hemsl.) Leroy, *Brosimum alicastrum, Hyperbaena jalcomulcensis* Pérez & Cast.-Campos, *Comocladia engleriana* Loes, *Ocotea* sp., *Protium copal* Engl., and *Psychotria erythrocarpa* Schltdl.

Wetlands Wetlands are dominated by aquatics like *Typha, Potamogeton, Cyperus,* and *Nymphaea.* These species are found in freshwater bodies at the coastal zone or in the flood plain close to river mouths (Castillo-Campos et al. 2011). The most common species in wetlands are *Suaeda linearis* (Elliott) Moq., *Salicornia* spp., *Monantochloe littoralis* Engelm, *Sesuvium portulacastrum* (L.) L., *Batis marítima* L., *Typha domingensis* Pers., *Pontederia sagittata* C Presl, *Thalia geniculata* L., *Cyperus articulatus* L., *C. digitatus* Roxb., *Sagittaria lancifolia* L., *Pistia stratiotes* L., and *Eleocharis mutata* (L.) Roem. and Schult (Castillo and Moreno-Casasola 1998; Martínez et al. 2014; Espejel et al. 2013, 2017).

Mangroves Mangroves are intermediate environments, where the sea encounters the freshwater discharge from rivers. *Avicennia germinans* (L.)L., *Conocarpus erectus* (L.), *Laguncularia racemosa* (L.) Gaertn, and *Rhizophora mangle* L. are the dominating species in this ecosystem.

Vegetation Response to Sea Level

Mangroves are good indicators of sea-level curves, because their survival depends on the balance between seawater and freshwater. This characteristic enables them to display rapid responses to hydrological, geomorphic, and climatic changes, colonizing and establishing in new geographical areas (Schaeffer-Novelli et al. 2002; Crigger et al. 2005; Day et al. 2008). During periods of sea regression, the low sea level caused a seaward migration of mangrove vegetation. Meanwhile during periods of sea transgression, mangroves move landward, as flooding of seawater covers the areas where estuaries dominated, leaving in the stratigraphy the imprints of the mangrove presence (Bocanegra-Ramírez et al. 2019).

Fluctuations of sea level can cause sudden mangrove death, leaving massive organic sediments during events originated either on land or marine areas (Bocanegra-Ramírez et al. 2019). During periods of increased precipitation in the highlands, the runoff increases, depositing large loads of sediment and water into the coastal zones. This freshwater input dilutes the water and changes the ionic concentration of the estuarine environment (Alongi 2008; Bocanegra-Ramírez et al. 2019). During sea events such as tsunamis or storms that cause sea regressions, there is an increase in the influence of marine processes (Ellison and Stoddart 1991). This sea intrusion generates mortality of the mangrove as the salt concentration increases.

A pollen sequence from the coastal plain of Veracruz showed a high sea stand 6000 years ago, which caused a landward displacement of the mangrove. The sea transgression caused a progradation of the inward land, which allowed the formation of estuarine ecosystems. Evidence of humid conditions in the Gulf of Mexico came from the abundance of pollen from tropical rainforest (Goman and Byrne 1998; Byrne and Horn 1989) characterized by Moraceae, Sapotaceae, Meliaceae, and Fabaceae. A significant drop in sea level which started around 5000 years ago increased the influence of inland processes that cause a change in the ionic conditions of the estuarine ecosystems, which were substituted by wetlands where *Potamogeton* was dominant, as the input of freshwater increased into the lagoon (Fig. 5.2) (Sluyter and Dominguez 2006).

The rearrangement of the estuarine environments, the humid condition that still prevailed in the Gulf of Mexico and other areas of Mesoamerica (Sluyter 1997; Solís-Castillo et al. 2013) during the middle Holocene, and the beginning of the late Holocene allowed the development of tropical forest in the lowland and montane forest in the middle elevations. The retreat of the sea around 5450 yr. BP in the Gulf of Mexico caused the rearrangement of landscapes. Mangroves invaded areas formerly occupied by taxa from the lowland rainforest.

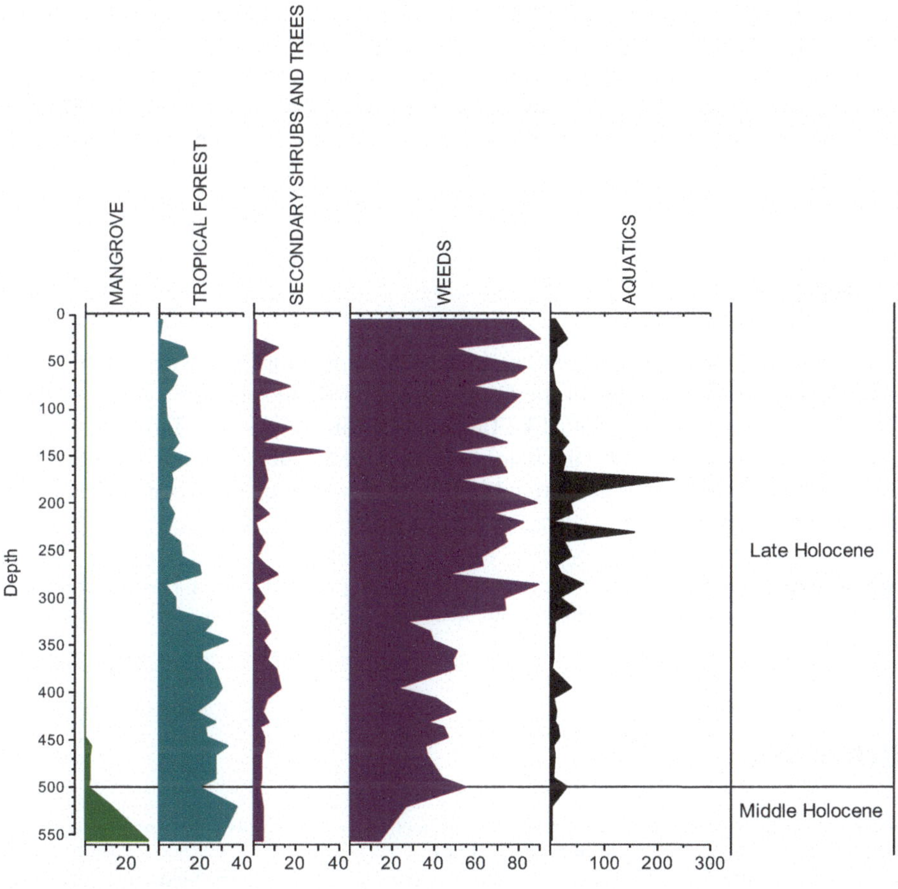

Fig. 5.2 Pollen record from Laguna, Santa Catarina, Veracruz. (Modified from Sluyter and Dominguez 2006)

Human Impact on Coastal Vegetation

There is no direct evidence (building, ceramics, burials) linked to early humans in the Gulf of Mexico during the early and middle Holocene. However, there is indirect evidence from pollen and phytoliths that indicate an impact in the environment and probably modification of landscape that can be related to an early human presence in the Gulf of Mexico (Pope et al. 2001; Sluyter 1997; Sluyter and Dominguez 2006; Torrescano and Islebe 2006; Pohl et al. 2007).

Pollen records evidence the presence of agriculture in the coastal plain of the Gulf of Mexico, as early as 7000 years ago (Pohl et al. 2007). Large *Zea* sp. pollen, typical of domesticated maize (*Zea mays*), appears with evidence of prehistoric anthropogenic forest clearance. Species characteristic of secondary habitats or fallows are frequently found in the Holocene record in neotropical lowlands (Sluyter 1997; Sluyter and Dominguez 2006) (Fig. 5.2). Forest logging for pasture, agriculture, or other human activities in the area created secondary habitats. Species with affinities to these habitats are heliophytes and early successional pioneers such as Asteraceae, Poaceae, Euphorbiaceae, Lamiaceae, *Piper*, *Crecropia*, *Croton*, and *Alchornea*. All these taxa appeared in a large proportion in the pollen record at ca. 7000 years B.P. in the Tabasco lowlands (Pohl et al. 2007) and Veracruz (Sluyter 1997; Sluyter and Dominguez 2006), suggesting an early Holocene prehistoric land use practices.

Conclusion

The reconstruction of ancient sea-level history and its impacts on vegetation distribution and on past societies in the Gulf of Mexico should include studies of past shorelines in areas with few data, like Tabasco and Tamaulipas. The Tabasco swamps and the flood plain of northern Chiapas should receive increased attention due the accelerated destruction of the wetlands and mangroves by channel construction and desiccation programs.

Acknowledgement We thank R. Dull, A. Sluyter, V. Castro, and V. Gómez for comments on the manuscript. Two anonymous reviewers are thanked for their comments that improved the manuscript.

References

Alongi DM (2008) Mangrove forest: resilience, protection from tsunamis and responses to global climate change. Estuar Coast Shelf Sci 76:1–13

Byrne R, Horn SP (1989) Prehistoric agriculture and forest clearance in the Sierra de los Tuxtlas, Veracruz, Mexico. Palynology 13:181–193

Blum MD, Sivers AE, Zayac T, Goble RJ (2003) Middle Holocene Sea-level and evolution of the Gulf of Mexico Coast. Gulf Coast Assoc Geol Soc Trans 53:64–77
Bocanegra-Ramírez DM, Li HC, Domínguez-Vázquez G et al (2019) Holocene climate change and sea level oscillations in the pacific coast of Mexico. Quat Int. https://doi.org/10.1016/j.quaint.2019.01.003
Castillo S, Moreno-Casasola P (1998) Análisis de la flora de dunas costeras del Golfo y Caribe de México. Acta Botánica Mexicana 45:55–80
Castillo-Campos G, Avendaño-Reyes S, Medina-Abreo ME (2011) Seccion IV. Diversidad de Ambientes. Ambientes terrestres. Flora y Vegetación. Conabio, pp. 159–179
Crigger DK, Graves GA, Fike DL (2005) Lake worth lagoon conceptual ecological model. Wetlands 25:943–954
Day JW, Christian RR, Boesch DM et al (2008) Consequences of climate change on the ecogeomorphology of coastal wetlands. Estuar Coasts 31:477–491
Ellison JC, Stoddart D (1991) Mangrove ecosystem collapse during predicted sea-level rise: Holocene analogues and implications. J Coast Res 7:151–165. https://www.jstor.org/stable/4297812
Espejel E, Peña-Garcillán P, Jiménez-Orocio O (2013) Flora de playas y dunas de México. Informe Técnico Final Conabio HJ007
Espejel I, Jiménez-Orocio O, Castillo-Campos G et al (2017) Flora en playas y dunas costeras de México. Acta Botánica Mexicana (121):39–81. https://doi.org/10.21829/abm121.2017.1290
Giattina JD, Altsman DT (1999) Gulf of Mexico Program: partnership with a purpose. In: Kumpf H, Steidinger K, Sherman K (eds) The Gulf of Mexico large marine ecosystem: assessment, sustainability and management. Blackwell Science, Inc, Malden, pp 3–13
Goman M, Byrne R (1998) A 5000-year record of agriculture and tropical forest clearance in the Tuxtlas, Veracruz, Mexico. The Holocene 8:83–89
Lot A (2004) Flora y vegetación de los humedales de agua dulce en la zona costera del Golfo de México. In: Caso M, Pisanty I, Ezcurra E (eds) Diagnóstico ambiental del Golfo de México. 1ª. Secretaría de Medio Ambiente y Recursos Naturales, Instituto Nacional de Ecología, Instituto de Ecología, A.C. and Harte Research Institute for Gulf of Mexico Studies. Distrito Federal, México, pp 521–553
Martínez ML, Moreno-Casasola P, Espejel et al (2014) Diagnostico General de las Dunas Costeras de México, Capitulo 5. Conafor, pp. 49–60
Miranda F, Hernández XE (1963) Los tipos de vegetación de México y su clasificación. Bol Soc Bot Méx 28:29–72
Moreno-Casasola P (1991) Sand dune studies on the eastern coast of Mexico. Proceedings Canadian Symposium on Coastal Dunes 1990. Guelph, ON, Canada, pp. 215–230
Moreno-Casasola P, Espejel I, Castillo S (1998) Flora de los ambientes arenosos y rocosos de las costas de México. In: Halffter G (ed) Biodiversidad en Iberoamérica, vol II. CYTED- Instituto de Ecología A.C, pp 177–258
Pohl MED, Piperno DR, Pope KO, Jones JG (2007) Microfossil evidence for pre-Columbian maize dispersals in the neotropics from San Andrés, Tabasco, Mexico. PNAS 104:6870–6875. https://doi.org/10.1073/pnas.0701425104
Pope KO, Pohl MED, Jones JG, Lentz DL, von Nagy C, Vega FJ, Quitmyer IR (2001) Origin and environmental setting of ancient agriculture in the lowlands of Mesoamerica. Science 292:1370–1373. https://doi.org/10.1126/science.292.5520.1370
Ramírez-Herrera MT, Lagos M, Huthchinson I et al (2012) Extreme wave deposits on the Pacific coast of Mexico: tsunamis or storms? — a multi-proxy approach. Geomorphology 139–140:360–371
Schaeffer-Novelli Y, Cintron-Molero G, Soares MLG (2002) Mangroves as indicators of sea level change in the muddy coasts of the world. Proc Mar Sci 4:245–262. https://doi.org/10.1016/S1568-2692(02)80083-3
SEMARNAP (1997) Programa de Manejo del Area de Protección de Flora y Fauna Laguna de Términos. SEMARNAP, México, p 167

Sluyter A (1997) Regional, Holocene records of the human dimension of global change: sea level and land-use change in prehistoric Mexico. Glob Planet Chang 14:127–146
Sluyter A, Dominguez G (2006) Early maize (Zea mays L.) cultivation in Mexico: dating sedimentary pollen records and its implications. Proc Nat Acad Sci U S A 103:1147–1151
Solís-Castillo B, Solleiro-Rebolledo E, Sedov S et al (2013) Paleoenvironment and Human Occupation in the Maya Lowlands of the Usumacinta River, Southern Mexico. Geoarchaeology 28:268–288
Torrescano N, Islebe GA (2006) Tropical forest and mangrove history from southeastern Mexico: a 5000 yr pollen record and implications for sea level rise. Veg Hist Archaeobotany 15:191–195

Chapter 6
Insights into the Holocene Environmental History of the Highlands of Central Mexico

Socorro Lozano-García, Margarita Caballero, Beatriz Ortega-Guerrero, and Susana Sosa-Nájera

Abstract With the aim of highlighting the present stage of development in the scenario of environmental and climate research during the Holocene in central Mexico, we present a synthesis of relevant events preserved in continental records. Climate conditions and intricate topography of this region led to the development of lake basins, whose records were the sources for several paleoecological studies. These records suggest higher moisture during the early Holocene, although the high insolation promoted higher evaporation. Toward the mid-Holocene, the southward displacement of the ITCZ led to drier conditions in the area. During the Late Holocene, environmental change and human activities are intertwined, with the latter expanding over the last 2000 years. A review of paleoecological signals such as cultural pollen taxa, microcharcoal, deforestation, and erosion reveals human activities during the Late Holocene. Understanding the past climate of central Mexico is important for predictions of global warming, as a large part of the Mexican population reside in this region. Additionally, changes in forest composition and hydrological conditions observed in this region provide important ecological and climatic information to characterize the Trans-Mexican Volcanic Belt.

Keywords Late Holocene · Trans-Mexican Volcanic Belt · Droughts · Pollen records

This chapter is dedicated to Dr. Jerzy Rzedowski

S. Lozano-García (✉) · S. Sosa-Nájera
Instituto de Geología, Universidad Nacional Autónoma de México, Ciudad de Mexico, Mexico
e-mail: mslozano@unam.mx

M. Caballero · B. Ortega-Guerrero
Instituto de Geofísica, Universidad Nacional Autónoma de México, Ciudad de Mexico, Mexico

N. Torrescano-Valle et al. (eds.), *The Holocene and Anthropocene Environmental History of Mexico*, https://doi.org/10.1007/978-3-030-31719-5_6

Introduction

Past landscape dynamics can be reconstructed by analyzing the terrestrial archives preserved in lacustrine sequences. These archives provide valuable information regarding limnological conditions, plant communities, and climate change. Environmental changes as well as their relationship with the archaeological history of central Mexico have been important research topics because of (a) a complex physiographic context due to the Trans-Mexican Volcanic Belt (TMVB) hosting important elevations and intermountain basins with sediments deposited over thousands of years which open the possibility of reconstructing past environments; (b) the fact that climate change has been an important driving force controlling species distribution and migration along the altitudinal gradient defined by the complex topography of the region; (c) the historical context of the region, as the TMVB is part of the core area for the development of the Mesoamerican cultures and was the cradle of maize domestication (Piperno et al. 2009).

Paleoecological data obtained from the analysis of lacustrine sediments allows the identification of ecosystem's responses to climate variability during different time frames. Geochemistry and diatom analysis are two proxies used in lake sediments that can give detailed information about changes in the hydrological balance, allowing the identification of either wetter or dryer intervals (ej. Metcalfe et al. 2010; Rodríguez-Ramírez et al. 2015; Bhattacharya et al. 2017). On the other hand, the palynological record has been used to reconstruct the vegetation history and the responses of plant communities to natural disturbances such as fires, volcanic activity, and climate change (Lozano-García et al. 1993, 2013, 2015; Lozano-García and Ortega-Guerrero 1994, 1998; Ortega-Guerrero et al. 2000; Torres-Rodriguez et al. 2015; Sedov et al. 2010; Correa-Metrio et al. 2012; Figueroa-Rangel et al. 2008; Castillo-Batista et al. 2016).

In this chapter, we present an overview of Holocene paleoecological records and available paleolimnological data from central Mexico to estimate past changes in temperature and episodes of increased aridity in the region. An emphasis is given to the review of previously published records providing evidences of changes in the vegetation communities and their altitudinal distribution that can be related with climatic changes or anthropogenic impacts.

Central Mexico

Geological Scenario

The Trans-Mexican Volcanic Belt (TMVB) (Fig. 6.1) extends between 19° and 21° N latitude from the Gulf of Mexico to the Pacific coast, with an E–W direction (Gomez-Tuena et al. 2005). This geological province is a continental magmatic arc that developed in the mid-Miocene (ca. 19 Ma) as result of the subduction of the Cocos and Rivera plates below the North America plate, and its activity continues

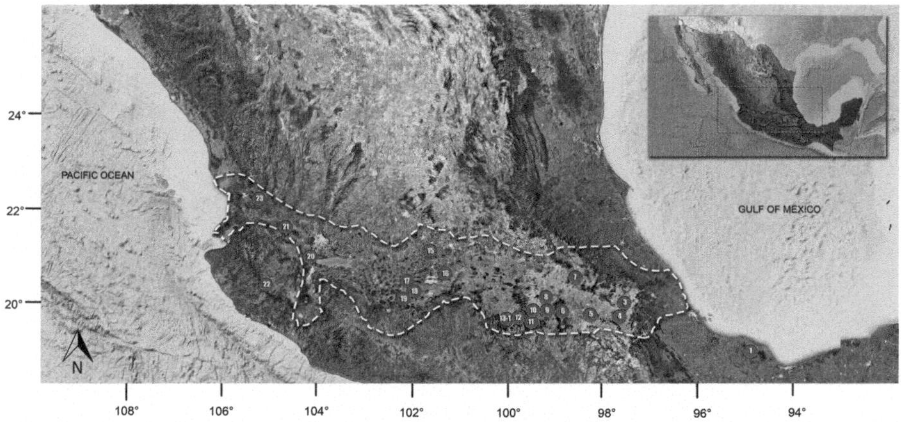

Fig. 6.1 The TMVB is marked with dotted line in the map of Mexico, and the numbers indicate the sites discussed in the text (Table 6.1)

up to modern times (Ferrari et al. 2012). Intense volcanic activity in the region led to the formation of the highest mountains like Iztaccihuatl (5286 m), Popocatepetl (5452 m), and Pico de Orizaba (5675 m). Extensional tectonic regimes during the Pliocene and the Quaternary developed several closed basins (Ferrari et al. 2012).

Topography is the most significant feature of this geological province, and large areas have altitudes of between 1000 and 2500 m asl. The regional altitudinal gradient is evident from an E-W profile (Fig. 6.2). The lower western sector has mean altitudes of <2000 m asl, and warmer climates and the higher central and eastern basins have higher altitudes, up to ~2600 m asl, and cooler climates (Castellano de Rosas 2007; Caballero et al. 2010). The hydrology of the region is closely related with the active volcanic and tectonic environments, which favors the presence of closed basins with the development of lacustrine systems. The lakes in the area vary from extensive and shallow systems like Texcoco (EdoMex) and Cuitzeo (Michoacán) to smaller and deeper crater lakes like Santa María del Oro (Nayarit) and Aljojuca (Puebla) (Sigala et al. 2017).

Climate Scenario

The climate is characterized by a summer rainy season with most of the precipitation concentrated between June and September. It occurs when the trade winds bring moisture from the Caribbean Sea and the Gulf of Mexico and the intertropical convergence zone (ITCZ) reaches its northerly position. Tropical storms and hurricanes in both the Atlantic and the Pacific Oceans are also an important source of summer moisture for the region. Besides, the western basins along the TMVB are part of the core area of the Mexican Monsoon system (Metcalfe et al. 2015), and therefore they also receive summer monsoon style precipitation from the Pacific

Table 6.1 Palaeoecological sites of central Mexico

Site name	Map key	Elevation (m)	Lat.	Long.	References
Lake Verde	1	100	18° 03′ N	95° 20′W	Lozano-García et al. (2010)
Cofre de Perote	2	3717	19°30′N	97°09′W	Caballero-Rodríguez et al. (2018)
Oriental Jalapasquillo	3	2400	19°13′N	97°25′W	Ohngemach (1977); Straka and Ohngemach (1989)
Aljojuca crater lake	4	2376	19°05'N	97°32'W	Bhattacharya et al. (2015)
Tlaloqua II crater lake	5	3100	19°13′N	98°02′W	Ohngemach and Straka (1983); Straka and Ohngemach (1989)
Valle Agua el Marrano	6	3860	19°12′N	98°39″W	Lozano-García and Vazquez-Selem (2005)
Lake Chalco CHAB	7	2240	19°15′N	19°15′N	Lozano-García et al. (1993), Caballero and Ortega (1998), Caballero et al. (2019)
Lake Chalco CHAD	7	2240	19°15′N	19°15′N	Lozano-García and Ortega-Guerrero (1998)
Lake Chalco CHAE	7	2240	19°15′N	19°15′N	Sosa-Nájera (2001), Correa-Metrio et al. (2013)
Lake Xochimilco	8	2240	19°13′N	99°08′W	Ortega-Guerrero et al. (2018)
Lake Texcoco	9	2210	19°28′N	99°58′W	Lozano-García and Ortega-Guerrero (1998), Sedov et al. (2010)
Lake Tecocomulco	10	2500	19°51′N	98°23′W	Caballero et al. (1999)
Lake Lerma SCA	11	2570	19°10′N	99°32′W	Caballero et al. (2002), Lozano-García et al. (2005)

(continued)

Table 6.1 (continued)

Site name	Map key	Elevation (m)	Lat.	Long.	References
Lake La Luna	12	4200	19°06′N	99°45′W	Cuna et al. (2014)
Lake Zempoala	13	2800	19°03′N	99°18′W	Almeida el al. (2005)
Quila	14	3100	19°04′N	99°19′W	Almeida el al. (2005)
Hoya Rincon de Parangueo, Hoya San Nicolas	15	1800	20°23′N	100°50′W-101°19′W	Park et al. (2010), Domínguez et al. (2019)
Lake Cuitzeo	16	1880	19°53′N-20°04′N	101°15′W	Israde et al. (2002, 2010)
Lake Zacapu	17	1970	19°50′N	101°40′W	Xclhuantzi-López (1994)
Lake Zirahuen	18	2075	19°26′N	101°44′W	Ortega-Guerrero et al. (2010), Lozano-García et al. (2015)
Lake Patzcuaro	19	2035	19°36′N	101°39′W	Bradbury (2000), Metcalfe et al. (2007)
Lake San Marcos	20	1352–1346	20°17′N-20°05′N	103°34′W-103°30′W	Castillo et al. (2017)
Etzatlan-Magdalena Basin	21	1360	20°47′N	104°39′W	Vázquez et al. (2017)
Sierra Manantlan Forest hollow	22	2570	21°23′N	103°52′W	Figueroa-Rangel et al. (2008)
Lake Santa Maria del Oro	23	750	19°42′N	104°35′W	Rodríguez-Ramírez et al. (2015)

Ocean. By the end of autumn and during winter, the ITCZ migrates southward, while the high-pressure cells over the North Pacific and North Atlantic cause dry conditions to most of Mexico.

Central Mexico is a sensitive region to recurrent atmosphere-ocean climatic oscillations like El Niño-Southern Oscillation (ENSO). The negative ENSO events (warmer tropical Pacific conditions during El Niño) are expressed as warmer and relatively dry summers and slightly cooler and wetter winters, while the positive ENSO events (cooler tropical Pacific conditions during La Niña) are related with higher than average summer precipitation (Pavia et al. 2006; Bravo et al. 2010; Magaña et al. 2003; Magaña 2004; Caballero et al. 2013)

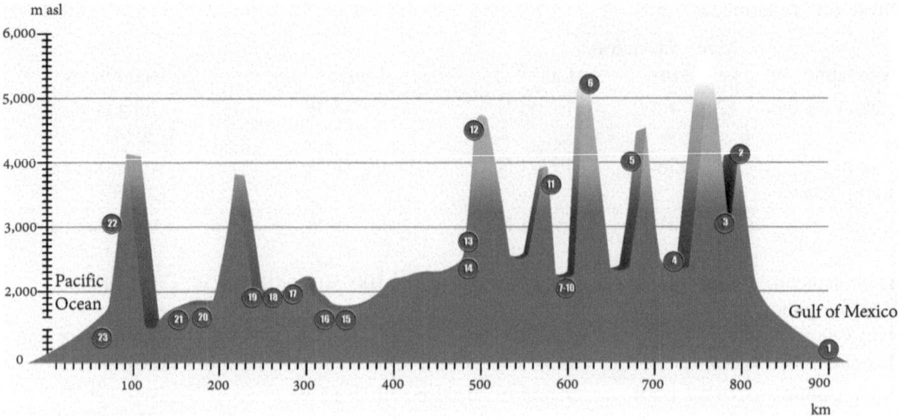

Fig. 6.2 Elevation profile of central Mexico showing the sites mentioned in the text (Table 6.1). Mountain name abbreviations from the east (E) to west (W) are as follows: CP Cofre de Perote, PO Pico de Orizaba, LM La Malinche, IZ Iztaccihuatl, AJ Ajusco, NT Nevado de Toluca, TA Tancitaro, NC Nevado de Colima

Plant Communities

Several vegetation types are present across these diverse landscapes. Alpine grasslands are generally found on the highest altitudes above the timberline and at lower elevations, and the montane forests, coniferous forests, and oak forests account for 60% of the vegetation (Luna et al. 2007). The timberline in the mountains of central Mexico varies in altitude from about 3800–4100 m asl, and the highest treeline in the world is at Pico de Orizaba, at 4200 m (Beaman 1962; Lauer 1978; Lauer and Klaus 1975). Below the treeline, the high elevation, sub-alpine *Pinus hartwegii* forest occurs from 3650 to 4200 m asl forming monospecific stands at the ecotone with alpine grasslands, under extreme temperatures (Perry 1991).

The coniferous forests are abundant as this kind of vegetation is favored by the acidic (pH 5–7), volcanic-derived soils common in the TMVB (Rzedowski 2006). *Pinus* forests (with ~48 species) are widely distributed, and except for *P. hartwegii*, they are present at elevations between 1500 and 3000 m asl, in temperate sub-humid climates, with annual average temperatures of 6–28 °C. The most common species in the area is *Pinus montezumae*, and it is substituted by *P. pseudostrobus* in less humid conditions and below 2000 m asl by *P. rudis* and *P. teocote*. *Pinus cembroides* (piñon pines) are generally found in the dryer areas with precipitation of less than 350 mm/yr. The volcanic history of central Mexico is linked with the evolutionary history of this genus, providing microhabitats that favored hybridization and adaptive radiation (Styles 1993). *Abies religiosa* forests in central Mexico are present at elevations from 2400 to 2600 m asl, with mean annual temperatures varying between 7 and 15 °C and precipitation of >1000 mm/yr. or in ravines protected from the high insolation (Rzedowski 2006).

Other prominent montane forests in the highlands of central Mexico are the *Quercus* forests, with 161 species, a highly diverse vegetation type. It occurs at elevations from 350 to 2000 m asl, with mean annual precipitation of 600–200 mm/yr. and annual average temperature of 10–26 °C, but more frequently in areas with temperature between 12 and 20 °C (Rzedowski 2006). The mesophytic or cloud forests develop in the coastal slopes at mid-elevations (1000–2500 m asl) above the lowland tropical vegetation and below the *Pinus*, *Pinus-Quercus*, and *Quercus* forests (Villaseñor 2010). Cloud forests are located at areas with mean annual precipitation higher than 1000 mm/yr., usually between 1500 and 3000 mm/yr., and the mean annual temperatures vary between 9 and 12 °C. This plant community presents high species diversity (Rzedowski 2006).

Holocene Environment

The present interglacial, the Holocene (11.7 cal ka BP to present), is characterized by an increase in global temperature after the cooler last glaciation. Based on estimations of terrestrial and marine temperature proxies, the beginning of the Holocene (11.7 cal ka BP) is marked by an increase in global temperature of 3–8 °C compared with the temperature of the last glacial maximum (21 ka cal BP) (Rehfeld et al. 2018). Besides, a lower global temperature variability led to relatively more stable climates over the Holocene. Temperature changes during this epoch promoted hydrological changes that affected plant community composition and distribution. The Holocene epoch is divided by two cold events at 8.2 ka cal BP and 4.2 ka cal BP (Walker et al. 2012) into the Greenlandian (11.7–8.2 cal ka BP), Northgrippian (8.2–4.2 cal ka BP), and Meghalayan (4.2 cal ka BP to today) ages.

Estimates of Past Temperature in Central Mexico

Important temperature changes have been reconstructed from different proxies during the transition from the Late Pleistocene to the Holocene. However, few of them provide quantitative estimates of these changes (Caballero et al. 2010, 2019). The temperature reconstruction based on equilibrium line altitudes (ELAs) of the mountain glaciers of central Mexico, particularly at the Iztaccihuatl Volcano, is one of them (Vazquez-Selem and Heine 2011). Other sources of quantitative estimates of the past temperature in the region are pollen- and diatom-based transfer functions, which allow estimating changes in temperature by calibrating fossil assemblages with modern ones present at different sites along a temperature gradient. Pollen- and diatom-based temperature transfer functions in central Mexico are available only for Lake Chalco, located at the southern part of the Basin of Mexico (Correa-Metrio et al. 2013; Caballero et al. 2019).

The lowest ELAs recorded between 20 and 14 cal ka BP infer that temperature decreased by 7.6–6.2 °C in central Mexico during the last glaciation (Vazquez-Selem and Heine 2011; Caballero et al. 2010). Pollen- and diatom-based transfer functions on the other hand suggest slightly lower maximum temperature decrease (4–5 °C, Correa-Metrio et al. 2013; Caballero et al. 2019). Deglaciation at the end of the Pleistocene generally led to the retreat of mountain glaciers, but several smaller re-advances are documented. The first one, dated at 12.0–10.0 cal ka BP, overlaps with the Younger Dryas cold event (12 cal ka BP) and the onset of the Holocene (11.7 cal ka BP), with an estimated temperature decrease of 4.4–5.4 °C. For the Younger Dryas, the pollen- and diatom-based transfer functions suggest smaller temperature decrease (about 1.5–2 °C). However, during the Greenlandian, they reconstructed important temperature increases, for example, warmer conditions were evident since ca. 11.5 ka cal BP from the diatom data, with a maximum warming (+3 °C) from 10.5 to 8.5 cal ka BP. The pollen data recorded this temperature increase at a slightly later date, that is, between 9 and 5 cal ka BP. Another glacial advance was recorded from 8.5 to 7.5 cal ka BP, contemporary with the 8.2 cal ka BP cold event that marks the transition between the Greenlandian and the Northgrippian. The ELAs estimated a temperature decrease of 2.6–3.3 °C during this event. This cold event is not clear in the pollen- and diatom-based transfer functions from Lake Chalco, which on the other hand show that temperatures reduced during the Northgrippian. Finally, a last glacial advance was recorded during the Meghalayan, which was correlated with the Little Ice Age (AD 1400–1900), but without a precise dating (Vazquez-Selem and Heine 2011).

Estimates of Past Moisture in Central Mexico

Paleolimnological records allow inferring past changes in available moisture during the Holocene. This kind of records can identify dry periods from intervals of reduced lake levels, increased in salinity, or reductions in the surface runoff. Along the TMVB, the lacustrine basins with paleolimnological records are distributed along an E-W altitudinal transect from the Gulf of Mexico to the Pacific Ocean (Fig. 6.2). The reviewed records for this section are also included in Table 6.1, where the corresponding bibliographic references are listed. For the sake of clarity, in this compilation, the names of the sites are mentioned in parenthesis, and an emphasis is given on the general trends.

During the Greenlandian (11.7–8.2 cal ka BP), many lacustrine records along the TMVB show sedimentary hiatuses (Tecocomulco, Texcoco, Zirahuen), saline environments (Chalco, Xochimilco, Cuitzeo), and trends to lower lacustrine levels (Upper Lerma, Patzcuaro, San Marcos, Etzatlan) that culminate during the 8.2 cal ka BP cold event that marks the end of this age. This trend to saline environments and generally low lake levels can be explained by the maximum values of summer insolation that promoted higher evaporation in the region. This trend is present in the lakes on the central and western sections of the TMVB. However, there is no paleolimnological information from the eastern sector of the TMVB for this interval.

The Northgrippian (8.2–4.2 cal ka BP) began, as explained in the previous paragraph, with generally low lake levels. Many of the records showed a recovery (Texcoco, Tecocomulco, Upper Lerma, Zirahuen, Etzatlan) or a transition from saline toward fresher conditions (Chalco, Xochimilco, Cuitzeo) by 6–5 cal ka BP. However, the records showed another dry period (Patzcuaro, Cuitzeo, Upper Lerma, Etzatlan) correlating with the cold event that marks the end of this age. Again for this period, there are no paleolimnological records from the eastern sector of the TMVB.

For the Meghalayan (4.2 cal ka BP to present), the number of paleolimnological records increased. These records covered the full extent of the TMVB for the last 2 cal ka BP. Some of these records (Upper Lerma, Patzcuaro, Etzatlan) showed that the dry spell that started by the end of the Northgrippian extended until about 3 to 2 ka cal BP. Lake levels recovered by that time, but soon they gave way to a new dry interval dating at around 1.5 cal ka BP (AD 600) to 0.8 cal ka BP (AD 1100). This dry period correlates with the demise of the Mesoamerican cultures during the Late Classic (AD 600–900) and is identified in nearly all the paleolimnological records from the region (Verde, Aljojuca, Upper Lerma, Patzcuaro, Zirahuen, Etzatlan, Juanacatlán, Santa María del Oro), from the Gulf Coast to the Pacific Ocean, representing the most important paleolimnological trend for the Meghalayan (Rodríguez-Ramírez. et al. 2015).

During the cooler climates of the Little Ice Age, several records showed evidences of dry conditions (La Luna, Juanacatlan, Santa Maria del Oro, Patzcuaro). However, the dates slightly differ between sites, and in some it shows a two-phase cooling pattern, that is, AD 1400 to 1600 and/or AD 1660 to 1770. Only at the lowland tropical site of Lago Verde, in Los Tuxtlas, a twofold increase in lake levels is evident. This two-phase behavior during the Little Ice Age that has also been recorded at other sites has been correlated with the Spörer (1450–1550) and Maunder (1650–1750) solar minima (Lozano et al. 2007).

Archaeological Background

Archaeological data regarding the peopling of the Americas is a controversial issue. Recent archaeological research suggests that Beringia could have been populated as early as 20 cal ka BP (Bourgeon et al. 2017). However, other archaeological data point to a later date, that is, 14 cal ka BP, during the deglacial. The early human migrations into America could have taken place through different routes, for example through an ice free corridor in North America between the Laurentide and Cordilleran ice sheets (Dixon 2001) or an alternative, that has become relevant in the recent decades, is the migration route through the Pacific coast, facilitated by the exploitation of coastal resources (Dixon 2001; Dillehay et al. 2008).

The transition from hunters-gatherers to agricultural villages was a gradual process that started during the Greenlandian, that is, 10 cal ka BP (Ranere et al. 2009), and consolidated during the so-called Archaic period (9–4 cal ka BP). Mexico has been identified as one of the main centers in the development of agriculture, with maize (*Zea mays* ssp. *mays*) as the most important crop in the region. Maize was domesticated from teosinte (*Zea mays* ssp. *parviglumis*) in the central Balsas region (Piperno et al. 2009; Ranere et al. 2009) by ca. 9 cal ka BP, and the molecular studies of maize landraces by Matsuoka et al. (2002) allowed to establish that there was a single domestication event after which diversification took place, extending from the Balsas region to the highlands of central Mexico, and from there, it spreads to the rest of Mesoamerica and surrounding regions.

Vegetation Paleorecord

Paleoecological research documented the vegetation history, lake levels, and erosion rates and related them to climate variability and other drivers. Some of these records cover the Last Glacial Maximum, as well as the transition from the Late Pleistocene to the Holocene (Caballero et al. 2010). Because of the region's intricate topography, climate change promoted variations in species distribution along the altitudinal gradient, and species migrations resulted in new configurations in plant communities. The climatic fluctuations over the Pleistocene, together with volcanic and tectonic activities, promoted topographic changes. A model of sky-island vegetation dynamics has been proposed to explain the biodiversity observed in the mountainous regions of Mexico (Mastretta-Yanes et al. 2015). Past changes in montane vegetation composition emerged from the various palynological studies carried out in the sedimentary sequences collected from lakes and bogs (Table 6.1).

According to studies of glacial advances in the mountains of central Mexico, most of the glaciers continued to have ELAs similar to those at the LGM until about 14 cal ka BP (Vazquez-Selem and Heine 2011; Caballero et al. 2010). This is consistent with the pollen record at the Tlaloqua II crater (3100 m asl), where the alpine grasslands were present during this period and the upper timberline was at ca. 3000 m asl, nearly 1000 m below its modern values (Ohngemach 1977; Straka and Ohngemach 1989). In the mid-elevations, pollen spectra from Lake Zirahuen (2075 m asl) showed changing vegetation during the deglacial, when dry and cold environments favored the dominance of *Pinus* forest. By the end of the Pleistocene (ca. 13.5 cal ka BP), the glaciers retreated (Vázquez-Selem and Heine 2011), and a rapid vegetation turnover was recorded (at 13.3 cal ka BP), with the presence of *Quercus* and *Alnus* and more diverse montane forests (Lozano-García et al. 2015). Similar palynological trends have also been documented in other records from mid-elevations (ca. 2000–2600 m asl). For example, increases in pollen percentages of *Quercus, Abies*, and *Fraxinus* are recorded at the western sector of central Mexico (lake Zacapu; Correa-Metrio et al. 2012), and pollen indicative of dry forest and shrublands was reported from Rincon de Parangueo (Dominguez-Vázquez et al. 2019).

For the Greenlandian (11.7–8.2 cal ka BP), insolation was the main forcing that affected precipitation and seasonality in central Mexico. The strength of summer insolation was at its highest during the beginning of the Holocene at ca. 12 cal ka BP. The high seasonality declined over the course of the last 10 ka, reaching its lowest values over the last 4 cal ka. Paleotemperature changes during the Late Pleistocene and Holocene were reconstructed using other proxies. Cooler conditions continued into the earliest Greenlandian (11.7–10 ka cal BP) according to the glacial history of the central volcanoes and the pollen-based transfer functions, even though diatom-based transfer functions recorded an earlier warming (since 11.5 cal ka cal BP). The high elevation pollen records showed periglacial conditions at the Agua El Marrano (3860 m asl) and Cofre de Perote (3717 m asl). However, alpine grasslands developed at the lower elevation site Tlaloqua II crater (3100 m asl) (Ohngemach 1977; Straka and Ohngemach 1989). The increase in temperature at ca. 10 cal ka BP, with higher than present summer insolation, promoted the establishment of alpine grasslands in all the high elevation sites. For example, the vegetation cover at the Agua El Marrano (3860 m asl) located in the NE slope of the Iztaccihuatl (Fig. 6.3) was alpine grasslands with the timberline at 500–700 m below its present position from 10.5 to 7 cal ka BP (Lozano-García and Vázquez-Selem 2005), and it continued up to 8.5 cal ka BP in the Tlaloqua II crater (Ohngemach 1977; Straka and Ohngemach 1989). At Lake Quila (3100 m asl), *Alnus* forest with low percentages of *Pinus* was present from ca. 11.6–10.6 cal ka

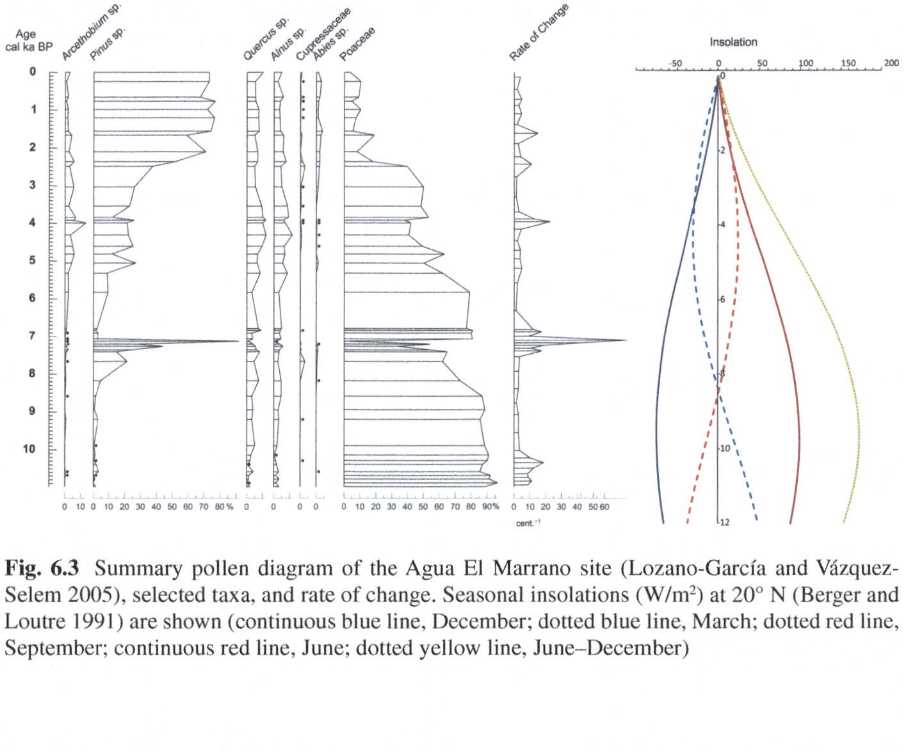

Fig. 6.3 Summary pollen diagram of the Agua El Marrano site (Lozano-García and Vázquez-Selem 2005), selected taxa, and rate of change. Seasonal insolations (W/m^2) at 20° N (Berger and Loutre 1991) are shown (continuous blue line, December; dotted blue line, March; dotted red line, September; continuous red line, June; dotted yellow line, June–December)

BP, and the contribution of *Pinus* forest increased afterward (Almeida-Leñero et al. 2005).

At mid-elevations sites, the forest composition changed between 12.5 and 9.5 cal ka BP. *Pinus* forests developed but the pollen spectra of Lake Patzcuaro, Lake Zirahuen, Lake Zacapu, Lake Chignahuapan, and Lake Chalco revealed increase in *Quercus, Alnus*, and *Abies* (Watts and Bradbury 1982; Xelhuantzin-López 1994; Lozano-García and Ortega-Guerrero 1998; Lozano-García et al. 2005, 1993; Torres-Rodríguez et al. 2012). This ecological turnover in vegetation is illustrated with the pollen diagram of Lake Zirahuen, where the rate of change remains below 20, then increases to 60, and becomes lower again (Fig. 6.4). By the end of the Greenlandian, at ca. 8 cal ka BP, *Pinus-Quercus* mix forests were present at several sites: Chalco, Zirahuen, Patzcuaro.

A significant palynological signal of warming during the Greenlandian was the disappearance of *Picea* (spruce) in pollen spectra at the high elevation basins (>2000 m asl), located in the eastern sector of the TMVB. The last record of *Picea* pollen at Tlaloqua II crater was at ca. 10 cal ka BP, and it was observed at ca. 7 cal ka BP and at ca. 10 cal ka BP in Lake Texcoco and Lake Chignahuapan, respectively. However, no *Picea* pollen has been reported in pollen sequences of the Late Pleistocene from the lower eastern basins (i.e., Patzcuaro, Zirahuen, Zacapu and Cuitzeo) that have records extending into the last glacial. This absence is indicative that the climate in these lower basins was not cold enough to sustain this conifer. Today, relict populations of *Picea chihuahuana* can be found at sites 700 km north from the Basin of Mexico, with mean annual temperatures from 9 to 12 °C and precipitations of 600 a 1300 mm/yr. It has been suggested that the high seasonality of the Greenlandian, with cold winters and warmer summers, compared with present conditions favored the ITCZ to shift toward a northern position carrying more

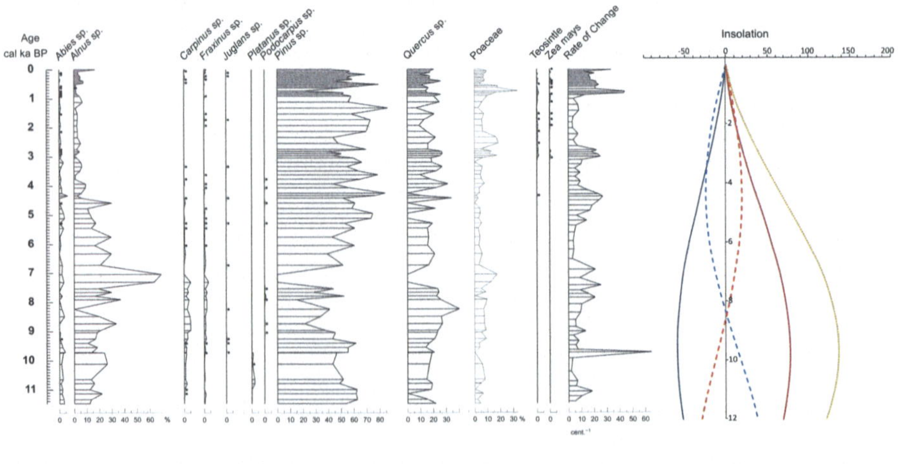

Fig. 6.4 Summary pollen diagram of Lake Zirahuen (Lozano-García et al. 2013), selected taxa, and rate of change. Seasonal insolations (W/m^2) at 20° N (Berger and Loutre 1991) are shown (continuous blue line, December; dotted blue line, March; dotted red line, September; continuous red line, June; dotted yellow line, June–December)

precipitation to the region (Metcalfe et al. 2015). However, the presence of hiatuses, low lake levels, and saline conditions is indicative of high evaporation that dominated the hydrological balance.

A transition from high to lower seasonality occurred during the Northgrippian, and it was translated into variable climatic conditions for central Mexico (Metcalfe et al. 2015). A compositional change in vegetation was recorded between 7.5 and 7.1 cal ka BP at Agua el Marrano (Fig. 6.3). The presence of a pollen assemblage similar to the modern pollen rain of *Pinus hartwegii* forest suggests that this community was established close to this site. The rate of change increased abruptly up to 70, and the pollen indicators of swampy conditions reduced, suggesting the establishment of drier conditions during this time. Afterward, from 7.1 to ca. 3 cal ka BP, the alpine grasslands reestablished in the site, and the rate of change values remained lower.

It was possible to detect significant changes in the forest composition from the high-resolution Zirahuen record, located at mid-elevations. A sudden collapse in *Alnus* was probably linked to the 8.2 ka cal BP cold oscillation. Another period of change was recorded between 7.3 and 7.1 cal ka BP with increases in *Alnus* pollen and reduction in other pollen trees such as *Pinus* and *Quercus*. An increase of the rate of change occurred between 7.9 and 7 cal ka BP indicating a compositional change in the community. A pulse of wetter conditions might explain this variation in the plant community. At other pollen records near Zirahuen, the increases in *Alnus* pollen were also recognizable. It occurred in Patzcuaro from ca. 7 to 6 cal ka BP (Watts and Bradbury 1982) and also in Zacapu (Xelhuanzti-López 1994), although the resolution in the last record was lower than Zirahuen.

During the Meghalayan, the last 4.2 cal ka cal, a general trend toward drier conditions was associated to the southward migration of the ITCZ (Metcalfe et al. 2015). The paleolimnological records for the region identify two long dry periods: the first one from 4.2 to 3 or 2 cal ka BP and the second one from 1.5 to 0.8 cal ka BP (AD 600 to 1100). Two shorter dry spells were observed during the Little Ice Age (i.e., AD 1400–1600 and 1660–1770). High-elevation sites showed the establishment of modern conditions at the Agua El Marrano. The record of alpine grasslands decreased, and *Pinus* pollen increased steadily until the *Pinus hartwegii* forest reached its modern range at ca. 3 cal ka BP. *Abies* forests were established at Quila, after a dry period from 5 to 2.5 cal ka BP, with mixed *Pinus-Quercus* forests (Almeida et al. 2005). A drought between 4.5 and 4.2 cal ka BP has been reported from Zirahuen, and it was related to the end of the orbital-controlled early Holocene warm period. Events of high erosion occurred after 4 cal ka BP with more unstable conditions linked to the ENSO forcing (Lozano-García et al. 2015). Dominance of *Pinus* forest, which is the main montane communities in the highlands of central Mexico, has been linked to drier climates. Evidence of *Pinus* forest expansion and its relationship with climatic changes has been documented in the pollen record of Manantlán, in particular with the arid intervals between 4.2–2.5 cal ka BP, 1.2–0.85 cal ka BP (AD 750–1100), and 0.5–0.2 cal ka BP (AD 1450–1750) (Figueroa-Rangel et al. 2008).

In central Mexico, the paleoecological records for the Meghalayan contained not only a climatic signal but also the imprint of human activities. This Late Holocene climatic signal was not evident in all the mid-elevation records, as some of them were perturbed by past human activities around the lakes. New hypothesis regarding the early distribution of gatherers and cultivators in Mexico suggests that agriculture and domestication dispersed inland from western Mexico through river basins such as the Santiago-Lerma and Balsas-Mezcala (Zizumbo-Villareal and Colunga 2008) and settled in numerous lakes of central Mexico with plentiful resources. The most distinctive evidences of human activity in the pollen records across central Mexico are a) the presence of maize pollen in the lacustrine sequences, b) reduction of tree pollen which marks deforestation events in the area, and c) abundant charcoal deposition probably related to the slash and burn practices characteristic of land clearing for agriculture.

Generally, the occurrence of cultigens has been reported, mainly maize, in the pollen records from central Mexico (Watts and Bradbury 1982; Goman and Byrne 1998; Sluyter and Domínguez-Vázquez 2006; Lozano-García et al. 2005; Lozano-García et al. 2007). These grains are not frequent in the lacustrine records because their large size (70–120 m) limits their dispersal. The identification of maize pollen is based on its size as well as its axis/pollen ratio; however, this criterion has been debated because *Zea mays* ssp. *parviglumis* and *Zea mays* subsp. *mays* overlap. An example of an early record of maize pollen comes from the crater lake of Rincon de Parangueo and at La Hoya San Nicolás. This record suggested small agriculture activities at ca. 5.7 cal ka BP, and these assumptions were based on the minor increases of *Zea* and Amaranthaceae (Park et al. 2010), with intensification of human activities at 2.4 cal ka BP. In other sites (e.g., Patzcuaro), the signal of agriculture was tracked by the presence of larger pollen grains attributable to *Zea* (at 3.6 cal ka BP; Watts and Bradbury 1982). In order to disentangle past land-use practices, a multiproxy approach with a combination of terrestrial pollen records including the sporadic presence of cultigen (maize), deforestation data, and charcoal records can be useful. For example, pollen of *Zea mays* ssp. *parviglumis* was found at ca. 6 cal ka BP at Lake Zirahuen. However, no other change in the pollen spectra indicated evidence of food production in the area. Extensive and intensive land use since the last 3 cal ka BP was inferred not only from the maize pollen but also by an increase in the herbaceous pollen assemblage (i.e., Amaranthaceae and Poaceae). All these, along with increases in magnetic susceptibility indicating higher erosion rates and abundant charcoal particles suggesting anthropogenic fires, provided evidence of land-use changes in the area.

In some paleoecological records that covered the last 2.8 cal ka, it was possible to unravel the climate variability imprint, in spite of the increasing intensity of anthropogenic impact. An example of this is the Lago Verde record in the eastern lowlands. A combination of proxies showed prehistoric human occupation with presence of maize and disturbance pollen such as Poaceae and Asteraceae throughout the mid-Classic (ca. BC 250 to 800 AD). Forest regeneration after 800 AD is inferred from the pollen spectra in response to abandonment with the highest tropical forest diversity during the Little Ice Age (Lozano-García et al. 2010;

Lozano-García et al. 2007). All these paleoecological research contributed to improve the understanding of changing environments in central Mexico, an area with a great proportion of the Mexican population. Multiple evidences of climate change, hydrological balance, vegetation history, and earlier human impacts provided indications of the environmental variability over the last 11,700 years. The past climate variability of central Mexico is relevant for understanding the environmental problems associated with the global warming.

References

Almeida-Leñero L, Hooghiemstra H, Cleef AM et al (2005) Holocene climatic and environmental change from pollen records of lakes Zempoala and Quila, central Mexican highlands. Rev Palaeobot Palynol 136:63–92

Beaman JH (1962) The tiemberlines of Iztaccihuatl and Popocatepetl, Mexico. Ecology 43:337–385

Berger A, Loutre MF (1991) Insolation values for the climate of the last 10 millon years. Quat Sci Rev 10:279–317

Bhattacharya T, Byrne R, Böhnel H et al (2015) Climate and cultural linkages in eastern Mexico. Proc Natl Acad Sci 112:1693–1698

Bhattacharya T, Chiang J C, Cheng W (2017) Ocean-atmosphere dynamics linked to 800–1050 CE drying in mesoamerica. Quaternary Science Reviews 169:263–277

Bourgeon L, Burke A, Higham T (2017) Earliest human in North America dated to the Last Glacial Maximum: new radiocarbon dates from Bluefish caves, Canada. PLoS One 12:e0169486

Bradbury JP (2000) Limnological history of Lago de Patzcuaro, Michoacan, Mexico for the past 48,000 years: impacts of climate and man. Palaeogeogr Palaeoclimatol Palaeoecol 163:65–95

Bravo-Cabrera JL, Azpra-Romero E, Zarraluqui Such V et al (2010) Significance tests for the relationship between "El Niño" phenomenon and precipitation in Mexico. Geofis Int 49:245–261

Caballero M, Lozano-García S, Ortega-Guerrero B, Correa-Metrio A (2019) Quantitative estimates of orbital and millennial scale climatic variability in central Mexico during the last~ 40,000 years. Quaternary Science Reviews 205:62–75

Caballero M, Lozano GS, Ortega B et al (1999) Environmental characteristics of lake Tecocomulco, northern basin of Mexico, for the last 50,000 years. J Paleolimnol 22:399–411

Caballero MM, Ortega B, Valadéz F et al (2002) Sta. Cruz Atizapán: a 22-ka lake level record and climatic implications for the late Holocene human occupation in the upper Lerma basin, central Mexico. Palaeogeogr Palaeoclimatol Palaeoecol 186:217–235

Caballero M, Lozano-García S, Vázquez-Selem L et al (2010) Evidencias de cambio climático y ambiental en registros glaciales y en cuencas lacustres del centro de México durante el último máximo glacial. Bol Soc Geol Mex 62:359–377

Caballero M, Rodríguez A, Vilaclara G et al (2013) Hydrochemistry, ostracods and diatoms in deep, tropical, crater lake in Western Mexico. J Limnol 72:512–523

Caballero-Miranda M, Ortega-Guerrero B (1998) Lake levels since about 40,000 years ago at Lake Chalco, near Mexico City. Quat Res 50:69–79

Caballero-Rodríguez D, Correa-Metrio A, Lozano-García S et al (2018) Late-Quaternary spatiotemporal dynamics of vegetation in Central Mexico. Rev Palaeobot Palynol 250:44–52

Castellano de Rosas E (2007) Reconocimiento espacial de los paisajes. In: Luna I, Monrroe JJ, Espinoza D (eds) Biodiversidad de la faja volcánica transmexicana. Universidad Nacional Autónoma de México, México, pp 39–55

Castillo M, Muñoz-Salinas E, Arce JL et al (2017) Early to present landscape dynamics of the tectonic lakes of west-central Mexico. J S Am Earth Sci 80:120–130

Castillo-Batista AP, Figueroa-Rangel BL, Lozano-García S et al (2016) Historia florística y ambiental del bosque mesófilo de montaña en el centro-occidente de México durante la pequeña edad de hielo. Revista Mexicana de Biodiversidad 87:216–229

Correa-Metrio A, Lozano-García MS, Sosa-Najera S et al (2012) Vegetation in Western Mexico during the last 50,000 years: modern analogs and climate in Zacapu Basin. J Quat Sci 27:509–518

Correa-Metrio A, Bush MB, Lozano-García S et al (2013) Millennial-scale temperature change velocity in the continental northern Neotropics. PLoS One 8:e81958

Cuna E, Zawisza E, Caballero M et al (2014) Environmental impacts of Little Ice Age cooling in central Mexico recorded in the sediments of a tropical alpine lake. J Paleolimnol 51:1–14

Dillehay TD, Ramírez C, Pino M et al (2008) Monte Verde: seaweed, food, medicine and the peopling of South America. Science 320:784–786

Dixon EJ (2001) Human colonization of the Americas: timing, technology and process. Quat Sci Rev 20:277–299

Domínguez-Vázquez G, Osuna-Vallejo V, Castro-López V, Israde-Alcántara I, Bischoff JA (2019) Changes in vegetation structure during the Pleistocene–Holocene transition in Guanajuato, central Mexico. Veg Hist Archaeobotany 28(1):81–91

Ferrari L, Orozco-Esquivel MT, Manea V et al (2012) The dynamic history of the Trans-Mexican Volcanic Belt and the Mexico subduction zone. Tectonophysics 522-523:122–149

Figueroa-Rangel BL, Willis K, Olvera-Vargas M (2008) 4200 years of pine dominated upland forest dynamics in west-central Mexico: human or natural legacy. Ecology 89:1893–19007

Goman M, Byrne R (1998) A 5000-year record of agricultural and tropical forest clearance in the Tuxtlas, Veracruz, Mexico. The Holocene 8:83–89

Gómez-Tuena A, Orozco-Esquivel MT, Ferrari L (2005) Petrogénesis ígnea de la Faja Volcánica Transmexicana. Boletín de la Sociedad Geológica Mexicana. Volumen Conmemorativo del Centenario. *Temas Selectos de la Geología Mexicana* LVII:227–285

Israde AI, Garduño-Monroy VH, Ortega MR (2002) Paleoambiente lacustre del Cuaternario tardío en el centro del lago de Cuitzeo. Hidrobiológica 12:61–78

Israde AI, Velázquez-Durán R, Lozano-García MS et al (2010) Evolución paleolimnologica del Lago de Cuitzeo, Michoacán durante el Pleistoceno-Holoceno. Bol Soc Geol Mex 62:345–357

Lauer W (1978) Timberline studies in Central Mexico. Arct Alp Res 102:383–396

Lauer W, Klaus D (1975) Geoecological investigations on the timberline of Pico de Orizaba. Arct Alp Res 74:315–330

Lozano-García S, Ortega-Guerrero B (1994) Palynological and magnetic susceptibility records of Chalco Lake, central Mexico. Palaeogeogr Palaeoclimatol Palaeoecol 109:177–191

Lozano-García MS, Ortega-Guerrero B (1998) Late Quaternary environmental changes of the central part of the Basin of Mexico. Correlation between Texcoco and Chalco basins. Rev Palaeobot Palynol 99:77–93

Lozano-Garcia MS, Vázquez-Selem L (2005) A high-elevation Holocene pollen record from Iztaccíhuatl volcano, central Mexico. The Holocene 15:329–338

Lozano-Garcia MS, Ortega-Guerrero B, Caballero M et al (1993) Late Pleistocene and Holocene paleoenvironments of Chalco lake. Quat Res 40:332–342

Lozano-García MS, Sosa-Nájera S, Sugiura Y et al (2005) 23, 000 years of vegetation history of the Upper Lerma, a tropical high altitude basin in central Mexico. Quat Res 64:70–82

Lozano-García MS, Caballero M, Ortega B et al (2007) Tracing the effects of the Little Ice Age in the tropical lowlands of eastern Mesoamerica. Proc Natl Acad Sci 104:16200–16203

Lozano-García S, Caballero M, Ortega B et al (2010) Late Holocene palaeoecology of Lago Verde: evidence of human impact and climate change in the northern limit of the neotropics during the late formative and classic periods. Veg Hist Archaeobotany 19:177–190

Lozano-García MS, Torres-Rodríguez E, Ortega B et al (2013) Ecosystem responses to climate and disturbances in western central Mexico during the late Pleistocene and Holocene. Palaeogeogr Palaeoclimatol Palaeoecol 370:184–195

Lozano-García S, Ortega B, Roy PD et al (2015) Climatic variability in the northern sector of the American tropics since the latest MIS 3. Quat Res 84:262–271

Luna I, Morrone JJ, Espinosa D (eds) (2007) Biodiversidad de la Faja Volcánica Transmexicana. Comisión Nacional para el Conocimiento y Uso de la Biodiversidad. Universidad Nacional Autónoma de México, Mexico City

Magaña V (ed) (2004) *Los impactos del niño en México*. Centro de Ciencias de la Atmósfera, Universidad Nacional Autónoma de México, Secretaría de Gobernación, México

Magaña VO, Vázquez JL, Pérez JL et al (2003) Impact el El Niño on precipitation in Mexico. Geofis Int 42:313–330

Mastretta-Yanes A, Moreno-Letelier A, Piñero D et al (2015) Biodiversity in the Mexican highlands and the interaction of geology, geography and climate within the Trans-Mexican Volcanic Belt. J Biogeogr 42:1586–1600

Matsuoka Y, Vigouroux Y, Goodman MM et al (2002) A single domestication for maize shown by multilocus microsatellite genotyping. Proc Natl Acad Sci 99:6080–6084

Metcalfe SE, Davies SJ, Braisby JD et al (2007) Long and short-term change in the Patzcuaro Basin, central México. Palaeogeogr Palaeoclimatol Palaeoecol 247:272–295

Metcalfe SE, Jones MD, Davies SJ et al (2010) Climate variability over the last two millennia in the North American monsoon region, recorded in laminated lake sediments from Laguna de Juanacatlán, Mexico. The Holocene 20:1195–1206

Metcalfe SE, Barron JA, Davies S (2015) The Holocene history of the North American monsoon: known knowns' and 'known unknowns in understanding its spatial and temporal complexity. Quat Sci Rev 120:1–27

Ohngemach D (1977) Pollen sequence of the Tlaloqua crater (La Malinche volcano, Tlaxcala, Mexico). Bol Soc Bot Méx 36:33–40

Ohngemach D, Straka H (1983) Beiträge zur Vegetations- und Klimageschichte im Gebiet von Puebla-Tlaxcala. Pollenanaysen im Mexiko-Projekt. Franz Steiner, Wiesbaden

Ortega B, Vázquez G, Caballero M et al (2010) Late Pleistocene: Holocene record of environmental changes in Lake Zirahuen, Central Mexico. J Paleolimnol 44:745–760

Ortega–Guerrero B, Thompson R, Urrutia-Fucugauchi J (2000) Magnetic properties of lake sediments from Lake Chalco, central Mexico, and their palaeoenvironmental implications. J Quat Sci 15:127–140

Ortega-Guerrero B, García LC, Linares-López C (2018) Tephrostratigraphy of the late Quaternary record from Lake Chalco, central México. Journal of South American Earth Sciences 81:122–140.

Park J, Byrne R, Bohnel H et al (2010) Holocene climate change and human impact, central Mexico: a record based on maar lake pollen and sediment chemistry. Quat Sci Rev 29:618–632

Pavia E, Graef F, Reyes J (2006) PDO-ENSO effects in the climate of Mexico. J Clim 19:6433–6438

Perry JP Jr (1991) The pines of Mexico and Central America. Timber Press, Portland

Piperno DR, Ranere AJ, Holst I et al (2009) Starch grain and phytolith evidence for early ninth millennium B.P. maize from the Central Balsas River Valley, Mexico. Proc Natl Acad Sci 106:5019–5024

Ranere AJ, Piperno DR, Holst I et al (2009) The cultural and chronological context of early Holocene maize and squash domestication in the Central Balsas River Valley, Mexico. Proc Natl Acad Sci 106:5014–5018

Rehfeld K, Münch T, Ho SL et al (2018) Global patterns of declining temperature variability from the Last Glacial Maximum to the Holocene. Nature 554:356–359

Rodríguez-Ramírez A, Caballero M, Roy P et al (2015) Climatic variability and human impact during the las 2000 years in western Mesoamerica: evidences of late classic (AD 600-900) and Little Ice Age drought events. Clim Past 11:1239–1248

Rzedowski J (2006) Vegetación de México. Comisión Nacional para el Conocimiento y Uso de la Biodiversidad, México

Sedov S, Lozano-García S, Solleiro-Rebolledo E et al (2010) Tepexpan revisited: a multiple proxy of local environmental changes in relation to human occupation from a paleolake shore section in Central Mexico. Geomorphology 122:309–322

Sigala I, Caballero-Miranda M, Correa-Metrio A et al (2017) Basic limnology of 30 continental waterbodies of the Trans-Mexican Volcanic Belt across climatic and environmental gradients. Boletín de la Sociedad Geológica de México 69:313–370

Sluyter A, Dominguez G (2006) Early maize (*Zea mays* L.) cultivation in Mexico: dating sedimentary pollen records and its implications. Proc Natl Acad Sci 103:1147–1151

Sosa-Nájera MS (2001) Registro palinológico del Pleistoceno tardío–Holoceno en el extremo meridional de la cuenca de México: paleoambientes e inferencias paleoclimáticas. Master Dissertation, Universidad Nacional Autónoma de México

Straka H, Ohngemach D (1989) Late Quaternary vegetation history of the Mexican highland. Plant Syst Evol 162:115–132

Styles BT (1993) Genus *Pinus*: a Mexican purview. In: Ramamoorthy TP, Bye R, Lot A (eds) Biological diversity of Mexico: origin and distribution. Oxford University Press, New York, pp 397–342

Torres-Rodríguez E, Lozano-García S, Figueroa-Rangel BL et al (2012) Cambio ambiental y respuestas de la vegetación de los últimos 17,000 años en el centro de México: el registro del lago de Zirahuén. Revista Mexicana de Ciencias Geológicas 29:764–778

Torres-Rodríguez E, Lozano-García S, Roy P et al (2015) Last Glacial droughts and fire regimes in the central Mexican highlands. J Quat Sci 30:88–99

Vázquez G, Roy PD, Solis B et al (2017) Holocene paleohydrology of the Etzatlán-Magdalena basin in western-central Mexico and evaluation of main atmospheric forcings. Palaeogeogr Palaeoclimatol Palaeoecol 487:149–157

Vázquez-Selem L, Heine K (2011) Late Quaternary glaciation in Mexico. Dev Quat Sci 15:849–861

Villaseñor JL (2010) El bosque húmedo de montaña en México y sus plantas vasculares: catálogo florístico-taxonómico. Comisión Nacional para el Conocimiento y Uso de la Biodiversidad-Universidad Nacional Autónoma de México, Mexico City

Walker MJ, Berkelhammer CM, Bjorck S et al (2012) Formal subdivision of the Holocene series/ epoch: a discussion paper by a working group of INTIMATE (integration of ice-core, marine and terrestrial records) and the Subcommission on Quaternary Stratigraphy (International Commission on Stratigraphy). J Quat Sci 27:649–659

Watts WA, Bradbury JP (1982) Paleoecological studies on the west central Mexican plateau and at Chalco in the basin of Mexico. Quat Res 17:56–70

Xelhuantzi–López MS (1994) Estudio palinológico de cuatro sitios ubicados en la Cuenca de Zacapu: fondo de ciénega, contacto Lomas–Ciénega, pantano interno y Loma Alta. In: Petrequin P (ed) 8000 años de la cuenca de Zacapu, evolución de los paisajes y primeros desmontes. Cuadernos de Estudios Michoacanos, Morelia, pp 81–93

Zizumbo-Villareal D, Colunga-García MP (2008) El origen agricultura, la domesticación de plantas y el establecimiento de corredores biológicos-culturales en Mesoamérica. Revista de Geografía Agrícola 41:85–113

Chapter 7
Integration of Landscape Approaches for the Spatial Reconstruction of Vegetation

Valerio Castro-López, Alejandro Velázquez, and Gabriela Domínguez-Vázquez

Abstract Reconstruction of ancient environmental conditions based on traditional analyses such as pollen and other proxies has provided understanding of climate variation and its influence on vegetation. These studies, however, are restricted to site-specific scales and have overseen the spatial understanding of past vegetation patterns. Hence, reconstruction of spatial distribution patterns of vegetation is yet a pending issue in paleoecological studies. The objective of this chapter is to develop a methodological integration of landscape approaches (paleoecological, bioclimatic, and geographical) to reconstruct spatial distribution patterns of vegetation in the Purepecha region in central Mexico. Correlation of climatic patterns, pollen rain, paleoecological data, and physical landscape components was jointly analyzed with the aid of a geographic information system. Result from this integrated approach indicated that during ca. 1000 B.C. (Preclassic period), climatic conditions were relatively more humid than the current climate. The dry climatic conditions were preferably dominant on valleys, plains, and footslopes. On the contrary, humid conditions were preferably distributed in hills and mountain landforms. Our outcomes provide a reproducible integrated methodology for reconstructing spatial patterns of vegetation. We, further, document for the first time the past spatial vegetation patterns in the core of the Purepecha culture.

Keywords Landscape approaches · Bioclimatology · Pollen rain · GIS · Paleoecological · Mexico

---

V. Castro-López (✉) · A. Velázquez
Centro de Investigaciones en Geografía Ambiental, Universidad Nacional Autónoma de México, Morelia, Michoacán, Mexico
e-mail: vcastro@pmip.unam.mx; alex@ciga.unam.mx

G. Domínguez-Vázquez
Facultad de Biología, Universidad Michacana de San Nicolas de Hidalgo, Morelia, Michoacán, Mexico
e-mail: gdoguez@yahoo.com.mx

N. Torrescano-Valle et al. (eds.), *The Holocene and Anthropocene Environmental History of Mexico*, https://doi.org/10.1007/978-3-030-31719-5_7

Introduction

Vegetation is a key component and one of the most recognizable elements on the Earth surface (Zonneveld 1989). Vegetation distribution and composition are determined by climatic, geomorphological, soil, and ecological factors, which interact simultaneously in long periods of time (Velázquez et al. 2016). While deep knowledge of the description and dynamics of local and regional vegetation is constantly rising (Rzedowski and Calderon 2013), there are still limited methodological contributions to reconstruct vegetation patterns, especially during the last millennium.

There are several approaches focusing at explaining the dynamic nature of vegetation. The paleoecological approach is based on the interpretation of fossil pollen and other proxies (e.g., diatoms, ostracods, charcoal), an objective view of climatic variations (Israde et al. 2010). The approach also provides valuable information on the impact of ancient societies (Islebe et al. 2016). The bioclimatological approach focuses on studying climate–vegetation relationship (Rivas-Martínez et al. 2011). This approach has been recently used as a tool for spatial reconstruction of the native vegetation (Castro-López and Velázquez 2018). The geographical approach relies upon landforms and soils and other physical attributes to delineate vegetation patterns. This approach uses geographic information systems (GIS) as a main integrative tool (Mas et al. 2004; Velázquez et al. 2016) enabling mapping and monitoring of a group of individuals, populations, and plant communities with different spatial resolution (Pedrotti 2013). Although there are numerous investigations to quantitatively reconstruct spatial patterns of vegetation, outcomes are yet far from being completed. Attempts to formalize the approaches have resulted in a great deal of interrelated and practical terminology under multiple forms (Reed et al. 2014). For this reason, the problem persists due to the lack of an integrative methodology that combines different approaches, namely, paleoecological, bioclimatic, and geographical, for contributing to the historical mapping of the vegetation. Any attempt to build a holistic approach will lead to a thorough understanding of the spatial distribution of vegetation. Some integrative methods have been proposed to reconstruct past scenarios from different perspectives. Some take as basis models to estimate the source area of the pollen and their deposits (Sugita et al. 2010). Others come to integrate landscape elements examined from a GIS (Etter et al. 2008; Brouwer 2013). In the same way, Caseldine and Fyfe (2006) integrate specific pollen parameters, vegetation mosaic, and landscape elements to simulate the distribution of pollen in a specific position within the landscape. Carrillo-Bastos et al. (2012) used fossil pollen interpolation methods to reconstruct the vegetation cover and infer precipitation variations. More comprehensive studies are reported by Gopar and Velázquez (2016) under a Boolean prediction framework giving emphasis to bioclimatic zoning. While the results are encouraging, they still offer limited outcomes. Reed et al. (2016) stated that the alternative of developing a holistic approach will tend toward providing detailed solutions to the problems of structure, composition, and spatial distribution of the past, present, and future vegetation.

The integration is paramount to the relevance of contribution to understand the spatial-specific impacts of climate change. However, this integration is still, to our knowledge, absent in scientific literature.

The objective of this chapter is to provide an integrative methodology of approaches (paleoecological, bioclimatic, and geographical) to express the changes of spatial distribution patterns of vegetation in different time periods. This chapter focuses on the Purepecha region between the boundaries of Michoacan and Guanajuato. The region was limited because it has high plant and climatic diversity (Rzedowski and Calderon 2013).

Methods

Study Area

The area of study is located in the central part of the Michoacan-Guanajuato volcanic field (MGVF) (Cano-Cruz and Carrasco-Núñez 2008; Fig. 7.1). The MGVF is controlled by the fault system with east–west direction (Garduño-Monrroy et al. 2011). Volcanic activity has had significant impact on landscape formation from

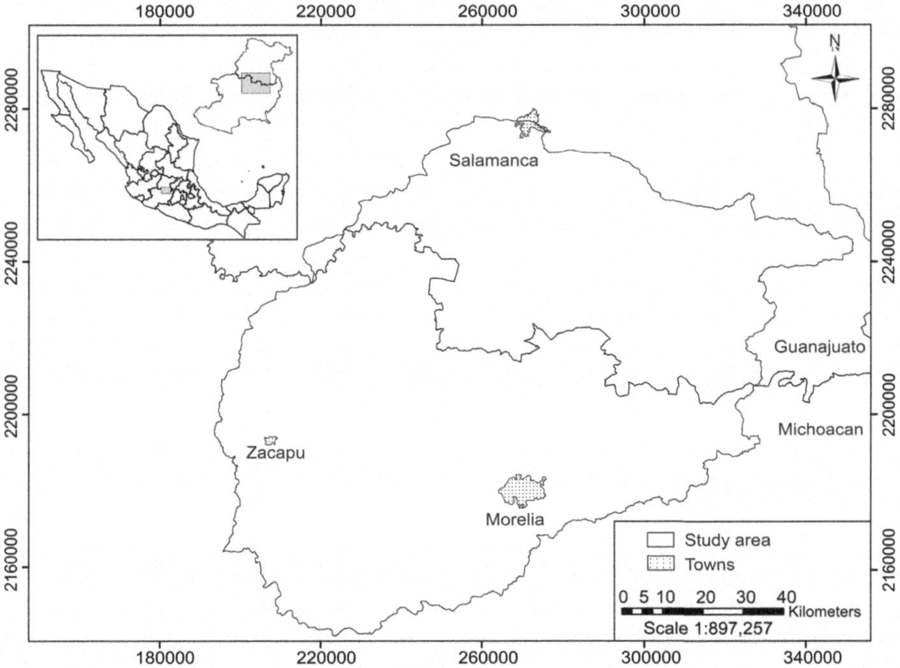

Fig. 7.1 Study area with the most important locations

Oligocene to Holocene (Gómez-Vasconcelos et al. 2015). Rocks that emerge on the surface are basalts, andesites, rhyolites, ignimbrites, and granites (Cano-Cruz and Carrasco-Núñez 2008; Gómez-Vasconcelos et al. 2015).

The altitudinal gradient oscillates from 1700 m.a.s.l. (meters above sea level) in the plains to 3400 m.a.s.l. in the mountain area. Precipitation is controlled by the West winds carrying high precipitation and is concentrated, mainly, in the mountain zone (Reyes et al. 1994). Precipitation varies from 1200 to 600 mm with an average temperature of 20 °C (INEGI 2010). In elevations above 2000 m, it is recurrent to find vegetation types such as *Abies*, *Pinus*, *Quercus*, *Arbutus*, *Alnus*, *Buddleja*, *Crataegus*, and *Clethra* (Pérez-Calix 1996). In elevations less than 2000 m, the predominant genera are *Acacia*, *Bursera*, *Cedrela*, *Condalia*, *Euphorbia*, *Erythrina*, *Eysenhardtia*, *Forestiera*, and *Yucca* and in the proximity of the rivers *Salix*, *Taxodium*, *Alnus*, and *Fraxinus* (Rzedowski and Calderon 2013). Over the last few decades, the plant diversity has experienced high degree of disturbance, especially on foot slopes, valleys, and plains (Villaseñor and Ortiz 2012).

Bioclimatic Analysis

The bioclimatic analysis was carried out using the databases of the Digital Climatic Atlas of Mexico (version 2.0) (http://uniatmos.atmosfera.unam.mx/ACDM/). The layers are in raster format with pixel of 1 km^2. The raster contains average temperature and precipitation information from 1902 to 2011. Raster layers were imported into the ArcGis to calculate bioclimatic indexes. The indexes used were the annual Ombrothermic Index $(Io = (Pp/Tp) * 10)$, Ombrothermic Index of the driest bimester of the driest quarter of the year $(Iod2 = (Ppd2/Tpd2) * 10)$, and Thermicity Index $(It = (T + M + m) * 10)$. The results of the indexes were compared with bioclimatic keys developed by Rivas-Martínez et al. (2011) to identifying bioclimates, thermotypes, and ombrotypes. The generated information was transformed to vector format to facilitate the editing of the data and to improve its spatial expression. The minimum cartographic area was 1 km^2 to maintaining consistency in the reading of the map. Our study only presents the thermotypes and ombrotypes to scale 1:250,000 with projection to NAD 1927 UTM 14 N Zone.

Pollen Rain Analysis

In total, 57 samples of moss and surface water were collected following a climatic gradient from dry tropical to wet temperate conditions. Samples were collected within different types of vegetation: temperate pine–oak forests and tropical dry forests. Pollen grains were extracted using the acetolysis technique of Erdtman (1952). Subsequently, the samples were mounted with glycerin for identification.

An optical microscope was used to identify pollen grains until reaching a minimum of 300 grains of arboreal taxa. Pollen taxa that presented more than 2% were used for canonical correspondence analysis (CCA) to examine whether they were related to climatic patterns. The CCA is a statistical technique for exploring relationships between two multivariate sets of variables (Gauch 1982). The analysis was done with the PC-ORD 5 Software (McCune and Mefford 2006).

Radiocarbon Data and Pollen Diagrams

The paleoecological approach was used to have a chronological framework in the lake sediments (Israde et al. 2010). Research outcomes often lack calibrations, so that dating uncertainties remain (Giesecke et al. 2014). We conducted a review of the paleoecological studies conducted in the study area for C^{14} dating. The database was calibrated to calendar years using Calib 14 (version 7.1) (Stuiver et al. 2018). All the lakes located within the study area were evaluated in the present survey. A first approximation ca. 1000 B.C. showed a greater relationship in the chronological dating. On the dating indicated, percentage data of pollen and the interpretations of the authors were collected for the model. In the same way, the minimum, average, and maximum values of the taxa were collected throughout the diagram.

Bioclimatic Approach

The isobioclimates were used to build a frame of reference in relation to the vegetation cover. The isobioclimates are the combination of bioclimates, thermotypes, and ombrotypes. All correlations were estimated in GIS. Landscape components were overlapped to find correlation patterns (Fig. 7.2). Subsequently, we placed the percentage values of pollen fossil in relation to the reference frame. In this test, CCA results were also used to validate the information. The isobioclimates that were empty subsequently were re-labeled based on the climatic gradient expressed by Gopar et al. (2015).

Statistical Analysis

Data under study that come from different approaches were subjected to canonical correspondence analysis to identify patterns of climatic, plant species assemblages and landscape components (Velázquez et al. 2016). Best ordinations axes with significant values to depict patterns were chosen, and from there plant species assemblages and its distribution were defined.

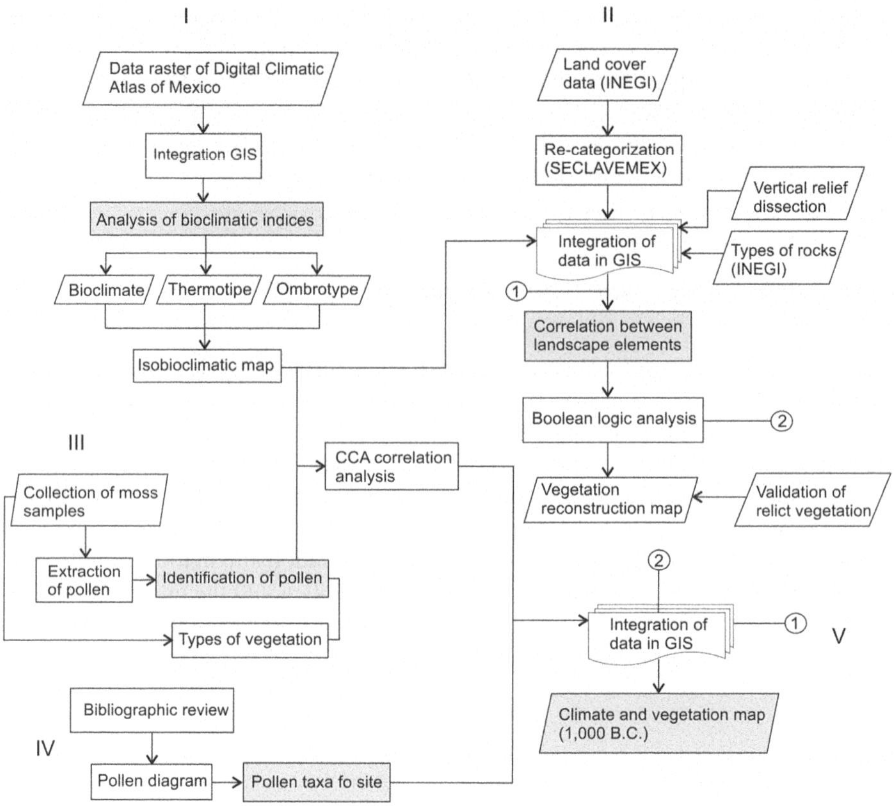

Fig. 7.2 Methodological processes for the reconstruction of the vegetal landscape. Each step must be done independently

Results

The integration of landscape approaches (paleoecological, bioclimatic, and geographical) under a baseline allowed the methodological construction for the reconstruction of climatic patterns and vegetation cover. The current bioclimatic analysis was determined by three thermotypes (temperature gradient) and three ombrotypes (precipitation gradient). The mesotropical thermotypes occupied the largest area, about 84%, whereas the thermotropical about 14%, and the supratropical about 2% (Fig. 7.3).

The ombrotypes is composed of the dry ombrotype with 49% of the total surface, followed by the subhumid ombrotype with 40% and the humid ombrotype around 11% (Fig. 7.4).

The CCA showed that the environmental variables showed a high relation with pollen (Fig. 7.5). The first axis contributed 11.2% with positive regression coefficient between humid climates to subhumid and temperate pollen. Dry climates

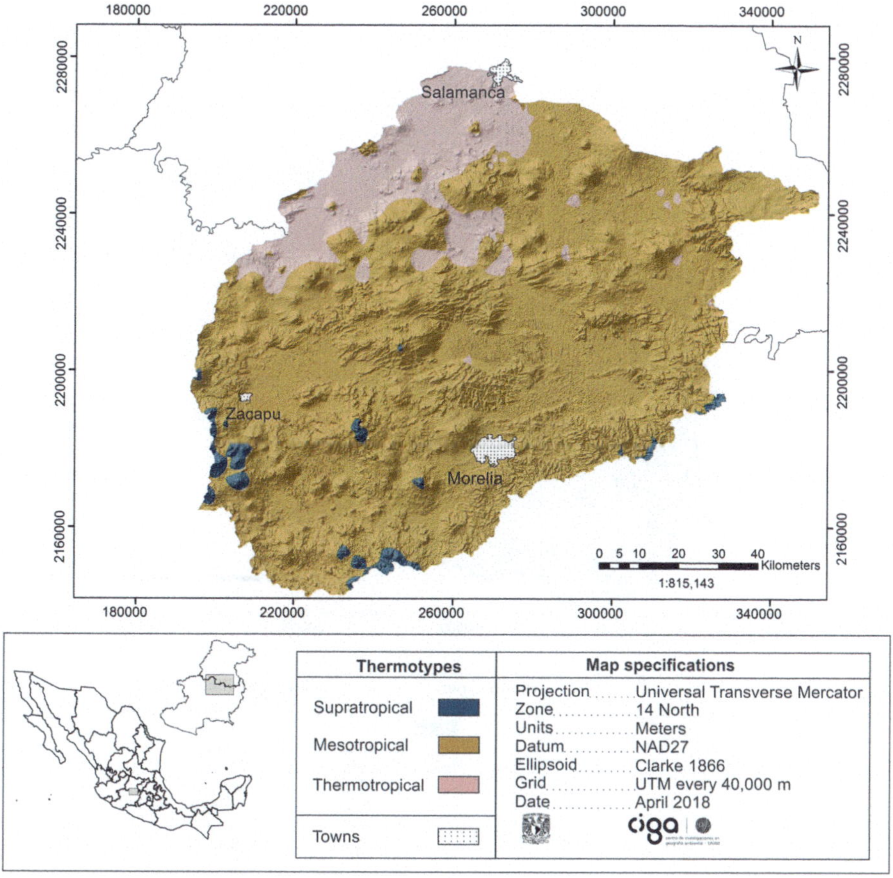

Fig. 7.3 Distribution of current thermotypes (temperature gradient)

showed negative regression in relation to tropical pollen and grasslands. The third axis contributed 6.4% of the variance explained. In this axis, the positive regression coefficient was associated with tropical and temperate taxa. On the other hand, pollen grasslands showed negative values. It should be also added that the Pearson correlation for axis 1 and axis 3 was 0.76 and 0.75 close to 1. Thus, our results indicate a good segregation of the data. The tropical pluviseasonal mesotropical humid (TPMH) isobioclimate presented a greater relation to *Pinus, Quercus*, and *Alnus*. The tropical pluviseasonal mesotropical subhumid (TPMSH) showed not relationship with any taxa specific.

The calibration of C^{14} dating revealed that six sites have corresponded to 1000 B.C., with a variation of ±200 years. Figure 7.6 compares the paleoecological records collected within the study area. The figure only shows the quantitative data of the proxies. For this dating, the study area was represented by mesotropical thermotype and humid ombrotype in most of the surface. On the other hand, a dry isobioclimate was observed (Fig. 7.7).

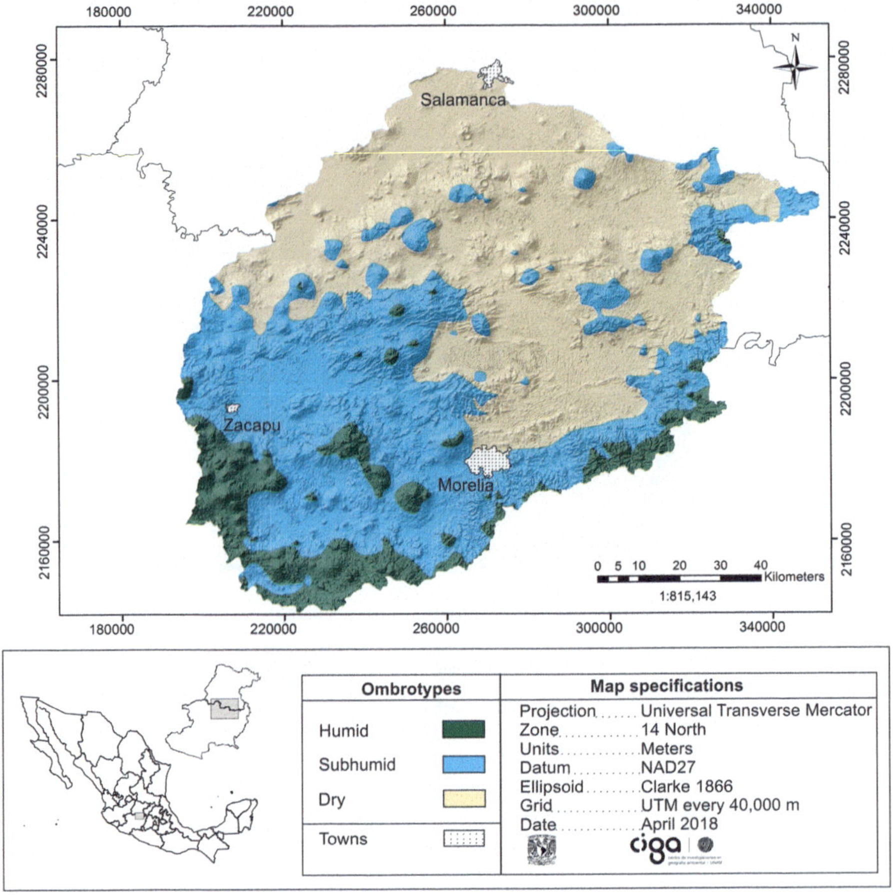

Fig. 7.4 Current distribution ranges of ombrotypes (precipitation gradient)

Discussion

The integration of landscape approaches provides a simple and useful tool for reconstructing past scenarios. A first approximation of ca. 1000 B.C. indicates that the humid ombrotype were located in the areas of hills and foot of mountains and dry ombrotype in the valleys. Torres-Rodríguez et al. (2012) report at the Lake Zirahuen humid conditions with elements of mesophilic forest before this data, and subsequent to this period, a decrease in humidity persists. Rivas-Martínez et al. (2011) argues that climates are positively associated with the altitudinal gradient, thermal and pluviometrical, where its regional distribution follows clear and uniform altitudinal gradients in systematic order. This would reinforce the argument that climatic strips showed a displacement following this climatic gradient.

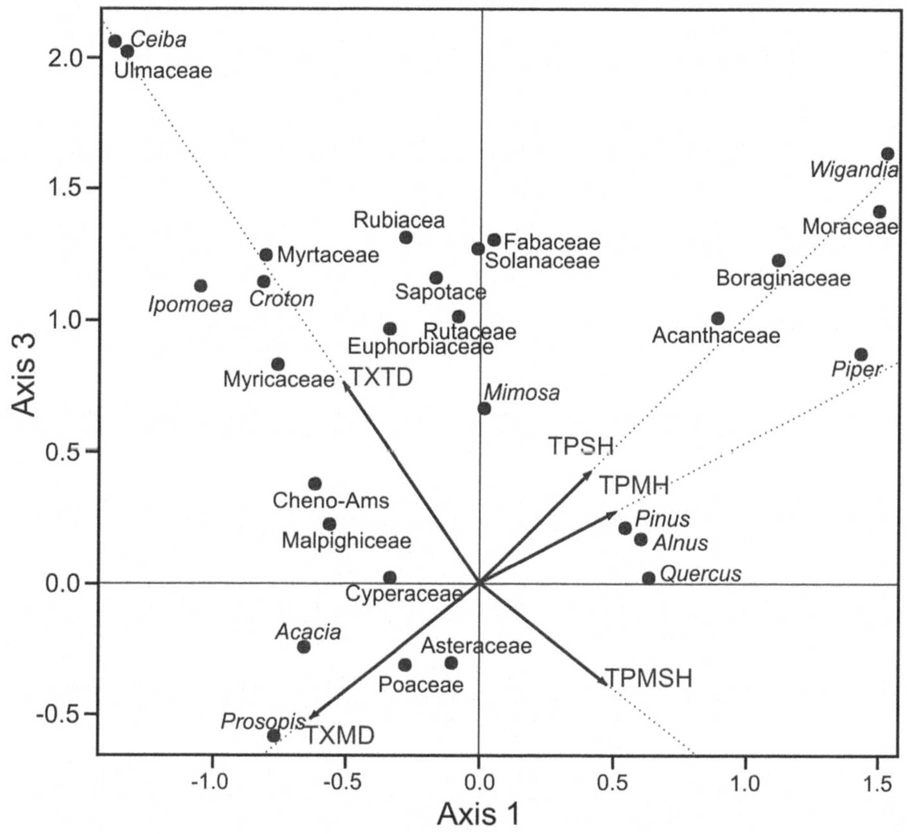

Fig. 7.5 Organizational chart of pollen taxa in relation to current isobioclimates. Environmental variables (isobioclimates) are displayed with arrows. TPSH tropical pluviseasonal supratropical humid, TPMH tropical pluviseasonal mesotropical humid, TPMSH tropical pluviseasonal mesotropical subhumid, TXMD tropical xeric mesotropical dry, and TXTD tropical xeric thermotropical dry

The methodological integration of landscape approaches (paleoecological, bioclimatic and geographical) enabled the recognition of the distribution of climatic patterns. This type of research is widely documented by paleoecological studies on a punctual scale. However, their understanding and spatial expression have been limited due to the scarcity of suitable sites. Therefore, they contribute to understanding part of the process and not the regional complexity. The integration of approaches gives a clear and concise view of the chorological changes in climatic patterns and plant communities in different temporal scales.

GIS is a powerful tool for integrating, analyzing, and managing input and output data from the implemented models (Devereux et al. 2004). Leverington et al. (2002) argue that the use of GIS lies in the rapidity to generate quantitative and georeferenced databases. It is used as an efficient tool to integrate local and spatial database

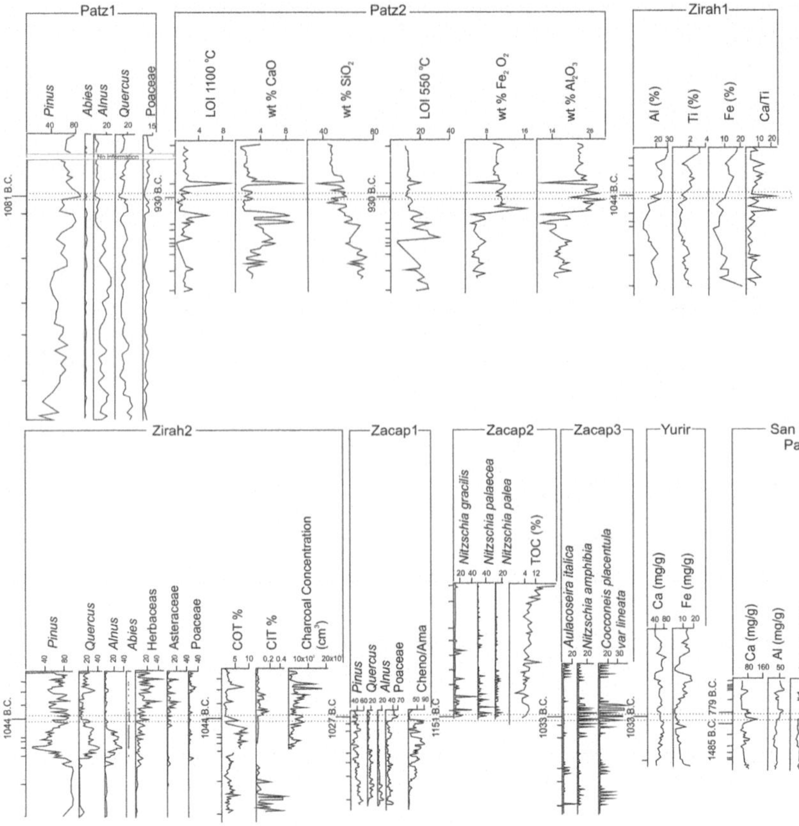

Fig. 7.6 Diagram of proxies located within the study area. The dotted horizontal line indicates the worked age. Patz1 (Watts and Bradbury 1982), Patz2 (Metcalfe et al. 2007), Zirah1 (Ortega et al. 2010), Zirah2 (Torres-Rodríguez et al. 2012), Zacap1 (Lozano-Garcia and Xelhuantzi-López 1997), Zacap2 (Leng et al. 2005), Zacap3 (Arnauld et al. 1997), Yurir (Metcalfe et al. 1989), and San Nicolas de Parangueo (Brown 1984) (*) interpolated data

and to express the scenarios of the historical landscape (Gaudin et al. 2008; Carrillo-Bastos et al. 2012; Brouwer 2013). It must be also added that an adequate integration of landscape components, hierarchically analyzed, will allow recognizing the different patterns that give subsistence to the vegetation types (Gopar and Velázquez 2016) and their different degrees of transformation (Etter et al. 2008) and provide sustainable natural resource solutions on multiple timescales (Reed et al. 2016). The results can be even more promising if more emphasis is given to bioclimatic zones. Gopar and Velázquez (2016) reported that bioclimatic analysis is an excellent tool to identify the climate-vegetation relationship, especially in regions with high geological, ecological, and human complexity. Currently, there is a rise in spatializing the historical landscape. However, the validation and calibration of historical maps has been hampered by the lack of more precise spatial data (Etter et al. 2008). By virtue of this, biophysical characteristics provide an alternative on the conditions

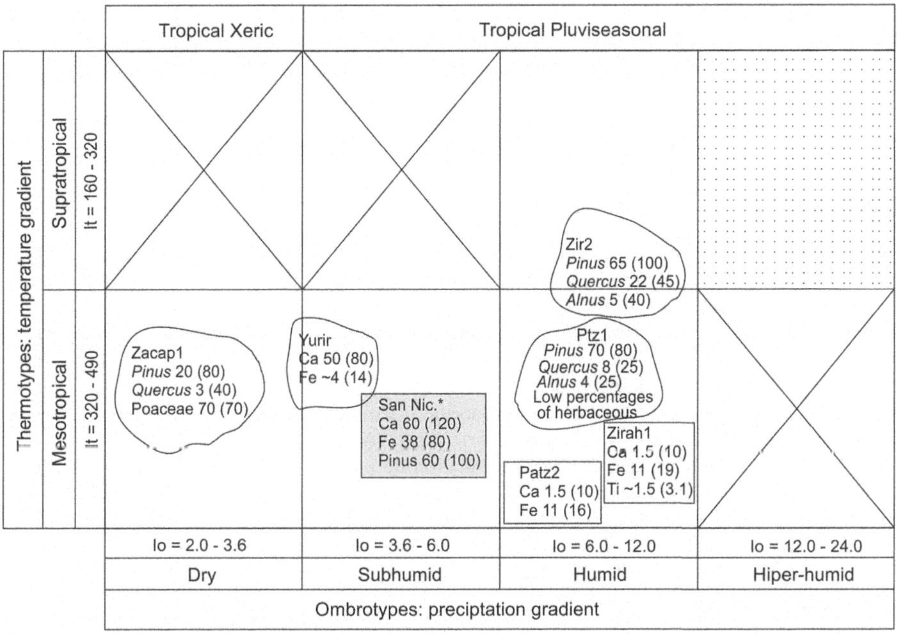

Fig. 7.7 Diversity climates in the area of study ca. 1000 B.C. Climate in "X" is not available in the study area. Climate of the hatched area is located at latitudes above 3100 m.a.s.l. * Data interpolated

of habitat, and their projection with the same physical characteristics reflected the complement of species associated to them (Simonson et al. 2018).

Overall, the analyses here used have proven to be an effective tool to help reconstructing vegetation distribution patterns. Nonetheless, pending uses remain, for example the association of the pollen spectrum with the climate and their relationship with vegetation patterns were not fully completed. It is necessary to recalibrate dates to have a complete chronological framework of the lake sediments. Furthermore, limitations of the study should be considered for future research; in particular, it can be used if topography has not changed significantly. García-Quintana et al. (2016) report that during the last 3000 years B.P., no major changes have occurred compared to what happened during the Pleistocene and the beginning of the Holocene. It should be also mentioned that the difference between lake sedimentation rates makes difficult having a linear chronological framework (Israde et al. 2010). In the same way, it is also necessary to incorporate the orientation of slopes to have a regional overview, since some types of vegetation are confined to certain slope orientations favoring their establishment and development (Rzedowski 2006).

To conclude, the integration of paleoecological, bioclimatic, and geographical approaches showed promising results for understanding past vegetation distribution patterns. This information may be a key to quantitatively assess in a region with a great cultural value.

Acknowledgments We would like to thank the Mexican National Council of Science and Technology (CONACYT) for providing a PhD scholarship to the first author. We thank Rocio Aguirre, Fernando Gopar, and Dulce Bocanegra for their invaluable help and assistance in the fieldwork.

References

Arnauld C, Metcalfe SE, Petrequin P (1997) Holocene climatic change in the Zacapu Lake Basin, Michoacan: synthesis of results. Quat Int 43(97):173–179

Brouwer M (2013) Reconstructing total paleo-landscapes for archaeological investigation: an example from the central Netherlands. J Archaeol Sci 40(5):2308–2320. https://doi.org/10.1016/j.jas.2013.01.008

Brown RB (1984) The paleoecology of the northern frontier of mesoamerica (pollen, mexico, archaeology). Dissertation, University of Arizona, USA

Cano-Cruz M, Carrasco-Núñez G (2008) Evolución de un cráter de explosión (maar) riolítico: Hoya de Estrada, Campo Volcánico Valle de Santiago, Guanajuato, México. Revista Mexicana de Ciencias Geológicas 25(3):549–564

Carrillo-Bastos A, Islebe GA, Torrescano-Valle N (2012) Geospatial analysis of pollen records from the Yucatán Peninsula, Mexico. Veg Hist Archaeobotany 21(6):429–437. https://doi.org/10.1007/s00334-270012-0355-1

Caseldine C, Fyfe R (2006) A modelling approach to locating and characterizing elm decline/landnam landscapes. Quat Sci Rev 25(5–6):632–644. https://doi.org/10.1016/j.quascirev.2005.07.015

Castro-López V, Velazquez A (2018) Reconstruction of native vegetation based upon integrated landscape approaches. Biodivers Conserv. https://doi.org/10.1007/s10531-018-1655-2

Devereux BJ, Devereux LS, Lindsay C (2004) Modelling the impact of traffic emissions on the urban environment: a new approach using remotely sensed data. In: Kelly REJ, Drake NA, Barr SL (eds) Spatial modelling of the terrestrial environment. Wiley, Chichester, pp 227–242

Erdtman OS (1952) Pollen morphology and plant taxonomy, angiosperms, (an introduction to palynology). Alqmus and Wiksell, Stockholm

Etter A, McAlpine C, Possingham H (2008) Historical patterns and drivers of landscape change in Colombia since 1500: a regionalized spatial approach. Ann Assoc Am Geogr 98(1):2–23. https://doi.org/10.1080/00045600701733911

García-Quintana A, Goguitchaichvili A, Morales J et al (2016) Datación magnética de rocas volcánicas formadas durante el Holoceno: caso de flujos de lava alrededor del Lago de Pátzcuaro (campo volcánico Michoacán-Guanajuato). Revista Mexicana de Ciencias Geológicas 33(2):209–220

Garduño-Monroy VH, Soria-Caballero DC, Israde-Alcántara I, Hernández-Madrigal VM, Rodríguez-Ramírez A, Ostroumov M, Rodríguez-Pascua MA, Chacon-Torres AC, Mora-Chaparro JC (2011) Evidence of tsunami events in the Paleolimnological record of Lake Pátzcuaro, Michoacán, México. Geofis Int 50(2):147–161

Gauch HG (1982) Multivariate analysis in community ecology. Cambridge University Press, Cambridge, NY

Gaudin L, Dominique M, Lanos P (2008) Correlation between spatial distributions of pollen data, archaeological records and physical parameters from North-Western France: a GIS and numerical analysis approach. Veg Hist Archaeobotany 17(5):585–595. https://doi.org/10.1007/s00334-008-0172-8

Giesecke T, Davis B, Brewer S et al (2014) Towards mapping the late quaternary vegetation change of Europe. Veg Hist Archaeobotany 23(1):75–86. https://doi.org/10.1007/s00334-012-0390-y

Gómez-Vasconcelos MG, Garduño-Monrroy VH, Macías JL et al (2015) The Sierra de Mil Cumbres, Michoacán, México: transitional volcanism between the Sierra Madre Occidental and the Trans-Mexican Volcanic Belt. J Volcanol Geotherm Res 301:128–147. https://doi.org/10.1016/j.jvolgeores.2015.05.005

Gopar-Merino LF, Velázquez A (2016) Componentes del paisaje como predictores de cubiertas de vegetación: estudio de caso del estado de Michoacán, México. Investigaciones Geográficas (90):75–88. https://doi.org/10.14350/rig.46688

Gopar-Merino LF, Velázquez A, Giménez de Azcárate J (2015) Bioclimatic mapping as a new method to assess effects of climatic change. Ecosphere 6(1):1–12. https://doi.org/10.1890/ES14-00138.1292

INEGI (2010) Compendio de información geográfica municipal, México http://www.inegi.org.mx/geo/contenidos/topografia/compendio.aspx. Accessed 1 July 2018

Islebe GA, Domínguez-Vázquez G, Espadas-Manrique C et al (2016) Cambio climático: contexto histórico, paleoecológico y paleoclimático. Tendencias actuales y perspectivas. In: Balvanera P, Arias-González JE, Rodríguez-Estrella R, Almeida-Leñero L, Schmitter-Soto JJ (eds) Una mirada al conocimiento de los ecosistemas de México. Universidad Nacional Autónoma de México (UNAM), Ciudad de México, pp 25–56

Israde-Alcántara I, Velázquez-Durán R, Lozano-García MS et al (2010) Evolución paleolimnológica del Lago Cuitzeo, Michoacán durante el Pleistoceno-Holoceno. Bol Soc Geol Mex 62(3):345–357. https://doi.org/10.1016/j.quaint.2004.10.022

Leng MJ, Metcalfe SE, Davies SJ (2005) Investigating Late Holocene climate variability in central Mexico using carbon isotope ratios in organic materials and oxygen isotope ratios from diatom silica within lacustrine sediments. J Paleolimnol 34(4):413–431. https://doi.org/10.1007/s10933-005-6748-8

Leverington DW, Teller JT, Mann JD (2002) A GIS method for reconstruction of late Quaternary landscapes from isobase data and modern topography. Comput Geosci 28(5):631–639. https://doi.org/10.1016/S0098-3004(01)00097-8

Lozano-García MS, Xelhuantzi-López MS (1997) Some problems in the Late Quaternary pollen records of Central Mexico: Basins of Mexico and Zacapu. Quat Int 43–44(97):117–123. https://doi.org/10.1016/S1040-6182(97)00027-X

Mas JF, Velázquez A, Díaz-Gallegosa JR et al (2004) Assessing land use/cover changes: a nationwide multidate spatial database for Mexico. Int J Appl Earth Obs Geoinf 5(4):249–261. https://doi.org/10.1016/j.jag.2004.06.002

McCune B, Mefford MJ (2006) PC-ORD, multivariate analysis of ecological data, version 5.31. MJM Software. Gleneden Beach, Oregon, EEUU

Metcalfe SE, Street-Perrott FA, Brown RB et al (1989) Late Holocene human impact on Lake basins in Central Mexico. Geoarchaeol Int J 4(2):119–141

Metcalfe SE, Davies SJ, Braisby JD et al (2007) Long and short-term change in the Pátzcuaro Basin, central Mexico. Palaeogeogr Palaeoclimatol Palaeoecol 247(3–4):272–295. https://doi.org/10.1016/j.palaeo.2006.10.018

Ortega B, Vázquez G, Caballero M et al (2010) Late Pleistocene: Holocene record of environmental changes in Lake Zirahuen, Central Mexico. J Paleolimnol 44(3):745–760. https://doi.org/10.1007/s10933-010-9449-x

Pedrotti F (2013) Plant and vegetation mapping. Springer, Berlin/Heridelberg

Pérez-Calix E (1996) Flora y vegetación de la cuenca del lago de Zirahuén Michoacán, México. Flora del Bajío y de regiones adyacentes, fascículo complementario XIII:1–73

Reed J, Deakin L, Sunderland T (2014) What are integrated landscape approaches and how effectively have they been implemented in the tropics: a systematic map protocol. Environ Evid 4(2):1–7. https://doi.org/10.1186/2047-2382-4-2

Reed J, Van Vianen J, Deakin EL, Barlow J, Sunderland T (2016) Integrated landscape approaches to managing social and environmental issues in the tropics: learning from the past to guide the future. Glob Chang Biol 22:2540–2554. https://doi.org/10.1111/gcb.13284

Reyes S, Douglas MW, Maddox RA (1994) El monzón del suroeste de Norteamérica (TRAVASON/SWAMP). Atmósfera 7:117–137

Rivas-Martínez S, Rivas-Sáenz S, Penas A (2011) Worldwide bioclimatic classification system. Glob Geobot 1:1–634. https://doi.org/10.5616/gg110001

Rzedowski J (2006) Vegetación de México. 1ª Edición digital, Comisión Nacional para el Conocimiento y Uso de la Biodiversidad. México. http://www.biodiversidad.gob.mx/publicaciones/librosDig/pdf/VegetacionMx_Cont.pdf. Accessed 1 May 2018

Rzedowski J, Calderón de Rzedowski G (2013) Datos para la apreciación de la flora fanerogámica del bosque tropical caducifolio de México. Acta Botánica Mexicana 102:1–23

Simonson WD, Allen HD, Parham E et al (2018) Modelling biodiversity trends in the montado (wood pasture) landscapes of the Alentejo, Portugal. Landsc Ecol 33(5):811–827. https://doi.org/10.1007/s10980-018-0627-y

Stuiver M, Reimer PJ, Reimer RW (2018) CALIB 7.1 [WWW program] at http://calib.org. Accessed 20 May 2018

Sugita S, Parshall T, Calcote R, Walker K (2010) Testing the landscape reconstruction algorithm for spatially explicit reconstruction of vegetation in northern Michigan and Wisconsin. Quat Res 74(2):289–300. https://doi.org/10.1016/j.yqres.2010.07.008

Torres-Rodríguez E, Lozano-García S, Figueroa-Rangel BL et al (2012) Cambio ambiental y respuestas de la vegetación de los últimos 17,000 años en el centro de México: el registro del lago de Zirahuén. Revista Mexicana de Ciencias Geológicas 29(3):764–778

Velázquez A, Medina-García C, Durán-Medina E et al (2016) Standardized hierarchical vegetation classification. Mexican and global patterns. Springer Nature, Basel

Villaseñor JL, Ortiz E (2012) La familia Asteraceae en la flora del Bajío y de regiones adyacentes. Acta Botánica Mexicana 100:259–291

Watts WA, Bradbury JP (1982) Paleoecological studies at Lake Patzcuaro on the west-central Mexican Plateau and at Chalco in the basin of Mexico. Quat Res 17(1):56–70. https://doi.org/10.1016/0033-5894(82)90045-X

Zonneveld IS (1989) The Land Unit a fundamental concept in landscape ecology, and its applications. Landsc Ecol 3(2):67–86

Chapter 8
Volcanic Activity in Mexico During the Holocene

José L. Macías and José L. Arce

Abstract This chapter presents an overview of the volcanic eruptions that have occurred in Mexico during the Holocene. Although volcanic regions are distributed all over the country, Holocene eruptions are mainly concentrated in the southern half of the country and, in particular, in the Trans-Mexican Volcanic Belt. Here, we summarize the details of the eruptions from the stratovolcanoes and monogenetic volcanoes, which have been extensively documented in the volcanological literature, and their radiometric or historical dates. Out of the 153 Holocene eruptions described so far, ~63.4% have occurred at active stratovolcanoes and calderas, while the remaining ~36.6% have occurred from vents within monogenetic volcanic fields. Surprisingly, it seems that volcanism increased through time from Early (~17.7%), Middle (~26.8%) to Late Holocene (~55.5%). These figures may be biased because younger deposits are better preserved than older ones, and the latter usually are eroded through time especially around active stratovolcanoes. Around 24 eruptions (~15.6%) have taken place in pre-Hispanic and historical time out of which 11 occurred during the Little Ice Age. These eruptions have posed a serious threat to the surrounding regions and their populations. Some stratovolcanoes have collapsed at least once, covering and destroying previous deposits and making it difficult to reconstruct past eruption records. Large, Plinian to sub-Plinian eruptions at stratovolcanoes are well recorded in the stratigraphy, but the small eruptions that did not produce widespread deposits are difficult to define. Eruptions from monogenetic volcanic fields were fed from central vents and fissures and mostly dominated by Strombolian activity and lava flows. The Holocene-collected data suggests that

J. L. Macías (✉)
Instituto de Geofísica, Unidad Michoacán, Universidad Nacional Autónoma de Mexico, Antigua Carretera a Pátzcuaro, Morelia, Michoacán, Mexico
e-mail: jlmacias@igeofisica.unam.mx

J. L. Arce
Instituto de Geología, Departamento de Procesos Litosféricos, Universidad Nacional Autónoma de México, Coyoacán, Mexico City, Mexico
e-mail: jlarce@geologia.unam.mx

N. Torrescano-Valle et al. (eds.), *The Holocene and Anthropocene Environmental History of Mexico*, https://doi.org/10.1007/978-3-030-31719-5_8

an eruption has taken place every ~65 years in Mexican territory during the past 10 ka. Monogenetic volcanoes should not be underestimated because at least 56 volcanoes have been created in Mexico during the Holocene, yielding an average recurrence of ~176 years. This chapter stresses the need to improve our knowledge of Holocene volcanism (e.g. still limited studies in some volcanoes) aimed to define average recurrence intervals and provide the data for probabilistic studies and hazard assessment to reduce future volcanic hazards.

Keywords Volcanism · Stratovolcanoes · Monogenetic fields · Geological · Historical · Modern records

Introduction

At first glance, Holocene volcanic areas in Mexico may appear in most parts of the territory (Fig. 8.1). However, this young volcanism is mainly concentrated in the Trans-Mexican Volcanic Belt (TMVB), the Chiapanecan Volcanic Arc (CVA),

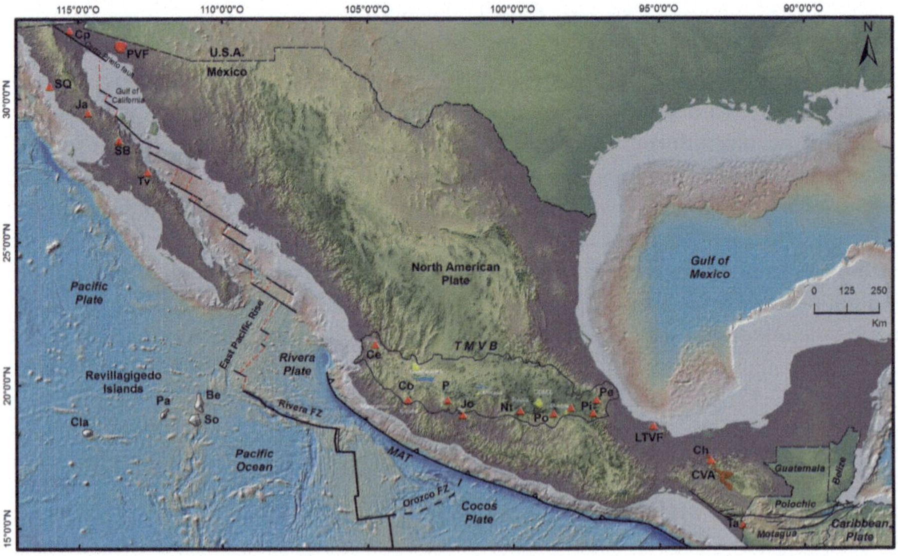

Fig. 8.1 Sketch map of tectonic settings in Mexico with boundaries of lithospheric plates and main volcanic regions. Abbreviations: TMVB, Trans-Mexican Volcanic Belt; CAVA, Central American Volcanic Arc; CVA, Chiapanecan Volcanic Arc; LTVF, Los Tuxtlas Volcanic Field; PVF, Pinacate Volcanic Field; SQ, San Quintin; TV, Tres Vírgenes Volcanic Complex; Cp, Cerro Prieto; Ja, Jaraguay; SB, San Borja; So, Socorro; Pa, Roca Partida; Be, San Benedicto; Cla, Clarión; Ce, Ceboruco; Co, Colima; P, Parícutin; Jo, Jorullo; Nt, Nevado de Toluca; Po, Popocatépetl; Pi, Pico de Orizaba; Pe, Cofre de Perote; Ch, Chichón; and Ta, Tacaná

the northwestern edge of the Central American Volcanic Arc (CAVA) and in random sites in Baja California (San Quintin volcanic fields), in El Pinacate Volcanic Field in Sonora (NW Mexico), and in the Revillagigedo Archipelago that is located at 700 km west of Manzanillo (see Fig. 8.1). Holocene volcanism has occurred from stratovolcanoes and monogenetic volcanic fields, and it has been explosive and effusive in nature (Schmincke 2004). Holocene volcanoes have experienced different types of eruptions including those associated with flank failure (i.e. Colima Volcano), Plinian to sub-Plinian columns (large magnitude in most stratovolcanoes as Popocatépetl and Nevado de Toluca), dome destruction (i.e. Colima), Strombolian activity (i.e. Paricutin Volcano), and lava flows (i.e. Pico de Orizaba and Ceboruco volcanoes; Alcalá-Reygosa et al. 2018; Nelson 1980). Stratovolcanoes form the biggest volcanic structures for instance Popocatépetl, Nevado de Toluca, and Tacaná volcanoes, constituted by an overlapping of lavas and pyroclastic deposits, recording several eruptive events intercalated by long quiescent periods (thousands of years) and recording important chemical changes through its volcanic history (Smith and Németh 2017). Meanwhile, monogenetic volcanoes are small structures that record a single eruptive event (i.e. Paricutin, Jorullo, and Xitle volcanoes) that can last for several years but also with small changes in chemical composition (Smith and Németh 2017). Clusters of volcanic vents formed by scoria cones and lava domes, maars and small shield volcanoes, whose products were erupted in a relatively short period (<2 Ma) with similar petrologic composition (mostly mafic) and are associated with tectonic features (e.g. basins, rifts, fractures or fault systems) are defined as volcanic fields (i.e. Connor and Conway 2000). Some examples in Mexico include the Pinacate, San Quintín, Jaraguay, San Borja, Michoacán-Guanajuato, Chichinautzin, Apan–Tezontepec, Serdan Oriental, Xalapa, and Tuxtla Volcanic Fields.

In this chapter, we summarize the current knowledge of Holocene volcanic activity in Mexico in the light of recent research in different areas. The boundary of the Pleistocene–Holocene has been defined in the Greenland NGRIP2 ice core at age 11.7 ka b2k (before AD 2000; Walker et al. 2008, 2009). In this interval of time, we were able to compile the date and age of 153 well-constrained eruptions that left traceable deposits or produced new monogenetic volcanoes. The chronology of these eruptions was identified by stratigraphic relationships, ^{14}C dates of paleosol and charcoal embedded in the deposits, some limited $^{40}Ar/^{39}Ar$ and exposure ^{36}Cl ages, ^{230}U-Th in zircons, and paleomagnetic determinations (Table 8.1). For the sake of simplicity, all dates mentioned in the text correspond to calibrated dates (cal. yr. BP) using the IntCal13 (Reimer et al. 2013) calibration curve, and 2σ range is included in Table 8.1. Some deposits of a single eruption (i.e. pink pumice of Popocatépetl; Siebe et al. 1996) have multiple ^{14}C results for which we used the best estimation considered by the authors. Two volcanoes have been so active, Colima Volcano, Popocatépetl, in historical times that we have combined all their modern activity into a single eruption, that is 1962–2019 and 1994–2019, respectively. Other volcanoes as Tacaná in Chiapas-Guatemala resumed phreatic activity in 1949 and 1986, however, we did not count these events as proper eruptions because they did not produce juvenile material or traceable deposits.

Table 8.1 Database of all Holocene eruptions reported for stratovolcanoes, calderas, and monogenetic volcanoes in México

	Location		Sample	Method	Conventional age	Calibrated age	
Volcanic landform	North	West	Type	(‰)	yr. BP.	2σ range (yr. BP)	Reference
Stratovolcanoes							
Everman							
Lomas Coloradas basalt			Charcoal in lacustrine sediments	^{14}C	4690 ± 270	6002–4628	1
1848			Historical record	N.A.			2
1896			Historical record	N.A.			2
May 22, 1951			Historical record	N.A.			2
January 29, 1993			Modern record	N.A.			2
Ceboruco							
Destiladero lava flow			Based on stratigraphic position	N.A.	>1088		3
Jala Pumice	21°10′42″	104°32′37″	Charcoal in paleosol below Jala Pumice	^{14}C	1060 ± 55	1197–1022	3
Copales lava flow			Based on stratigraphic position	N.A.	<1060		3
Cajón lava flow			Based on stratigraphic position	N.A.	<1060		3
Coapan I lava flow			Based on stratigraphic position	N.A.	<1060		3
El Norte lava flow			Based on stratigraphic position	N.A.	<1528		3
Ceboruco lava flow			Based on stratigraphic position	N.A.	<1528		3
1870–1875 lava flow			Historical record	N.A.			
Colima							
Paleofuego							
Unit H				^{14}C	9770 ± 60	11,305–10,882	4
Mesa Yerbabuena DAD	n.d.	n.d.	Charcoal in DAD	^{14}C	9671 ± 88	11,231–10,756	5
Unit J	n.d.	n.d.	Charcoal in pyroclastic deposit	^{14}C	7750 ± 60	8510–8177	4
Unit K				^{14}C	7530 ± 80	8510–8176	4
Unit L				^{14}C	7520 ± 50	8409–8202	4

Unit M				^{14}C	7070 ± 60	8008–7761	4
Alvarez–Coquimatlan DAD	n.d.	n.d.	Charcoal in DAD	^{14}C	7040 ± 160	8175–7590	6
Unit N				^{14}C	6950 ± 50	7925–7679	4
Unit O				^{14}C	6150 ± 40	7164–6943	4
Unit P				^{14}C	5980 ± 50	6942–6678	4
Unit Q				^{14}C	5850 ± 60	6791–6497	4
Unit R				^{14}C	5550 ± 80	6498–6189	4
Unit S				^{14}C	5430 ± 50	6312–6021	4
Unit T	n.d.	n.d.	Charcoal in pyroclastic deposit	^{14}C	4810 ± 50	5645–5333	4
Unit U				^{14}C	4740 ± 40	5586–5326	4
Unit V	n.d.	n.d.	Charcoal in pyroclastic deposit	^{14}C	4540 ± 60	5445–4975	4
Unit W				^{14}C	4480 ± 60	5309–4889	4
Unit Y	n.d.	n.d.	Charcoal in pyroclastic deposit	^{14}C	4460 ± 40	5295–4894	4
Unit X				^{14}C	4380 ± 80	5286–4834	4
Fuego de Colima							
Los Ganchos DAD	n.d.	n.d.	Charcoal in DAD	^{14}C	3600 ± 120	4245–3587	7
El Remate DAD	n.d.	n.d.	Charcoal in DAD	^{14}C	2690 ± 40	2861–2748	8
Unit Z	n.d.	n.d.	Charcoal in pyroclastic deposit	^{14}C	360 ± 60 4240 ± 110		4
February 15, 1818			Historical record	N.A.			9
January 20, 1913			Historical record	N.A.			10
1962-present			Modern record	N.A.			11
Jocotitlán							
Sector collapse	19°45′46″	99°41′31″	Charcoal in pyroclastic surge deposit	^{14}C	9690 ± 85	11,239–10,769	12
Pumiceous block-and-ash flow deposit	19°44′52″	99°46′24″	Charcoal in block-and-ash flow deposit	^{14}C	680 ± 80	757–527	13

(continued)

Table 8.1 (continued)

	Location		Sample	Method	Conventional age	Calibrated age	
Volcanic landform	North	West	Type	(‰)	yr. BP.	2σ range (yr. BP)	Reference
Nevado de Toluca							
Upper Toluca Pumice	19°17′30″	99°21′00″	Charcoal in pyroclastic flow deposit	^{14}C	10,445 ± 95	12,621–12,025	14
Obligo dome	19°06′30″	99°45′17″	Whole rock	Cl-36	9100 ± 500		14
Ash flow deposit	19°13′26″	99°47′22″	Charcoal in ash flow deposit	^{14}C	3435 ± 50	3580–3831	15
Popocatépetl							
Lower pre-ceramic eruption I	n.d.	n.d.	Charcoal in pyroclastic surge deposit	^{14}C	10,700 ± 200	13,037–12,060	13
Intermediate pre-ceramic eruption I	n.d.	n.d.	Charcoal in pyroclastic surge deposit	^{14}C	9100 ± 200	10,775–9564	13
Intermediate pre-ceramic eruption II	n.d.	n.d.	Charcoal in pyroclastic surge deposit	^{14}C	7100 ± 200	8328–7591	13
Upper pre-ceramic Plinian eruptive sequence (ochre pumice)	n.d.	n.d.	Charcoal in pyroclastic surge deposit	^{14}C	5000 ± 200	6262–5318	16
Lower ceramic Plinian eruptive sequence (Lorenzo pumice)	n.d.	n.d.	Charcoal in paleosol below fallout	^{14}C	2150 ± 200	2713–1715	16
Intermediate ceramic eruptive sequence	n.d.	n.d.	Charcoal in pyroclastic surge deposit	^{14}C	1700 ± 200	2120–1266	13
Upper ceramic Plinian eruptive sequence (pink pumice)	n.d.	n.d.	Charcoal in pyroclastic surge deposit	^{14}C	1200 ± 200	1522–733	16
1540			Historical record	N.A.			17
1592			Historical record	N.A.			17
1664			Historical record	N.A.			17

1919–1927			Modern record	N.A.			17
December 21, 1994-present			Modern record	N.A.			17
La Malinche							
Malinche Pumice II	19°15′40″	98°04′20″	Charcoal in ash flow deposit	^{14}C	9030 ± 85	10,419–9901	18
Altamira pumice	19°15′44″	97°58′17″	Charcoal in pumice-and-ash flow deposit	^{14}C	7430 + 200/−195	7838–8631	18
Ash fall/flow	19°16′55″	98°00′23″	Charcoal in ash flow covered by modern soil	^{14}C	3115 ± 55	3450–3181	18
Pico de Orizaba							
Ash-and-scoria flow deposit (Coscomatepec eruptive episode)			Carbonized roots	^{14}C	9400 ± 170	10,248–11,142	19
Chichimeco dome complex			Charcoal in scoria flow deposit	^{14}C	8630 ± 90	9474–9900	20
Citlaltépetl pumice	2,108,242	695,553	Charcoal in pyroclastic deposit	^{14}C	8505 ± 50	9546–9448	20
Ash-and-scoria flow deposit (Loma Grande eruptive event)			Carbonized roots	^{14}C	7020 ± 120	7608–8148	19
Block-and-ash flow deposit			Charcoal in pyroclastic deposit	^{14}C	4040 ± 80	4297–4823	21
Dacitic tephra T-IV (Jacal eruptive event)			Carbonized roots	^{14}C	3450 ± 70	3514–3897	19
Lava flow A	18.99°	97.30°	Whole rock	Cl-36	3040 ± 660		22
Block-and-ash flow deposit (Texmola eruptive event)			Carbonized roots	^{14}C	1860 ± 40	1707–1885	19
Lava flow B	19.01°	97.27°	Whole rock	Cl-36	1028 ± 220		22
Lava flows (Excola erupive episode)			Historical record		AD 1537–1687		
San Martin Tuxtla							
Ash layer			Brown paleosol under altered lapilli beds	^{14}C	4080 ± 55	4816–4431	23

(continued)

Table 8.1 (continued)

	Location		Sample	Method	Conventional age	Calibrated age	
Volcanic landform	North	West	Type	(‰)	yr. BP.	2σ range (yr. BP)	Reference
March 2, 1793	n.d.	n.d.	Historical record	N.A.			23
El Chichón							
L	n.d.	n.d.	Charcoal in pyroclastic deposit	^{14}C	7740 ± 50	8597–8420	24
K	n.d.	n.d.	Charcoal in pyroclastic deposit	^{14}C	3675 + 80/−75	4246–3730	24
J	n.d.	n.d.	Charcoal in pyroclastic deposit	^{14}C	3105 ± 70	3468–3081	24
I	n.d.	n.d.	Charcoal in pyroclastic deposit	^{14}C	2470 ± 50	2718–2365	24
H	n.d.	n.d.	Charcoal in pyroclastic deposit	^{14}C	2205 ± 60	2345–2060	24
G	n.d.	n.d.	Charcoal in pyroclastic deposit	^{14}C	1885 + 75/−70	1994–1622	24
F	n.d.	n.d.	Charcoal in pyroclastic deposit	^{14}C	1695 ± 65	1805–1414	24
E	n.d.	n.d.	Charcoal in pyroclastic deposit	^{14}C	1490 ± 45	1521–1302	24
D	n.d.	n.d.	Charcoal in pyroclastic deposit	^{14}C	1250 ± 70	1299–999	25
C	n.d.	n.d.	Charcoal in pyroclastic deposit	^{14}C	845 ± 75	916–674	24
B	n.d.	n.d.	Charcoal in pyroclastic deposit	^{14}C	550 ± 60	655–506	26
A			Modern record	N.A.		AD 1982	24
Tacana							
Once de Abril ash flow deposit	15°03′25″	92°08′42″	Charcoal in pyroclastic deposit	^{14}C	9960 ± 50	11,620–11,244	27
Papales ash flow deposit 1	15°07′03.6″	92°06′03.4″	Charcoal in pyroclastic deposit	^{14}C	7630 ± 115	8683–8180	28
Chocab flow deposit	15°06′30″	92°04′40″	Charcoal in pyroclastic deposit	^{14}C	6910 ± 95	7933–7595	28
Pyroclastic surge-flow deposit	n.d.	n.d.	Charcoal in pyroclastic deposit	^{14}C	5860 ± 125	6998–6400	27
Moat pyroclastic surge deposit	15°08′3.5″	92°06′27.4″	Paleosol	^{14}C	2660 + 55/−50	2920–2715	28
Mixcun block-and-ash flow deposit	n.d.	n.d.	Charcoal in pyroclastic deposit	^{14}C	1950 ± 50	2003–1740	27
Scoria flow deposit	15°09′21″	92°09′04″	Charcoal in pyroclastic deposit	^{14}C	920 ± 35	924–748	28
Pumice fall deposit (Pfd)	15°07′43.3″	92°06′24.9″	Charcoal in pyroclastic deposit	^{14}C	810 + 110/−105	935–748	28

Yellow pyroclastic flow deposit	15°09′10″	92°06′23.1″	Charcoal in pyroclastic deposit	^{14}C	370 + 80/−75	536–157	28
Papales ash flow deposit 2	15°07′05″	92°06′05″	Charcoal in pyroclastic deposit	^{14}C	280 ± 60	496 to present	28
White fall-surge deposit	15°06′21″	92°06′27.2″	Charcoal in pyroclastic deposit	^{14}C	150 ± 40	285 to present	28
Calderas							
La primavera caldera							
Tajo dome	n.d.	n.d.	Whole rock	K-Ar	5 + 8/−14 ka		29
Los Humeros caldera							
San Antonio-Las Chapas lava flows	n.d.	n.d.	Charcoal in lower paleosol	^{14}C	8910 ± 30	10,180–9915	30
Cuicuiltic member	n.d.	n.d.	Charcoal in pyroclastic deposit	^{14}C	7355 ± 105	8374–7982	30
Olivine-bearing lava flows	n.d.	n.d.	Underlying paleosol	^{14}C	3870 ± 130	4798–3905	30
El Pajaro lava flows	n.d.	n.d.	Charcoal at the base of the flow	^{14}C	2860 ± 30	3067–2878	30
Volcanic fields							
Barcena							
August 1, 1952			Modern record	N.A.			31
Pinacate							
The Ives lava flow	n.d.	n.d.	Whole rock	$^{40}Ar/^{39}Ar$	13 ± 3 ka		32
La Laja cinder cone	n.d.	n.d.	Whole rock	$^{40}Ar/^{39}Ar$	12 ± 4 ka		33
Ceboruco-San Pedro graben							
Molcajete cone	21°03′05″	104°24′22″	Paleosol below scoriae fallout	^{14}C	9220 + 170/−165	11,072–9918	3
Pedregoso dome	21°04′33″	104°26′51″	Paleosol below Pedregoso fallout	^{14}C	3550 ± 110	4150–3571	3
Potrerillo II cone	21°10′28″	104°34′49″	Charcoal in pyroclastic flow deposit	^{14}C	2430 + 50/−45	2705–2352	3
Pochetero dome	21°04′33″	104°26′51″	Paleosol below Pochetero sequence	^{14}C	2355 ± 110	2728–2150	3

(continued)

Table 8.1 (continued)

	Location		Sample	Method	Conventional age	Calibrated age	
Volcanic landform	North	West	Type	(‰)	yr. BP.	2σ range (yr. BP)	Reference
Pichancha coulée			Based on stratigraphic position		>1060		3
Michoacán-Guanajuato							
La Taza	213,688	2,161,002	Paleosol	^{14}C	9300 ± 40 BP	10,649–10,300	34
Hoya El Huanillo	187,070	2,179,244	Charcoal in the soil under ash and lapilli layers	^{14}C	9180 ± 250	11,130–9688	35
C. Grande	225,798	2,121,831	Paleosol	^{14}C	8715 + 145/−140	10,173–9502	36
La Mina	243,949	2,182,924	Paleosol	^{14}C	7325 ± 135	8400–7879	37
Los Caballos	243,719	2,182,084	Paleosol	^{14}C	7045 + 125/−120	8159–7657	37
Cerro Amarillo	213,394	2,161,540	Paleosol	^{14}C	6970 ± 40 years	7927–7696	34
Cerro Colorado	213,688	2,161,002	Paleosol	^{14}C	6970 ± 40 BP	7927–7696	34
Tandeparacua	243,452	2,200,628	Paleosol	^{14}C	6105 ± 45	7159–6860	37
Los Lobos	n.d.	n.d.	Paleosol	^{14}C	6015 ± 200	7318–6413	38
C. La Tinaja	236,823	2,123,678	Rock	Paleomagnetic	5325 ± 130		39
El Metate	185,950	2,163,628	Paleosol	^{14}C	4700 ± 200	5888–4868	40
C. El Jabali	803,169	2,153,069	Charcoal in the soil under ash and lapilli layers	^{14}C	3830 ± 150	4801–3834	35
Hoya Urutzen	191,378	2,167,030	Charcoal	^{14}C	3650 ± 30	4084–3888	41
C. El Zoyate	222,434	2,116,155	Paleosol	^{14}C	3505 + 115/−110	4090–3481	36
La Palma	n.d.	n.d.	Rock	Paleomagnetic	3220–2880		39
Las Vigas scoria cone (El Infiernillo lava low)	201,246	2,198,563	Paleosol	^{14}C	3200 ± 30	3475–3366	35
Non associated lava flow	19.61°	102.07°		^{14}C	2277 ± 25	2165–2349	42
Mesa La Muerta	n.d.	n.d.	Rock	Paleomagnetic	2240–2070		39
Cicapien	19.57°	102.09°		^{14}C	1806 ± 24	1631–1819	42
Cutzaróndiro Malpais	n.d.	n.d.	Rock	Paleomagnetic	410–320		39

El Tecolote	198,873	2,192,241	Volcanic glass and plagioclase	$^{40}Ar/^{39}Ar$	0.002 + −0.041		34
El Jorullo	213,839	2,099,957	Historical record	N.A.		AD 1759–1776	38
El Parícutin	788,557	2,157,661	Historical record	N.A.		AD 1943–1952	43
Chichinautzin							
Mezontepec	n.d.	n.d.	Whole rock	$^{40}Ar/^{39}Ar$	11,000 ± 3000		34
Pelado	n.d.	n.d.	Charcoal in paleosol below a lava flow	^{14}C	10,270 ± 190	12,571–11,331	41
Tenango	n.d.	n.d.	Charcoal in paleosol below the lava flow	^{14}C	8390 ± 130	9579–9022	44
Tres Cruces	n.d.	n.d.	Paleosol below ash	^{14}C	8390 ± 100	9541–9130	45
Oyameyo	482,109	2,115,486	Charcoal in sandy-silty ash flow	^{14}C	8315 ± 155	9584–8782	41
Cuauhtzin	488,112	2,118,930	Charcoal in block-and-ash flow deposit	^{14}C	7360 ± 120	8387–7968	46
Tabaquillo	n.d.	n.d.	Whole rock	$^{40}Ar/^{39}Ar$	7000 ± 9000		34
Tlaloc	498,656	2,112,503	Paleosol	^{14}C	6200 ± 85	7302–6885	46
Guespalapa	480,839	2,110,410	Paleosol	^{14}C	2835 ± 75	3159–2782	41
Pelagatos	n.d.	n.d.	Reworked charcoal against lava flow	^{14}C	2520 ± 105	2836–2346	35
Jumento	19°12′01″	99°18′26″	Charcoal in surge deposit	^{14}C	2010 ± 30	2041–1885	37
PL- 1 Tecuautzi	n.d.	n.d.	Whole rock	$^{40}Ar/^{39}Ar$	2000 ± 56,000		34
Chichinautzin	480,839	2,110,410	Charcoal in ash fallout	^{14}C	1835 ± 55	1889–1618	41
Xitle	2,134,121	480,972	Charcoal in ash	^{14}C	1675 ± 40	1700–1424	47
La Cima	n.d.	n.d.	Paleosol	^{14}C	1160 ± 70	1260–937	48
Apan-Tezontrepec							
El Cuello scoria cone	n.d.	n.d.		^{14}C	9645 + 283/−1143	12,928–9153	49

(continued)

Table 8.1 (continued)

Volcanic landform	Location		Sample	Method	Conventional age	Calibrated age	Reference
	North	West	Type	(‰)	yr. BP.	2σ range (yr. BP)	
Serdán Oriental							
Alchichica				^{14}C	9750 ± 30	11,234–11,153	50
Tecuitlapa				^{14}C	8040 ± 30	9021–8778	50
Atexcac				^{14}C	7480 ± 30	8376–8021	50
Tepextitl				^{14}C	5540 ± 30	6300–6204	50
Aljojuca				^{14}C	2430 ± 30	2698–2354	50
Las Derrumbadas	n.d.	n.d.		^{14}C	1970 ± 30	1933–1866	50
Xalapa							
Coacotzintla lava flow	n.d.	n.d.		^{14}C	2980 ± 55	3340–2980	
Volcancito scoria cone	n.d.	n.d.		^{14}C	850 ± 40	905–685	51
Tuxtla volcanic field							
Cerro Mono Blanco	n.d.	n.d.	Charcoal below fallout deposit	^{14}C	3270 ± 250	4147–2865	52
Puntiagudo	n.d.	n.d.	Charcoal below fallout deposit	^{14}C	1835 ± 195	2301–1346	53

The youngest ^{14}C are reported when multiple dated samples are available, and in some case, we used the preferred age by the authors. Calibrated dates (cal. yr. BP) were obtained with the IntCal13 (Reimer et al. 2013) calibration curve and 2σ range

References are: (1) Farmer et al. (1993), (2) Siebe et al. (1995a), (3) Sieron and Siebe (2008), (4) Luhr et al. (2010), (5) Komorowski et al. (1997), (6) Cortés et al. (2005), (7) Navarro et al. (2004), (8) Siebe et al. (1993), (9) Macías et al. (2017), (10) Saucedo et al. (2010), (11) Martín Del Pozzo et al. (1987), (12) Siebe et al. (1992), (13) Siebe and Macías (2006), (14) Arce et al. (2003), (15) Macías et al. (1997a), (16) Siebe et al. (1996), (17) Martin del Pozzo et al. (2016), (18) Castro-Govea and Siebe (2007), (19) Höskuldsson and Robin (1993), (20) Rossotti and Carrasco-Núñez (2004), (21) Siebe et al. (1993), (22) Alcalá-Reygosa et al. (2018), (23) Espíndola et al. (2010), (24) Espíndola et al. (2000), (25) Duffield et al. (1984), (26) Macías et al. (1993), Tilling et al. (1984), (27) Macías et al. (2010), (28) Macías et al. (2015), (29) Mahood and Drake (1982), (30) Carrasco-Núñez et al. (2017), (31) Williams (1952), (32) Turrin et al. (2008), (33) Gutmann et al. (2000), (34) Jaimes-Viera et al. (2018), (35) Guilbaud (2009), (36) Hasenaka and Carmichael (1985a), (37) Arce et al. (2015), (38) Osorio-Ocampo et al. (2018), (39) Mahgoub et al. (2017), (40) Chevrel et al. (2016a), (41) Siebe et al. (2004), (42) Mahgoub et al. (2019), (43) Luhr and Simkin (1993), (44) Bloomfield (1973), (45) Bloomfield (1975), (46) Siebe et al. (2005), (47) Siebe (2000), (48) Kirianov et al. (1990), (49) Hernández-Javier (2008), (50) Chédeville et al. (2019), (51) Siebert and Carrasco-Núñez (2002), (52) Espíndola et al. (2016), (53) Reinhardt (1991)

Abbreviations: *n.d.* not determined either because ages represent average of several samples or because they were not given; *n.a.* not applicable. Calibrated 14C using IntCal13, with 95.4% probability

Distribution of Volcanoes in Mexico

The tectonic setting of Mexico is dominated by three main tectonic environments: (a) the subduction of the Cocos and Rivera plates beneath the North American plate at the Middle America Trench (Pardo and Suárez 1995), (b) the extensional environment of the Gulf of California and its continuation towards the San Andreas Fault (Atwater 1989; Stock and Hodges 1989), (c) and post-abandonment magmatism of the Revillagigedo Islands along a failed mid-ocean ridge (Siebe et al. 1995a) (Fig. 8.1). All these tectonic environments have been the focus of intense submarine and subaerial volcanism that have extended in many cases into the Holocene.

The oblique subduction (16°) of the Rivera and Cocos slabs occurs beneath the North American and the Caribbean plates along the Middle America Trench (Demant 1978). This subduction generates three main segment areas of volcanism which from west to east are the Trans-Mexican Volcanic Belt (TMVB), an active continental volcanic arc with over 8500 volcanic structures that transect central Mexico (Fig. 8.2); the Los Tuxtlas Volcanic Field (LTVF) composed of 5 large volcanoes, 353 monogenetic cones, and 42 maar structures (Espíndola et al. 2016); and the Chiapanecan Volcanic Arc (CVA) composed of scattered extrusions of dioritic bodies and trachy–andesitic–dacitic domes and rare volcanoes (Mora et al. 2007, 2012). In Southeastern Mexico, the Cocos slab plunges underneath the tectonic boundary between the North America and the Caribbean plates and forms a diffuse triple-point junction. In Central America, this subduction process produces the Central American Volcanic Arc (CAVA) that runs from western Panama to the Tacaná volcanic complex in Southeastern Mexico (DeMets et al. 1990; Rebollar et al. 1999).

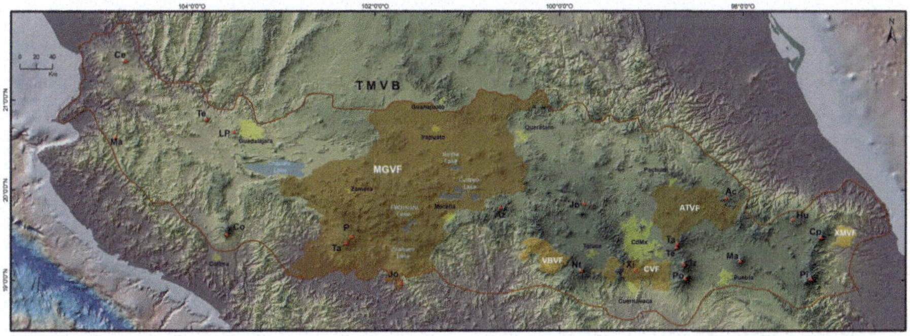

Fig. 8.2 Shaded relief model of the Trans-Mexican Volcanic Belt in central Mexico and the location of main active volcanoes, calderas, and volcanic fields such as Chichinautzin, CVF; Michoacán–Guanajuato, MGVF; Mascota, MVF; Valle de Bravo, VBVF; Jalapa, XMVF; and Apan–Tezontepec, APVF. Abbreviations: Ce, Ceboruco; Ma, Mascota; Te, Tequila; LP, La Primavera; Co, Colima; Ta, Tancítaro; P, Parícutin; Jo, Jorullo; Az, Azufres; J, Jocotitlán; Nt, Nevado de Toluca; Xi, Xitle; Ta, Tlaloc; Te, Telapón; Iz, Iztaccihuatl; Po, Popocatépetl; Ma, Malinche; Ac, Acoculco caldera; Hu, Humeros caldera; Cp, Cofre de Perote; and Pi, Pico de Orizaba

The extensional tectonics of the Gulf of California began ~5 Ma ago (Klitgord and Mammerickx 1982). This tectonism has generated a diverse range of volcanic centres in Baja California and the San Quintin, Jaraguay, San Borja, and La Purisima (Tres Vírgenes) volcanic fields that represent intraplate magmatism. These three volcanic fields are constituted by scoria cones and lava flows (Luhr et al. 1995) with OIB-type (San Quintin) and calc-alkaline magmatism (Storey et al. 1989; Luhr et al. 1995). Other isolated volcanic centres such as Cerro Prieto in Mexico (García-Sánchez et al. 2017) and Salton Buttes in the USA occurred along the NW–SE rift basins, and embryonic oceanic spreading centres are interconnected by NW–SE-oriented transform fault systems (Lonsdale 1989; Oskin and Stock 2003). In northwestern Sonora in mainland Mexico, the Pinacate Volcanic Field is found, and it has been associated to the basin and range tectonics (Gutmann 2002, 2007). Most of these Pinacate rocks are alkaline to transitional basalt in composition (Lynch and Gutmann 1987) with ages ranging from 1.5 Ma through to the Holocene (Turrin et al. 2008). The Revillagigedo Archipelago is located 650 km southwest of the shores of mainland Mexico. It is constituted by the Socorro, Clarión, San Benedicto, and Roca Partida islands that along with several seamounts form the northern trace of the abandoned north-trending Mathematician Ridge (Mammerickx et al. 1988).

Holocene Eruptions from Volcanoes in Mexico

Volcanic activity in Mexico has indistinctly occurred in all tectonic environments mentioned above. In addition, there has been submarine activity but it is not documented in this chapter. In mainland Mexico, the Holocene eruptions have been concentrated in active volcanoes and monogenetic fields, although scattered eruptions have also occurred in other environments (e.g. Revillagigedo Archipelago). With this overall perspective, we describe well-documented Holocene eruptions that occurred in polygenetic volcanoes and volcanic fields. All elevations are given in meters above sea level.

Active Stratovolcanoes and Calderas

Revillagigedo Archipelago

Everman Volcano (1050 m)

Everman is a basaltic shield volcano that rises ~3000 m from the seafloor and reaches an altitude of 1050 m. The volcano eruptive activity has not yet been studied in detail but historic accounts are listed in Siebe et al. (1995a) with eruptions in 1848 and 1896 and on May 22, 1951. The last event was described as a short-lived steam explosion from a cinder cone west of Lomas Coloradas on the southeast side of the island (Crowe and Crowe 1955). Additionally, there is a basaltic lava flow at Lomas Coloradas that overlies lacustrine sediments dated at cal. yr. BP 6002–4628 (Farmer et al. 1993). The best documented event from Everman Volcano is a submarine eruption that occurred on January 29, 1993. The eruption occurred from submarine shallow vents located at 30 and 210 m depth, ~4 km offshore to the west of Socorro Island (Siebe et al. 1995a), that expelled basaltic scoriaceous blocks and steam plumes.

Bárcena Volcano (332 m)

Bárcena, located 53 km north of Socorro Island, is a tuff cone that was born on August 1, 1952. The volcano formed through successive hydromagmatic explosions that generated a 400 m high tuff cone in 2 weeks (Williams 1952). The eruption continued with the emplacement of a central trachyte dome that breached the base of the cone forming a fanlike lava flow (Richards 1957). By February 1953, the volcanic activity at Bárcena had ceased with just fumarolic activity remaining.

Ceboruco Volcano (2164 m)

Ceboruco is a late Quaternary volcano located in the western part of the TMVB in the State of Nayarit, which has recorded intense Holocene activity (Nelson 1980; Sieron and Siebe 2008). At least two explosive events have occurred at Ceboruco during Late Holocene: the Marquesado block-and-ash flow deposit dated at 1500 ± 300 yr. BP (Nelson 1980) and the Plinian eruption that emplaced the Jala pumice (Gardner and Tait 2000) at cal. yr. BP 990–1020 (Sieron and Siebe 2008). After the Jala eruption, the latter authors mentioned that at least six lava flows have been erupted from Ceboruco (oldest to youngest): Copales, Cajón, Coapan I, El Norte, Ceboruco, and the AD 1870–1875 lavas. They did not report specific dates for this eruption but concluded that the El Norte and Ceboruco lavas were younger than year AD 1528. The youngest event of Ceboruco is the well-known AD 1870–1875 lava eruption that has been widely documented by several authors (e.g. Caravantes 1870; Sieron and Siebe 2008 and references therein).

La Primavera Caldera (2270 m)

La Primavera Caldera is located just west of the City of Guadalajara. It is a rhyolitic complex of domes and pyroclastic deposits (Mahood 1977). The caldera formed at ~95 ka with an explosive eruption that emplaced the Toba Tala ignimbrite that generated an 11-km-wide crater (Mahood 1980). Post-caldera activity (<95 ka Holocene) emplaced a complex sequence of pyroclastic deposits that cover the caldera and extruded domes along the ring-fault and on faults inside and outside of the caldera (Mahood 1980, 1981; Walker et al. 1981). Dates of these events suggested that at least one eruption from the southeastern Tajo dome could have occurred during the Holocene at 5 + 8/−14 ka, although the K-Ar age has a large error (Mahood and Drake 1982).

Colima Volcano (3860 m)

Colima Volcano is located 30 km north of the City of Colima, and it represents the southern active volcano of the N–S Colima volcanic complex chain that formed Cántaro (1.6–1.0 Ma), Nevado de Colima (0.53–0.03 Ma), Paleofuego (0.04 Ma), and Colima volcanoes (Recent) (Robin et al. 1987; Komorowski et al. 1994; Cortés et al. 2005, 2019). Cortés et al. (2005) reported that Paleofuego Volcano collapsed at least five times, emplacing the same number of debris avalanches. Two of these debris avalanches occurred during the Holocene at cal. yr. BP 11,231–10,756 (Mesa Yerbabuena debris avalanche) and at 8175–7590 (Alvarez-Coquimatlán debris avalanche) (Cortés et al. 2019). After the last collapse of Paleofuego, the volcanic activity migrated to the south to form modern Colima Volcano. This volcano has collapsed at least twice at 4245–3587 (Los Ganchos debris avalanche) (Cortés 2002) and at 2861–2748 (El Remate debris avalanche) (Komorowski et al. 1997).

The complex activity of Paleofuego and Colima volcanoes has made the reconstruction of their Holocene activity difficult. Luhr et al. (2010) described units A–Z ranging in age from 15,960 to 1400 years BP; however, only a few units have a distinctive stratigraphy that can be traced across several sections and were supported by ^{14}C determinations, to resolvable the separate units. These authors presented specific stratigraphy and age determinations of four units J, T, V, and Y. Crummy et al. (2019a) refined this stratigraphy describing 10 units that later on expanded to 17 units within more specific time frames (Crummy et al. 2019b). These units are the H tephra (11,305–10,882), the J tephra (8510–8177), the K interbedded PDC and fall deposits (8510–8176), the L tephra (8409–8202), the M ash-rich deposit with lenses of tephra (8008–7761), the N tephra (7925–7679), the O surge and ash fall deposits (7164–6943), the P tephra (6942–6678), the Q tephra separated by surge deposits (6791–6497), the R surge deposit (6498–6189), the S tephra interbedded with PDC deposits (6312–6021), the T tephra bounded by surge deposits (5645–5333), the U tephra (5586–5326), the V surge deposit (5445–4975), the W tephra (5309–4889), the X surge deposit (5286–4834), the Y tephra (5295–4894), and unit Z layer that is made of massive ash, surges, and fallouts with an unresolvable age varying from 4240 ± 110 to 360 ± 60 yr. BP.

According to the chronology of the collapses described by Cortés et al. (2019) and Holocene stratigraphy presented by Crummy et al. (2019b), at least six volcanic units (H–M) should be related to Paleofuego Volcano. Because Colima Volcano was already formed prior to Los Ganchos collapse at 4245–3587 (Cortés 2002), eruptions N to Y may be attributable either to Paleofuego or Colima Volcano. Colima Volcano is composed of interlayered andesitic lava flows and pyroclastic deposits (fall, surge, and flow) (Luhr and Carmichael 1990a, b). After its last collapse at 2861–2748 (El Remate debris avalanche), its eruptive chronology has not been described in detail because of its persistent activity that has filled and eroded its surrounding gullies.

During the last 400 years, Colima Volcano has experienced ca. 43 eruptions that placed it among the most active volcanoes in North America (De la Cruz-Reyna 1993; Saucedo et al. 2005). The best known historical Plinian eruptions of Colima occurred in February 15, 1818 (Macías et al. 2017) and January 20, 1913 (Saucedo et al. 2010). After the 1913 eruption, the volcano restarted explosive activity in 1962 with the construction and destructions of domes and the generation of PDC deposits and associated lavas (Martín Del Pozzo et al. 1987; Rodríguez-Elizarrarás et al. 1991; Navarro-Ochoa et al. 2002; Saucedo et al. 2002; Saucedo et al. 2005; Macías et al. 2006). The most recent event took place in June 2015, emplacing a 10-km-long block-and-ash flow to the south of the volcano (Capra et al. 2016; Macorps et al. 2018).

Jocotitlán Volcano (3950 m)

This volcano is located 51 km to the north of the City of Toluca. It is a composite volcano made of calc-alkaline andesites and dacites and rests on an older Tertiary complex of dacite domes and andesitic scoria cones (Siebe et al. 1992; Siebe and Macías 2006). Jocotitlán collapsed towards the north at cal. yr. BP 11,239–10,769 and produced a typical debris avalanche deposit that entered a lake and deformed the lacustrine sediments. The deposit consists of conical-shaped hills made of shattered rocks of the previous cones. Siebe et al. (1992) also reported a series of pyroclastic surge deposits exposed on the upper flanks of the volcano that are dated at cal. yr. BP 757–527.

Nevado de Toluca Volcano (4560 m)

Nevado is located about 80 km to the SW of Mexico City and 21 km to the SW of the City of Toluca. It is a compound volcano built on Miocene volcanic rocks and the Guerrero terrain basement rocks (García-Palomo et al. 2002a). The volcanic history of Nevado started 1.5 Ma ago with the emplacement of andesitic and dacitic lava flows, followed by explosive eruptions that accompanied pyroclastic deposits at around 0.8 Ma (García-Palomo et al. 2002a; Torres-Orozco et al. 2017). The volcano summit is made of several lava dome remnants with relative young ages (40–50 ka; Torres-Orozco et al. 2017). Some of these domes produced block-and-ash flow deposits that were emplaced around the volcano during the Late Pleistocene, when a couple of Plinian eruptions also took place at 21.7 and 12.1 ka, depositing the pumice fallout named Lower Toluca Pumice and Middle Toluca Pumice, respectively (Bloomfield and Valastro 1974; Arce et al. 2005; Capra et al. 2006). Nevado de Toluca produced one of the most powerful Plinian eruptions from central Mexico in the Holocene, the Upper Toluca Pumice deposit at cal. yr. BP 12,621–12,025 (Arce et al. 2003). Around 8 km^3 of magma was erupted that provoked environmental changes during the Younger Dryas. The pumice fallout blanketed the Lerma and Mexico basins with 1 m (lapilli sized pumice) and 40 cm (coarse ash), respectively (Bloomfield and Valastro 1977; Arce et al. 2003). The large volume ejected could have produced the current collapse structure at the summit of the volcano that is 2 by 1.5 km in diameter. This huge explosive event was followed by an effusive eruption that took place in the central crater, emplacing the Ombligo lava dome dated by ^{36}Cl exposure dating to 9.7 ka (Arce et al. 2003). The last volcanic activity at Nevado de Toluca corresponds to an explosive eruption that occurred at cal. yr. BP 3580–3831 (Macías et al. 1997b). This eruption produced a small ash flow, distributed only towards the north of the volcano, up to distances of 14 km from the summit, charring vegetation.

Popocatépetl Volcano (5552 m)

During Early and Middle Holocene, the volcano has produced at least four eruptions dated at cal. yr. BP 13,037–12,060, 10,775–9564, 8328–7591, and 6262–5318 (Siebe et al. 1997). During Late Holocene, the volcano produced three main eruptions dated at cal. yr. BP 2713–1715, 2120–1266, and 1522–733 (Siebe et al. 1996; Panfil et al. 1999; Siebe and Macías 2006). Next, we summarized the larger magnitude Plinian eruptions which occurred at the volcano during the past 5000 years.

The cal. yr. BP 6262–5318 eruption (Ochre pumice) produced wet pyroclastic surges followed by a Plinian tephra fallout dispersed to the north and subsequent pyroclastic flow deposits (Siebe et al. 1996; Arana-Salinas et al. 2010). This eruption coincides the beginning of the Mesoamerican calendar of the pre-Hispanic cultures in central Mexico. The cal. yr. BP 2713–1715 eruption (Lorenzo pumice) produced a yellow pumice fallout that was deposited on the NE flank of the volcano. The fallout buried pre-Hispanic settlements in the region (Seele 1973) that later on were called Tetimpa by archaeologists (Panfil et al. 1999; Plunket and Uruñuela 1998). This eruption coincides with the transition from the Pre-Classic to the Classic of the Mexican archaeology. The cal. yr. BP 1522–733 eruption (pink pumice) produced wet pyroclastic surges followed by three Plinian fallouts dispersed to the NW–N–NE and pyroclastic flow deposits. These eruptions blocked the hydrological network of Popocatépetl, and Iztaccíhuatl volcano located to the north. Springs and meteoric waters saturated the unconsolidated material produced by the pyroclastic flow-generated lahars that flooded the basin of Puebla. These lahars severely impacted pre-Hispanic populations and caused the abandonment of surrounding Classic ceremonial centres as Cholula, Cacaxtla, and Xochitécatl (Siebe et al. 1996). The eruption coincides (likely AC 823) with the Classic–Post-Classic period transition.

During historical times, at least four eruptions have been well recorded: AD 1540, 1592, 1664, and 1919–1927 (De la Cruz Reyna et al. 2008; Martin Del Pozzo et al. 2016). Finally, the most recent eruption started on December 21, 1994, that still continues today. It has extruded ~85 domes inside the crater and subsequently destroyed them with Vulcanian-type explosions (Gómez-Vázquez et al. 2016). This ongoing eruption had filled the volcano's crater with lava domes since year 2000 (Macías and Siebe 2005), which poses a threat for the surrounding population.

La Malinche Volcano (4461 m)

La Malinche is a Quaternary stratovolcano located 25 km to the northeast of the City of Puebla. It lacks a summit crater because its top is occupied by several dacitic domes (Castro-Govea and Siebe 2007). These authors divided the evolution of the volcano in two main stages: pre-Malinche (>45.8 ka) and Malinche stage (<45.8 to recent times). The Malinche stage included period 1 (~45.8–23.2 ka), period 2 (~21.4–15.9 ka), and period 3 (12–3.1 ka). During period 3, these authors documented at least four eruptions during the Holocene:

1. The Malinche Pumice II, a fallout deposit overlain by ash flow deposits on the NE slope of the volcano with a maximum age of 10,419–9901 cal. yr. BP.
2. A dome destruction event that emplaced block-and-ash flow deposits followed by lahars and pumice-and-ash flow deposits emplaced towards the northeast of the volcano dated at 7838–8631 cal. yr. BP (Castro-Govea and Siebe 2007).
3. The Altamira pumice-rich pyroclastic fall dated at cal. yr. BP 8980–8420, described on the northeastern flank of the volcano.
4. On top of the Malinche's sequence is a widespread ash fall deposit that is exposed on the N, W, and S flanks of the volcano. This layer is rich in coarse pumice ash, crystals, and lithics that were dated at least at cal. yr. BP 3450–3181 (Castro-Govea and Siebe 2007).

Pico de Orizaba (5675 m)

Pico de Orizaba is a Quaternary volcano built upon Cretaceous limestones and claystone rocks (Yáñez-García and García-Durán 1982). Its eruptive history shows several episodes of construction and destruction of ancient edifices (Robin and Cantagrel 1982; Höskuldsson 1992; Carrasco-Núñez and Ban 1994; Carrasco-Núñez 2000). The current volcanic edifice has been built during four eruptive phases from oldest to youngest: the Torrecillas cone, Espolón de Oro cone, the silicic peripheral domes, and Citlaltépetl cone.

The Holocene activity (Citlaltépetl cone) occurred in six eruptive stages dubbed Coscomatepec, Xilomich, Loma Grande, Avalos, Jacal, and Texmola by Höskuldsson and Robin (1993) that involved effusive and explosive eruptions. The oldest ash-and-scoria flow deposit of the Coscomatepec episode occurred at cal. yr. BP 10,248–11,142, followed by the emplacement of the Chichimeco dome complex made of amphibole-bearing andesitic lavas dated at cal. yr. BP 9474–9900 (Carrasco-Núñez 1993). At cal. yr. BP 9546–9448, occurred the Citlaltépetl eruption that emplaced pumice fallout and pyroclastic flow deposits (Carrasco-Núñez and Rose 1995; Rossotti and Carrasco-Núñez, 2004) that in age correspond to the Xilomich eruptive stage of Höskuldsson and Robin (1993). Afterwards, dome destruction events emplaced block-and-ash flow deposits on the western and southeastern flanks of the volcano at around cal. yr. BP 4297–4823 (Siebe et al. 1993; Carrasco-Núñez 1999) corresponding with the Avalos eruptive stage of Höskuldsson and Robin (1993). These authors also reported eruptions at cal. yr. BP 3514–3897 (Jacal eruptive period) and cal. yr. BP 1707–1885 (Texmola eruptive period). The Excola episode began with a Plinian eruption followed by seven effusive eruptions that occurred between AD 1537 and 1687 (Höskuldsson and Robin 1993). Recently, Alcalá-Reygosa et al. (2018) calculated ^{36}Cl exposure ages of two lava flows named "A" (3.03 ± 0.70 ka), which correlates with the Jacal episode of Höskuldsson and Robin (1993), and lava flow "B" (1.45 ± 0.35 ka) that correlates with the Texmola episode.

Los Humeros Caldera (3150 m)

Los Humeros Caldera is located some 20 km to the northwest of the town of Perote on the eastern part of the TMVB (Fig. 8.2). The caldera formed during the explosive eruption that emplaced the Xaltipan ignimbrite some 164 ± 4.2 ka (zircon ^{238}U-Th and plagioclase $^{40}Ar/^{39}Ar$ dating), and it was explosively reactivated during the formation of the nested Los Potreros Caldera at ~77 ka that emplaced the Zaragoza ignimbrite (Carrasco-Núñez et al. 2018). The eruptive activity of the caldera has alternated between episodes of effusive and explosive eruptions since the Pleistocene to the Holocene (Ferriz and Mahood 1984; Carrasco-Núñez et al. 2017). During the Holocene, studies have shown that Los Humeros extruded trachyandesitic lava flows (San Antonio-Las Chapas) at cal. yr. BP 10,180–9915 and experienced explosive activity (Cuicuiltic member) at 8374–7982 with the emplacement of an alternated succession of trachydacite pumice fall with strombolian basaltic andesite scoria fall deposits (Dávila-Harris and Carrasco-Núñez 2014), and these were followed by two events from the inner part of the caldera that emitted olivine-bearing basaltic lava flows at 4798–3905 and the El Pájaro trachytic lava flows at 3067–2878 (Carrasco-Núñez et al. 2017).

San Martín Tuxtla Volcano (1659 m)

It is a basaltic stratovolcano located in southern Veracruz near the Gulf of Mexico. San Martín is the highest peak of the Tuxtlas Volcanic Field (TVF) that consists of hundreds of scoria cones and maars (Espíndola et al. 2010). The volcano has a massive structure with a 1-km-wide crater in which there are two cinder cones. According to Nelson and Gonzalez-Caver (1992), the construction of the volcano may have started around 0.8 Ma with the emission of basanitic lavas. No other lavas forming the structure of the volcano have been dated. A widely distributed ash layer overlies a paleosol dated at cal. yr. BP 4816–4431 (Espíndola et al. 2010). These authors indicated that there were historic sources that indicated the volcano resumed activity in AD 1534 and 1664; however, these deposits are not traceable around the volcano. The March 2, 1793, eruption was instead very well preserved and documented by Moziño (1870) and García (1835) and summarized by Espíndola et al. (2010). The Smithsonian Volcanoes of the World (Simkin and Siebert 1994) considered it a VEI = 4 eruption that produced a volume of 10^8 m^3 of tephra. Espíndola et al. (2010) concluded that the eruption began with phreatomagmatic explosions followed by Strombolian activity from March to October 1793 and effusive activity that lasted ca. 2 years. The eruption formed two cinder cones inside the crater dubbed East and West by García (1835). Espindola et al. (2010) calculated a total volume of magma of 2×10^7 m^3 for this eruption.

Chichón Volcano (1100 m)

El Chichón is located in the NW portion of the State of Chiapas, at about 75 km to the SW of the City of Villahermosa, State of Tabasco. El Chichón is the northwestern most active volcano of the Chiapanecan Volcanic Arc (CVA) (Damon and Montesinos 1978). It has a 1.5 × 2-km-wide Somma crater with an inner 1-km-wide crater, which formed during its 1982 eruption. During the last 8000 years, the volcano has produced at least 11 explosive eruption units (A–L) separated by paleosols (Tilling et al. 1984; Espíndola et al. 2000; Scolamacchia and Capra 2015). The oldest Holocene eruption recorded at the volcano produced a pyroclastic flow deposit that was dated at cal. yr. BP 8597–8420 (unit L). The following eruption occurred 4246–3730 cal. yr. BP producing widespread pyroclastic flow deposits (Unit K). At about 3468–3081 cal. yr. BP, the volcano produced pyroclastic flows and surge deposits (Unit J). Around 2718–2365 cal. yr. BP, another eruption produced a pyroclastic flow deposit that incorporated pottery shards (unit I). A hydromagmatic eruption dispersed wet pyroclastic surges that were dated at 2345–2060 cal. yr. BP (unit H). Two eruptions occurred in a short period of time producing wet pyroclastic surges at 1994–1622 cal. yr. BP (unit G) and pyroclastic flow deposits at 1805–1414 cal. yr. BP (unit F). A dome destruction event generated a pyroclastic flow deposit at around 1521–1302 cal. yr. BP (unit E), and this eruption has been linked to the disruption of the Maya civilization during the Late Classic period (Nooren et al. 2017). The volcano reactivated at circa 1299–999 cal. yr. BP producing widespread pyroclastic flow deposits (unit D); the paleosol on top of this unit contains abundant pottery shards. A unique eruption of El Chichón generated a pumice-rich pyroclastic flow at around 916–674 cal. yr. BP (unit C). A Plinian eruption dispersed a widespread fallout layer dated at 655–506 cal. yr. BP (unit B). After 550 years of quiescence, the volcano reactivated in March 28, 1982, through a complex series of explosions that produced three Plinian fall layers and associated pyroclastic flow and surge deposits (Sigurdsson et al. 1984; Macías et al.

1997a; Scolamacchia and Macías 2005) whose deposits were collectively called unit A (Tilling et al. 1984).

Tacaná Volcano (4060 m)

Tacaná is a volcanic complex consisting of four NE–SW trending edifices which from oldest to youngest are Chichuj (3800 m), Tacaná (4060 m), Las Ardillas dome (3782 m), and San Antonio (3700 m) (García-Palomo et al. 2006). The complex started its formation ~225 ka with the emission of lavas and construction of Chichuj volcano. At ~50 ka, east of Chichuj, a new volcano started to grow, Tacaná, with the emission of lava flows building a semi-conical summit (Macías 2007; Macías et al. 2015). At around 20 ka, another volcano, San Antonio, started its construction southwest of Tacaná, with the emission of lava flows and a summit dome. Lastly, a lava dome Ardillas (unknown age) was extruded between San Antonio and Tacaná. At around 15 ka, the NW flank of Tacaná failed emplacing a debris avalanche followed by block-and-ash flows and subsequent lahars. The collapse left a 600-m-wide summit horseshoe-shaped crater that was open to the SW, which was subsequently filled by lava flows and two summit domes of unknown ages.

At the beginning of the Holocene, Tacaná collapsed to the north producing the Tuimanj debris avalanche at 10 ka (Macías et al. 2010; Limón 2011). This collapse might be related to the explosive eruption that produced the Once de Abril pyroclastic flows and surges that reached the base of the cone dated at cal. yr. BP 11,620–11,244 (Macías et al. 2000). After this event, nine small eruptions (VEI or Volcanic Explosivity Index = 2 of Newhall and Self, 1982) have taken place at Tacaná: the Papales ash flow deposit 1 that overlies a paleosol dated at 8683–8180 cal. yr. BP, the Chocabj pumice flow with charcoal embedded at 8683–8180 cal. yr. BP (García-Palomo et al. 2006), a pyroclastic sequence of flows and surges between paleosols dated at 6998–6400 cal. yr. BP, a pyroclastic surge sequence with an underlying paleosol dated at 2920–2715 cal. yr. BP, a scoria block and ash flow with charcoal dated at 924–748 cal. yr. BP, and a sub-Plinian fall and wet pyroclastic surges at 935–748 cal. yr. BP. These deposits cover the two summit domes that in turn must have been emplaced prior to these eruptions. Another three explosions that need to be further analysed have produced a yellow pyroclastic flow deposit dated at 536–157 cal. yr. BP (Limón 2011), the Papales ash flow deposit at cal. yr. BP 496 to present, and the white surge pyroclastic surge deposit at cal. yr. BP 285 to present (Macías et al. 2015). At around cal. yr. BP 2003–1740, the San Antonio Volcano collapsed to the SW, and this collapse was accompanied by pyroclastic flows (VEI = 4) and followed by lahars (Macías et al. 2000). The horseshoe-shaped crater left by this event at San Antonio was then filled by lava flows and a summit dome. Tacaná has been active during the last two centuries with phreatic explosions occurred in 1881 and 1949 (Müllerried 1951) and 1986 (De la Cruz-Reyna et al. 1989).

Monogenetic Volcanic Fields

Most of the Holocene activity documented in the Mexican volcanic fields comes from the Chichinautzin and Michoacán-Guanajuato areas with minor activity documented in the Apan–Tezontepec, Tuxtla, Jalapa, Serdan Oriental, Pinacate, and San Quintin volcanic fields (Figs. 8.1 and 8.2).

Baja California Volcanic Fields

These volcanic fields are constituted by three restricted volcanic areas, known as San Quintin, Jaraguay, and La Purisima (southern part of Baja California Norte), and San Borja (in the limit of Baja California Norte and Sur) (Fig. 8.1). Since the cessation of the subduction of the Farallon Plate beneath the North American plate 12 Ma ago, volcanism in this region of Mexico shifted to an extensional regime (Sawlan and Smith 1984). San Quintín is known because of the mantle xenoliths hosted in Ti-rich tholeiites, whereas Jaraguay, San Borja, and La Purísima (including the Tres Vírgenes Volcanic Complex) are calc-alkaline in nature. From these four volcanic areas, only San Quintín, Jaraguay, and San Borja have been considered as Holocene in age (Gorsline and Stewart 1962; Rogers et al. 1985; Storey et al. 1989), although radiocarbon data is only available for San Quintin. Gorsline and Stewart (1962) dated sediments in the San Quintin bay between 5 and 6 ka, overlaid by volcanic deposits of the monogenetic cones, suggesting a Holocene age for some structures from this volcanic field. Based on morphologic characteristics and comparisons with San Quintin volcanic structures, some authors have suggested that Jaraguay and San Borja could be of Holocene age (Rogers et al. 1985; Luhr et al. 1995).

Pinacate Volcanic Field

Pinacate Volcanic Field is located in the State of Sonora, close to the Mexico–US border (Fig. 8.1). It has been defined as Pleistocene–Holocene volcanic field, consisting of more than 465 eruptive centres that include cinder cones, maars, shield volcanoes, and lava flows. Chemical compositions of material which erupted from this volcanic field are heterogeneous, and basanites, trachytes, alkali-basalts, and tholeiites have been recognized (Lynch et al. 1993). Because of the arid environment in this region, radiocarbon data is not available, and instead some studies have attempted to obtain $^{40}Ar/^{39}Ar$ ages from products of El Pinacate Volcanic Field even though the rocks are K-poor. The two $^{40}Ar/^{39}Ar$ ages are at the limit of the Holocene, with an age of 12 ± 4 ka for the La Laja cinder cone (Gutmann et al. 2000) and 13 ± 3 ka for the Ives lava flow (Turrin et al. 2008). It is possible that some other voclanoes are younger than 10 ka and could be dated using other methods or techniques that have been recently improved to date Holocene rocks, such as ^{238}U-Th in zircons (Bernal et al. 2014) or ^{10}Be and ^{36}Cl (Alcalá-Reygosa et al. 2018).

Ceboruco–San Pedro Graben

Sieron and Siebe (2008) reported Holocene monogenetic activity around Ceboruco Volcano as follows: an eruption of the Molcajete scoria cone located 10 km SE of the volcano (cal. yr. BP 11,072–9918), Potrerillo II phreatomagmatic eruption (cal. yr. BP 2705–2352) on the flanks of Cerro Grande dome complex, the Pedregoso (cal. yr. BP 4150–3571), Pochetero (cal. yr. BP 2728–2150) domes, and the Pichancha coulée (>1060 yr. BP). The presence of some other monogenetic volcanoes of Holocene age in this area could not be discarded.

The Michoacán–Guanajuato Volcanic Field (MGVF)

This volcanic field consists of 1040 volcanoes that cover an approximated area of 40,000 km^2 (Hasenaka and Carmichael 1985b). Cinder cones, lava domes, shield volcanoes (Ban et al. 1992; Hasenaka 1994; Hasenaka and Carmichael 1987), and fissural lava flows (Hasenaka and Carmichael 1985a) have been reported. Volcanic activity in the field began during Early Pliocene, ~5 Ma (Guilbaud et al. 2012) or ~6 Ma (Osorio-Ocampo et al. 2018), and includes two historic volcanic structures, Jorullo (AD 1759–1776) and Parícutin (AD 1943–1952) (Fries Jr. 1953; Luhr and Carmichael 1985; Luhr and Simkin 1993; Guilbaud et al. 2011). Recent studies which have made significant advances to the understanding of the eruptive history of the volcanic field are those that focus on the Tancítaro stratovolcano (Ownby et al. 2007, 2011), Tacámbaro–Puruarán (Guilbaud et al. 2012; Mahgoub et al. 2017), Zacapu (Siebe et al. 2014; Kshirsagar et al. 2015, 2016), El Metate shield volcano (Chevrel et al. 2016a, b), Jorullo (Guilbaud et al. 2011), El Estribo volcano (Pola et al. 2014, 2015), Zacapu (Reyes-Guzmán et al. 2018), Parícutin (Larrea et al. 2017), and Pátzcuaro (Osorio-Ocampo et al. 2018). So far, these studies have documented 23 Holocene eruptions, which are distributed towards the central and mostly the southern part of the field (Fig. 8.3). According to these data, these eruptions correspond to the El Huanillo (cal. yr. BP 11,130–9688), La Taza (10,649–10,300), Cerro Grande (10,173–9502), La Mina (8400–7879), Los Caballos (8159–7657), Cerro Amarillo (7927–7696), Tendeparacua (7159–6860), Cerro Colorado (7927–7696), Los Lobos (7318–6413; personal communication, C. Siebe), La Tinaja (5325 ± 130 paleomagnetic), El Metate shield (5888–4868), El Jabali (4801–3834), El Zoyate (4090–3481), Hoya Urutzen (4084–3888), La Palma (3220–2880 BC), el Infiernillo lava flow (3475–3366), unamed lava flow (2165–2349), La Muerta (2240–2270 BC), Cicapien (1631–1819), Malpaís de Cutzarondiro (420–320 BC), El Tecolote (0.002 + −0.041 ^{40}Ar/^{39}Ar), Jorullo (AD 1759–1776), and Parícutin (AD 1943–1952).

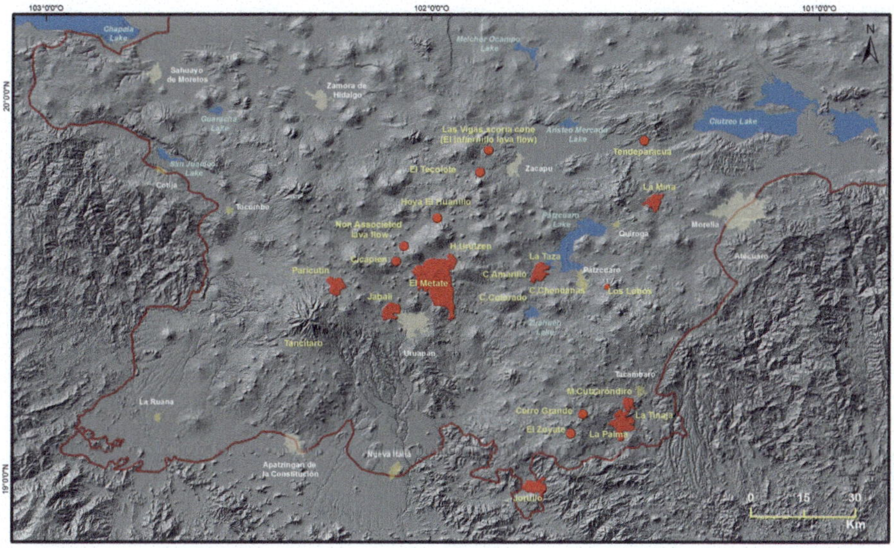

Fig. 8.3 Location of the 23 documented monogenetic volcanoes formed during the Holocene in the Michoacán–Guanajuato Volcanic Field. Most of these volcanoes are located towards the central and southern parts of the volcanic field (i.e. Jorullo and Parícutin)

Chichinautzin Volcanic Field

This volcanic field is located in the southern portion of the Mexico basin and extends from the slopes of Popocatépetl up to the slopes of Nevado de Toluca volcanoes and is oriented E–W (Fig. 8.4). This volcanic field is one of the most studied areas in the Trans-Mexican Volcanic Belt due to its heterogeneous composition and because of its proximity to Mexico City and other densely inhabited areas. It is composed of more than 220 volcanic structures (Bloomfield 1975; Martin Del Pozzo 1982; Siebe et al. 2004, 2005; Agustin-Flores et al. 2011). Radiocarbon dates published of this volcanic field (Table 8.1) indicate that at least 15 eruptions have taken place during the Holocene (summarized in García-Palomo et al. 2002a; Siebe et al. 2004; Meriggi et al. 2008; Agustin-Flores et al. 2011; Jaimes-Viera et al. 2018). These Holocene volcanoes are distributed across the entire Chichinautzin Volcanic Field. According to these data, the oldest volcanoes do not have precise ages; however, they are constrained at 11,000 ± 3000 (Mezontepec, $^{40}Ar/^{39}Ar$), cal. yr. BP 12,571–11,331 (Pelado), and 1160 ± 70 (La Cima, $^{40}Ar/^{39}Ar$) followed by six volcanic structures that include Oyameyo (9584–8782), Tenango (9579–9022), Tres Cruces (9541–9130), Cuauhtzin (8387–7968), Tabaquillo (7000 ± 9000; $^{40}Ar/^{39}Ar$), and Tláloc (7302–6885) (Table 8.1; Fig. 8.4). The younger volcanic structures range between 2.8 and 1.6 ka and are represented by Guespalapa (3159–2782), Pelagatos (2836–2346), Jumento (2041–1885), Tecuautzin (2000 ± 56,000; $^{40}Ar/^{39}Ar$), Chichinautzin (1889–1618), and Xitle volcanoes (1700–1424). There are other volcanic structures with a very

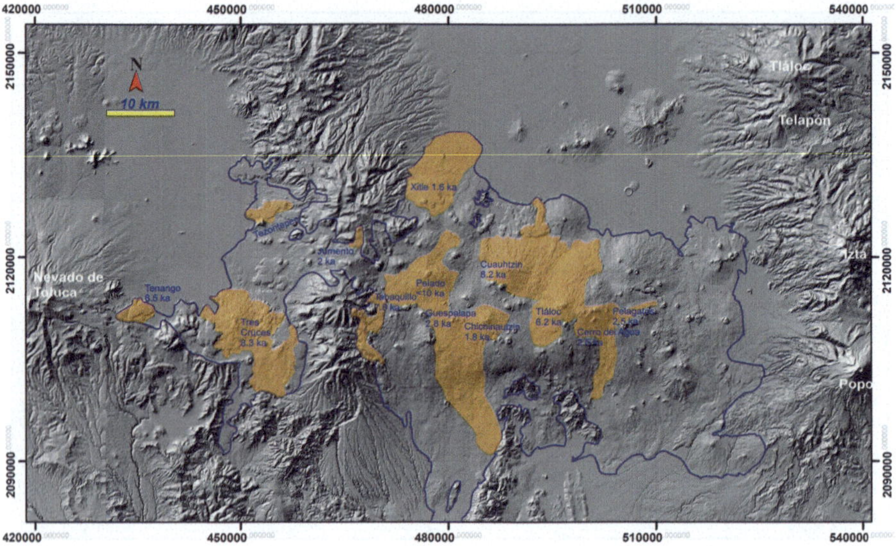

Fig. 8.4 Digital Elevation Model of the Chichinautzin Volcanic Field located showing in light-brown the location of 15 Holocene monogenetic landforms including the 1.6 ka Xitle Volcano (Siebe 2000) located to the southwestern part of Mexico City. The volcanic field is bounded to the west by Nevado de Toluca Volcano and to the east by Popocatépetl Volcano

well-preserved morphology suggesting Holocene ages (i.e. Tezontepec cone and Xicomulco lava flow). Xitle experienced a Strombolian-type eruption, starting with explosive events that deposited ash and scoria fallout, which are intercalated with seven main lava flows dispersed towards the north. These lava flows travelled about 12 km from the crater, covering the southwestern part of Mexico City. Additionally, these lavas disrupted the Cuicuilco archaeological site provoking its complete abandonment (Siebe 2000). Furthermore, it has not been discarded that volcanoes younger than Xitle could be dated in the future.

The Apan–Tezontepec Volcanic Field

This volcanic field is located in the northern part of the Mexico basin (30 km to the NE of Mexico City), including the States of Mexico, Hidalgo, and Tlaxcala (Fig. 8.2). It consists of about 140 volcanic structures including shield volcanoes, scoria cones, and dacitic lava domes that are aligned in two main directions (NE–SW and NW–SE) (García-Palomo et al. 2002b) according to fault systems with the same orientations. The only Holocene volcano in this region is called El Cuello volcanic complex made of four scoria cones with associated lava flows dated cal. yr. BP 12,928–9153 (Hernández-Javier 2008). Even if the Holocene volcanic activity is not important in the Apan region, seismic activity has been recorded in modern times, apparently related to movement along the NE–SW and NW–SE fault systems (Lermo-Samaniego et al. 2001; García-Palomo et al. 2018).

Serdan Oriental Volcanic Field

This volcanic area is located in the States of Puebla and Tlaxcala, in the eastern sector of the TMVB. It is a basin where a series of monogenetic volcanoes occur, including maars, cinder cones, and domes (Siebe et al. 1995b; Carrasco-Núñez et al. 2007; Ort and Carrasco-Núñez 2009). Just recently, Chédeville et al. (2019) described five new Holocene structures, among these volcanoes with cal. yr. BP at 11,234–11,153 (Alchichica maar), 9021–8778 (Tecuitlapa), 8376–8021 (Atexcac), 6300–6204 (Tepextitl), and 2698–2354 (Aljojuca). Las Derrumbadas twin rhyolitic domes (Siebe and Verma 1988) are the youngest structures in the region that yielded an isochron zircon age of 4.7 + 1.3/−1.2 ka (2 s, $n = 17$, MSWD = 1.18, $p = 0.28$) (Bernal et al. 2014) but a younger age at cal. yr. BP 1933–1866 (Chédeville et al. 2019).

The Xalapa Monogenetic Volcanic Field

The XMVF sits in the eastern part of the TMVB, on the east flank of the Cofre de Perote volcano (CP), around the City of Xalapa in the State of Veracruz, Mexico (Rodríguez-Elizarrarás et al. 2010). The XMVF contains over 50 monogenetic Quaternary volcanoes that consist of abundant scoria cones, small shield volcanoes, and tuff rings. The oldest age of these vents dates back to 2.59 ± 0.05 Ma (El Estropajo scoria cone). In this volcanic field, there are at least two radiocarbon-dated Holocene volcanoes, the Coacotzintla lava flow at cal. yr. BP 3340–2980 and the Volcancito scoria cone at 905–685 (mean weighted) that were both dated by Siebert and Carrasco-Núñez (2002). It seems that other volcanoes in the XMVF may be of Holocene age; however, there are no dates to support this yet.

The Tuxtla Volcanic Field

This is an isolated basaltic volcanic field, located in the State of Veracruz (Fig. 8.1) and separated from the TMVB and the CVA (Espíndola et al. 2016). This volcanic field is roughly NW–SE oriented, transected by two main faults called Catemaco and Sontecomapan faults (Andreani et al. 2008; Espíndola et al. 2016). The youngest volcanoes range in age from 0.8 Ma to the Recent, with a historical eruption of the composite San Martín Tuxtla Volcano recorded in 1793 (Nelson and Gonzalez-Caver 1992; Espíndola et al. 2010). During the last few decades, some studies have described Holocene eruptions of some of the monogenetic cones (Espíndola et al. 2016). A charcoal fragment sampled beneath the Cerro Mono Blanco ash fallout exposed in the Matacapan archaeological site yielded an age of cal. yr. BP 4147–2865. This volcanic event was responsible for the decline of the Olmecs that inhabited the area at that time (Santley 1992). Another cone called Cerro Puntiagudo emitted ash fallout and lava flows, and a charcoal fragment from beneath its ash fall yielded an age of cal. yr. BP 2301–1346 (Reinhardt 1991). It is possible that more eruptions from the Tuxtla Volcanic Field have taken place during the Holocene, but they have not been documented yet.

Outlook of Holocene Volcanism

Volcanic activity during the Holocene has taken place all over Mexican territory although the magnitudes and the recurrence intervals vary. Since 11.7 ka b2k (Walker et al. 2008) after the Younger Dryas (Broecker et al. 2010), we were able to compile the date and age of 153 well-constrained eruptions that left traceable deposits or the born of new monogenetic volcanoes. Following the internal subdivisions of the Holocene proposed by Walker et al. (2012), with the Early–Middle Holocene boundary at 8.2 ka and Middle–Late Holocene boundary at 4.2 ka, we account for the number of eruptions that occurred in Mexico. Based on the record outlined above, 17.7% of the eruptions occurred in the Early Holocene, 26.8% in the Middle Holocene, and surprisingly 55.5% during Late Holocene (Fig. 8.5a, b). Within the Late Holocene, around 15.6% of the eruptions occurred during pre-Hispanic and historical times. Eleven of these eruptions coincide with the Little Ice Age period (Porter 1986). In addition, during Late Holocene, around 79% of these eruptions have taken place within the Trans-Mexican Volcanic Belt either in active stratovolcanoes or as newly formed monogenetic volcanoes, followed by ~15% of volcanic activity in the State of Chiapas and <6% elsewhere. These percentages are a good approximation of the Holocene eruptions in Mexico; however, these figures may be biased because of two main reasons: (1) there could be volcanoes and eruptions which have not been dated yet or deposits may not be preserved (i.e. volcanic fields in Baja California or the Pinacate in Sonora State), and (2) the apparent increase in Late Holocene volcanism may due to the fact that young eruptions are easier to recognize, track, and date than older deposits that are partially covered by subsequent deposits or eroded through time. In active stratovolcanoes, large flank collapses (i.e. Colima Volcano, Cortés et al. 2019) disrupted the edifice and destroyed or covered previous deposits making it difficult to correlate deposits around the volcano and to link to a single eruption (Luhr et al. 2010). A recent reinterpretation of previous data aided with new fieldwork, new ^{14}C data, and correlation with previous studies has improved the eruptive record of Colima (Crummy et al. 2019a, b) that is no yet completely understood. In other cases, small-volume eruptions from stratovolcanoes produced deposits with limited distributions that are easily eroded by post-dispositional processes or covered by subsequent eruptions making their precise identification in space and time very difficult (e.g. Pico de Orizaba, Tacaná Volcanic Complex). On the contrary, volcanic stratigraphy has been improved over the last two decades with new studies concerning glacial chronology (Vázquez-Selem 2000; Vázquez-Selem and Heine 2011; Vázquez-Selem and Lachiniet 2017), tephrostratigraphy and palaeoenvironmental reconstruction in the lakes of central Mexico (Caballero et al. 1999, 2019; Israde-Alcántara and Garduño-Monroy 1999; Israde-Alcántara et al. 2005, 2010, 2018; Ortega-Guerrero et al. 2010, 2018a, b), pedogenesis (Sedov et al. 2003; Solleiro-Rebolledo et al. 2004, 2007, 2015, 2016; Ibarra-Arzave et al. 2019), statistical methods trying to link geological and historical eruption time series in polygenetic volcanoes (Mendoza-Rosas and De la Cruz-Reyna 2008, 2010), and dendrochronology (Biondi et al. 2003; Franco-Ramos et al. 2019).

In the near future, new and improved dating techniques ($^{40}Ar/^{39}Ar$, ^{36}Cl and ^{10}Be exposed ages, paleomagnetism, and ^{238}Th-U in zircons) in addition to the traditional ^{14}C of organic matter, often preserved in the deposits or in the bracketing paleosols, would provide new chronological constraints.

Despite these limitations, our Holocene database of volcanic activity gives a good idea of the volcanic hazards in Mexican territory (Fig. 8.5a). Out of the 97 eruptions recorded so far in stratovolcanoes, 15 of them have produced large explosive events (VEI > 4). Most of them have occurred within the TMVB: Popocatépetl with at least three Plinian eruptions (cal. yr. BP 6262–5318, 2713–1715, and 1522–733 eruptions), Nevado de Toluca with one Plinian event (cal. yr. BP 12,621–12,025 at the boundary of the Younger Dryas), Ceboruco with a Plinian eruption (cal. yr. BP 1197–1022), Fuego de Colima with four flank failures (11,231–10,756, 8175–7590, 4245–3587, 2861–2748), and at least two Plinian events (AD 1818 and 1913), El Chichón with three events cal. yr. BP 1521–1302, 655–506, and AD 1982 Plinian eruptions, Pico de Orizaba with the Citlaltépetl Plinian-sub-Plinian eruption occurred at cal. yr. BP 9546–9448, and Tacaná with at least one eruption occurring at cal. yr. BP 2003–1740 that destroyed part of San Antonio Volcano producing pyroclastic density currents and subsequent floods that affected the ceremonial centre of Izapa causing its abandonment (Macías et al. 2000, 2018). All these eruptions have affected large areas producing environmental changes (i.e. El Chichón AD 1982) and altering the development of Mesoamerican civilization in central Mexico (Siebe et al. 1996; Plunket and Uruñuela 1998; Siebe 2000) and southern Chiapas (Macías et al. 2000) or the disruption of the Maya civilization in the Central Maya Lowlands (Nooren et al. 2017).

On the other hand, monogenetic volcanism should not be ignored, considering that at least 56 volcanoes have been created during the Holocene, yielding an average recurrence of ~176 years (Fig. 8.5b). Of course, these recurrence intervals are different for each volcanic field and represent rough figures for the birth of a new volcano. For example, at Chichinautzin Volcanic Field, a new volcano is created every ~667 years, while at the Michoacán–Guanajuato Volcanic Field, a new volcano forms every ~435 years, according to our database. Two of these 56 eruptions produced some of the largest eruptions that took place during Late Holocene: these were the Las Derrumbadas twin rhyolitic domes in the Serdan Oriental Volcanic Field (~11 km^3 dense-rock equivalent, DRE) and at El Metate andesite shield in the MGVF (~9.2 km^3 DRE). The construction of these two monogenetic volcanoes occurred surprisingly during pre-Hispanic time at cal. yr. BP 5888–4868 (Metate, Chevrel et al. 2016a, b) and 1933–1866 (Las Derrumbadas, Chédeville et al. 2019). At around cal. yr. BP 1700–1424, Xitle Volcano erupted ca. 1 km^3 of magma disrupting the lives of inhabitants of the ceremonial centre of Cuicuilco located in the southern part of Mexico City (Siebe 2000). Therefore, even a small eruption like Xitle's will disrupt the life of more than 30 million people of the cities of Mexico and Cuernavaca posing diverse problems to infrastructure and communications in central Mexico.

The database presented here gives a clear indication of which volcanoes should be considered active, with eruptions recorded in the Holocene time. This includes

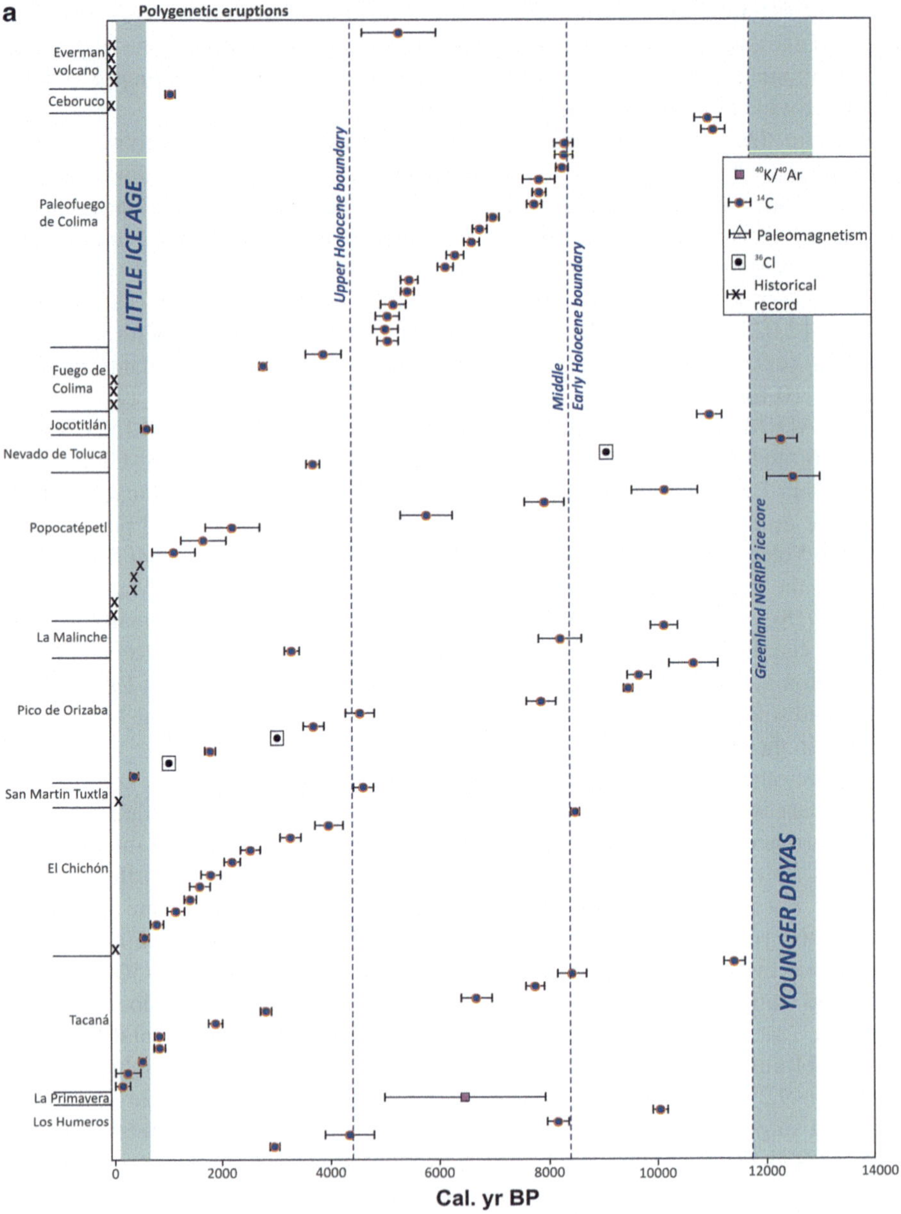

Fig. 8.5 Time versus age of volcanic eruptions that occurred during the Holocene in Mexico: (**a**) polygenetic volcanoes (stratovolcanoes and calderas) and (**b**) volcanic fields (scoria cones, maars, domes, and fissural lava flows). The limits of the Early, Middle, and Late Holocene are based on Walker et al. (2008), Younger Dryas on Broecker et al. (2010), and the Little Ice Age on Porter (1986)

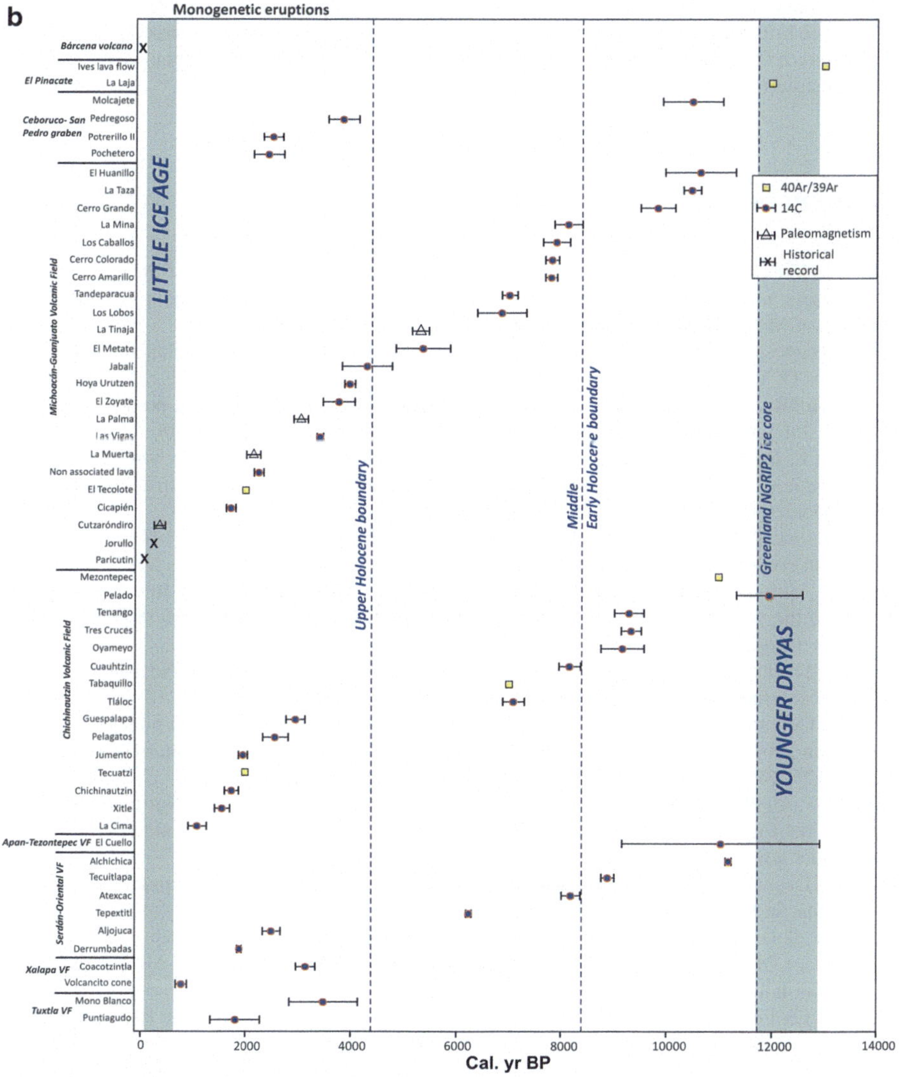

Fig. 8.5 (continued)

13 volcanoes of which 2 are calderas and 8 volcanic fields. Calderas are commonly associated with large explosive eruptions, involving several cubic kilometres of magma, and are considered the most catastrophic volcanic events, but such large events have long recurrence intervals compared to eruptions from stratovolcanoes (Lipman 2000). The monogenetic volcanic fields also pose serious hazards to the society because new volcanoes may erupt anywhere within the field.

Acknowledgement This work was supported by CONACyT PN522 and FC2016-01-2406 projects to José Luis Macías and PAPIIT IN102317 project to José Luis Arce. We appreciate the technical support of Guillermo Cisneros. We are indebted to Vicky Smith and an anonymous reviewer for their constructive comments of this chapter.

References

Agustin-Flores J, Siebe C, Guilbaud MN (2011) Geology and geochemistry of Pelagatos, Cerro del Agua, and Dos Cerros monogenetic volcanoes in the Sierra Chichinautzin volcanic field, south of México City. J Volcanol Geotherm Res 201:143–162

Alcalá-Reygosa J, Palacios D, Schimmelpfennig I, Vázquez-Selem L, García-Sancho L, Franco-Ramos O, Villanueva J, Zamorano JJ, Aumaître G, Bourlès D, Keddadouche K (2018) Dating late Holocene lava flows in Pico de Orizaba (Mexico) by means of in situ-produced cosmogenic 36Cl, lichenometry and dendrochronology. Quat Geochronol 47:93–106

Andreani L, Rangin C, Martínez-Reyes J, Le Roy C, Aranda-García M, Le Pichon X, Peterson-Rodriguez R (2008) The Neogene Veracruz fault: evidences for left-lateral slip along the southern Mexico block. Bull Soc Géol Fr 179:195–208

Arana-Salinas L, Siebe C, Macías JL (2010) Dynamics of the ca. 4,965 yr. B.P. "Ochre Pumice" Plinian eruption of Popocatépetl Volcano, México. J Volcanol Geotherm Res 192:212–231

Arce JL, Muñoz-Salinas E, Castillo M, Salinas I (2015) The ~ 2000 yr BP Jumento volcano, one of the youngest edifices of the Chichinautzin Volcanic Field, Central Mexico. J Volcanol Geotherm Res 308:30–38

Arce JL, Macías JL, Vázquez-Selem L (2003) The 10.5 ka Plinian eruption of Nevado de Toluca, Mexico: stratigraphy and hazard implications. Geol Soc Am Bull 115:230–248

Arce JL, Cervantes KE, Macías JL, Mora JC (2005) The 12.1 ka middle Toluca pumice: a dacitic Plinian-subplian eruption of Nevado de Toluca in central Mexico. J Volcanol Geotherm Res 147:125–143

Atwater T (1989) Plate tectonic history of the northeast Pacific and western North America. In: Winterer EL, Hussong DM, Decker RW (eds) The eastern Pacific Ocean and Hawaii. GSA, Boulder, pp 21–72

Ban M, Hasenaka T, Delgado-Granados H (1992) K-Ar ages of lavas from shield volcanoes in Michoacán-Guanajuato, Mexico. Geofis Int 31:467–473

Bernal JP, Solari LA, Gómez-Tuena A, Ortega-Obregón C, Mori L, Vega-González M, Espinosa-Arbeláez DG (2014) In-situ 230Th/U dating of Quaternary zircons using LA-MCICPMS. Quat Geochronol 23:46–55

Biondi F, Galindo I, Gavilanes JC, Elizalde A (2003) Tree growth response to the 1913 eruption of Volcán de Fuego de Colima, Mexico. Quat Res 59:293–299

Bloomfield K (1973) The age and significance of the Tenango Basalt, Central Mexico. Bull Volcanol 37:586–595

Bloomfield K, Valastro SJr (1977) Late Quaternary tephracronology of the Nevado de Toluca, Central Mexico: Over Geol Min Res 46, 15 p

Bloomfield K (1975) A late-Quaternary monogenetic volcano field in central Mexico. Geol Rundsch 64:476–497

Bloomfield K, Valastro S (1974) Late Pleistocene eruptive history of Nevado de Toluca, central Mexico. Geol Soc Am Bull 85:901–906

Bloomfield K, Sánchez-Rubio G, Wilson L (1977) Plinian eruptions of Nevado de Toluca Volcano. Geol Rundsch 66:120–146

Broecker WS, Denton GH, Edwards RL, Cheng H, Alley RB, Putnam AE (2010) Putting the Yunger Dryas cold event into context. Quat Sci Rev 29:1078–1081

Caballero M, Lozano-García S, Ortega-Guerrero B, Urrutia-Fucugauchi J, Macías JL (1999) Environmental characteristics of Lake Tecocomulco, northern basin of Mexico, for the last ca. 50,000 years. J Paleolimnol 22:399–411

Caballero M, Lozano-García S, Ortega-Guerrero B, Correa-Metrio A (2019) Quantitative estimates of orbital and millennial scale climatic variability in central Mexico during the last ~40,000 years. Quat Sci Rev 205:62–75

Capra L, Carreras LM, Arce JL, Macías JL (2006) The lower Toluca Pumice: a ca. 21,700 yr B.P. Plinian eruption of Nevado de Toluca volcano, Mexico. Spec Pap Geol Soc Am 402:155–173

Capra L, Macías JL, Cortés A, Saucedo R, Osorio-Ocampo S, Dávila N, Arce JL, Gavilanes-Ruíz JC, Corona-Chávez P, García-Sánchez L, Sosa-Ceballos G, Vázquez R (2016) Preliminary report on the July 10–11, 2015 eruption at Volcán de Colima: pyroclastic density currents with exceptional runouts and volumes. J Volcanol Geotherm Res 310:39–49

Caravantes A (1870) El Ceboruco. La Naturaleza, Periódico Científico de la Sociedad Mexicana de Historia Natural, Tomo 1. Imprenta de Ignacio Escalante, Mexico, pp 248–252

Carrasco-Núñez G (1993) Structure, eruptive history, and some major hazardous events of Citlaltépetl volcano (Pico de Orizaba), Mexico. Ph.D. thesis, Michigan Technological University, USA, 182 p

Carrasco-Núñez G (1999) Holocene block-and-ash flows from summit dome activity of Citlaltépetl volcano. J Volcanol Geotherm Res 88(1–2):67–75

Carrasco-Núñez G (2000) Structure and proximal stratigraphy of Citlaltépetl volcano (Pico de Orizaba), Mexico. Spec Pap Geol Soc Am 334:247–262

Carrasco-Núñez G, Ban M (1994) Geologic map and structure sections of the summit area of Citlaltepetl volcano, Mexico, with summary of the geology of the Citlaltépetl volcano summit area. Cartas Geológicas y Mineras # 9, Instituto de Geología, UNAM

Carrasco-Núñez G, Rose WI (1995) Eruption of a major Holocene pyroclastic flow at Citlaltépetl volcano (Pico de Orizaba), México, 8.5–9.0 ka. J Volcanol Geotherm Res 69:197–215

Carrasco-Núñez G, Ort MH, Romero C (2007) Evolution and hydrological conditions of a maar volcano (Atexcac crater, eastern Mexico). J Volcanol Geotherm Res 159:179–197

Carrasco-Núñez G, Hernández J, De León L, Dávila P, Norini G, Bernal JP, Jicha B, Navarro M, López-Quiroz P (2017) Geologic map of Los Humeros volcanic complex and geothermal field, eastern trans-Mexican Volcanic Belt. Terra Digit 1(2):1–11

Carrasco-Núñez G, Bernal JP, Davila P, Jicha B, Giordano G, Hernández J (2018) Reappraisal of Los Humeros volcanic complex by new U/Th zircon and 40Ar/39Ar dating: implications for greater geothermal potential. G-Cubed 19:132–149

Castro-Govea R, Siebe C (2007) Late Pleistocene–Holocene stratigraphy and radiocarbon dating of La Malinche volcano, central Mexico. J Volcanol Geotherm Res 162:20–42

Chédeville C, Guilbaud MN, Siebe C (2019) Stratigraphy and radiocarbon ages of Late Holocene Las Derrumbadas rhyolitic domes and surrounding vents in the Serdán-Oriental basin (Mexico): Implications for archaeology, biology, and hazard assessment. The Holocene, (In press)

Chevrel MO, Guilbaud MN, Siebe C (2016a) The AD 1250 effusive eruption of el Metate shield volcano (Michoacán, Mexico): magma source, crustal storage, eruptive dynamics, and lava rheology. Bull Volcanol 78:32. https://doi.org/10.1007/s00445-016-1020-9

Chevrel MO, Siebe C, Guilbaud MN, Salinas S (2016b) The AD 1250 el Metate shield (Michoacán): Mexico's most voluminous Holocene eruption and its significance for archaeology and hazards. The Holocene 26:471–488

Connor C, Conway, F (2000) Basaltic volcanic field. In: Sigurdsson, H. (Ed.), Encyclopedia of Volcanoes. Academic Press, USA, pp. 331–343

Cortés A (2002) Depósitos de Avalancha y Flujos de escombros originados hace 3,300 años por el Colapso del Sector Suroeste del Volcán de Colima. Ph.D. thesis, UNAM, México

Cortés A, Garduño-Monroy VH, Navarro-Ochoa C, Komorowski JC, Saucedo R, Macías JL, Gavilanes JC (2005) Cartas Geológicas y Mineras 10. Carta Geológica del Complejo Volcánico de Colima, con Geología del Complejo Volcánico de Colima: México D.F., Universidad Nacional Autónoma de México, Instituto de Geología, escala 1:10,000, mapa con texto explicativo 37 p., 15 figures, 2 tables

Cortés A, Komorowski JC, Macías JL, Capra L, Layer WP (2019) Late Pleistocene-Holocene debris avalanche deposits from Volcán de Colima, Mexico (Chapter 4). In: Varley N et al (eds)

Volcán de Colima, active volcanoes of the World. Springer, Berlin, Heidelberg, pp 55–80. https://doi.org/10.1007/978-3-642-25911-1_4
Crausaz W (1994) Pico de Orizaba o Citlaltépetl: geology, archaeology, history, natural history and mountaineering routes. Geopress International, Amherst, 594 p
Crowe B, Crowe P (1955) Heaven, hell and salt water. Adlard Coles Ltd., London, 264 p
Crummy JM, Savov IP, Navarro-Ochoa C, Morgan DJ (2019a) Holocene eruption history and magmatic evolution of the Colima volcanic complex (Chapter 1). In: Varley N et al (eds) Volcán de Colima, active volcanoes of the World. Springer, Berlin, Heidelberg, pp 1–25. https://doi.org/10.1007/978-3-642-25911-1_4
Crummy JM, Savov IP, Loughlin SC, Connor CB, Connor L, Navarro-Ochoa C (2019b) Challenges of determining frequency and magnitudes of explosive eruptions even with an unprecedented stratigraphy. J Appl Volcanol 8:3. https://doi.org/10.1186/s13617-019-0083-7
Dávila-Harris P, Carrasco-Núñez G (2014) An unusual syn-eruptive bimodal eruption: the Holocene Cuicuiltic member at Los Humeros caldera, Mexico. J Volcanol Geotherm Res 271:24–42
Damon P, Montesinos E (1978) Late Cenozoic volcanism and metallogenesis over an active Benioff Zone in Chiapas, Mexico. Ar Geol Soc Dig 11:155–168
De la Cruz Reyna S, Yokoyama I, Martínez Bringas A, Ramos E (2008) Precursory seismicity of the 1994 eruption of Popocatépetl volcano, central Mexico. Bull Volcanol 70:753–767
De la Cruz-Reyna S (1993) Random patterns of occurrence of explosive eruptions at Colima Volcano, Mexico. J Volcanol Geotherm Res 55:51–68
De la Cruz-Reyna S, Armienta MA, Zamora V, Juárez F (1989) Chemical changes in spring waters at Tacaná Volcano, Chiapas, México. J Volcanol Geoth Res 38:345–353. https://doi.org/10.1016/0377-0273(89)90047-4
Demant A (1978) Características del Eje Neovolcánico Transmexicano y sus problemas de interpretación: Instituto de Geología, UNAM: Revista 2, pp 172–187
DeMets C, Gordon RG, Argus DF, Stein S (1990) Current plate motions. Geophys J Int 101:425–478
Duffield WA, Tilling RI, Canul R (1984) Geology of El Chichón volcano, Chiapas, Mexico. J Volcanol Geotherm Res 20:117–132
Espíndola JM, Macías JL, Tilling RI, Sheridan MF (2000) Volcanic history of El Chichón volcano (Chiapas, Mexico) during the Holocene, and its impacts on human activity. Bull Volcanol 62:90–104
Espíndola JM, Zamora-Camacho A, Godinez ML, Schaaf P, Rodriguez-Elizarraras S (2010) The 1793 eruption of San Martin Tuxtla volcano, Veracruz, México. J Volcanol Geotherm Res 197:188–208
Espíndola JM, Lopez-Loera H, Mena M, Zamora-Camacho A (2016) Internal architecture of the Tuxtla volcanic field, Veracruz, Mexico, inferred from gravity and magnetic data. J Volcanol Geotherm Res 324:15–27
Farmer JD, Farmer MC, Berger R (1993) Radiocarbon ages of lacustrine deposits in volcanic sequences of the Lomas Coloradas area, Socorro Island, Mexico. Radiocarbon 35:253–262
Ferriz H, Mahood G (1984) Eruption rates and compositional trends at Los Humeros volcanic center, Puebla, Mexico. J Geophys Res 89:8511–8524
Franco-Ramos O, Vázquez-Selem L, Stoffel M, Cerano-Paredes J, Villanueva-Díaz J (2019) Tree-rings based analysis of the 2001 pyroclastic flow and post-eruptive tree colonization on Popocatépetl volcano, Mexico. Catena 179:149–159
Fries C Jr (1953) Volumes and weights of pyroclastic material, lava, and water erupted by Parícutin volcano, Michoacán, Mexico. EOS Trans Am Geophys Union 34:603–616
García JA (1835) Eruptionen des Vulkanes on Tustla in den Jahren 1664 und 1793. El Constitucional (Jalapa 8 Dic. 1830). Neues Jahrbuch für Mineralogie, 6, 40–456
García-Palomo A, Macias JL, Arce JL, Capra L, Garduño VH, Espíndola JM (2002a) Geology of Nevado de Toluca volcano and surrounding areas, central Mexico, Geological Society of America map and chart series MCH089. Geological Society of America, Boulder, pp 1–26
García-Palomo A, Macías JL, Tolson G, Valdez G, Mora JC (2002b) Volcanic stratigraphy and geological evolution of the Apan region, east-central sector of the trans-Mexican Volcanic Belt. Geofis Int 41:133–150

García-Palomo A, Macías JL, Arce JL, Mora JC, Hughes S, Saucedo R, Espíndola JM, Escobar R, Layer P (2006) Geological evolution of the Tacaná volcanic complex, México-Guatemala. In: Rose WI, Bluth GJS, Carr MJ, Ewert JW, Patino LC, Vallance JW (eds) Natural hazards in Central America, Geological Society of America special paper 412. Geological Society of America, Boulder, pp 39–57

García-Palomo A, Macías JL, Jiménez A, Tolson G, Mena M, Sánchez-Núñez JM, Arce JL, Layer PW, Santoyo MA, Lermo-Samaniego J (2018) NW-SE Pliocene-Quaternary extension in the Apan-Acoculco region, eastern trans-Mexican Volcanic Belt. J Volcanol Geotherm Res 349:240–255

García-Sánchez L, Macías JL, Arce JL, Sosa-Ceballos G, Garduño-Monroy VH, Saucedo R, Avellán D, Rangel E, Layer PW, Lopez-Loera H, Rocha VS, Cisneros G, Reyes-Agustín G, Jiménez A (2017) Genesis and evolution of the Cerro Prieto volcanic complex, Baja California, Mexico. Bull Volcanol 79:44. https://doi.org/10.1007/s00445-017-1126-8

Gardner JE, Tait S (2000) The caldera-forming eruption of Volcán Ceboruco, Mexico. Bull Volcanol 62:20–33

Gómez- Vázquez A, De la Cruz-Reyna S, Mendoza-Rosas A.T. (2016) The ongoing dome emplacement and destruction cyclic process at Popocatépetl volcano, Central Mexico. Bull Volcanol 78:1–15

Gorsline DS, Stewart RA (1962) Benthic marine exploration of Bahia San Quintin, Baja California, 1960-1961, marine and Quaternary geology. Pac Nat 3:282–319

Guilbaud MN, Siebe C, Layer P, Salinas S, Castro-Govea R, Garduño-Monroy VH, Le Corvec N (2011) Geology, geochronology, and tectonic setting of the Jorullo Volcano region, Michoacán, México. J Volcanol Geotherm Res 201:97–112

Guilbaud MN, Siebe C, Layer P, Salinas S (2012) Reconstruction of the volcanic history of the Tacámbaro-Puruarán area (Michoacán, México) reveals high frequency of Holocene monogenetic eruptions. Bull Volcanol 74:1187–1211

Guilbaud MN, Siebe C, Agustin-Flores J (2009) Eruptive style of the young high-Mg basaltic -andesite Pelagatos scoria cone, southeast of Mexico City. Bull Volcanol 71:859, https://doi.org/10.1007/s00445-009-0271-0

Gutmann JT (2002) Strombolian and effusive activity as precursors to phreatomagmatism: eruptive sequence at maars of the Pinacate volcanic field, Sonora, Mexico. J Volcanol Geotherm Res 113:345–356

Gutmann JT (2007) Geologic studies in the Pinacate volcanic field. J Southwest 49:189–243

Gutmann JT, Turrin BD, Dohrenwend JC (2000) Basaltic rocks from the Pinacate volcanic field yield notably young 40Ar/39Ar ages. EOS Trans Am Geophys Union 81:33–37

Hasenaka T (1994) Size, distribution, and magma output rate for shield volcanoes of the Michoacán-Guanajuato volcanic field, central Mexico. J Volcanol Geotherm Res 63:13–31

Hasenaka T, Carmichael ISE (1985a) The cinder cones of Michoacán-Guanajuato, central Mexico: their age, volume and distribution, and magma discharge rate. J Volcanol Geotherm Res 25:105–124

Hasenaka T, Carmichael ISE (1985b) Compilation of location, size, and geomorphological parameters of volcanoes of the Michoacán-Guanajuato volcanic field, central Mexico. Geofis Int 24:577–607

Hasenaka T, Carmichael I (1987) The cinder cones of Michoacán-Guanajuato, central México – petrology and chemistry. J Petrol 28:241–269

Heine K (1975) Studien zur Jungquartären glazialmorphologie mexikanischer Vulkane mit einem Ausblick auf die Klimaentwicklung. In: Lauer W (ed) Das Mexiko-Projekt der Deutschen Forschungsgemeinschaft VII. Franz Steiner Verlag GmbH, Wiesbaden, 178 p

Hernández-Javier I (2008) Geología y geomorfología volcánica de la región de los yacimientos de obsidiana en Otumba en el sector norte de la Sierra Nevada de México. Bachelor thesis, Universidad Nacional Autónoma de México, México, D.F., 124 p

Höskuldsson A (1992) Le complexe volcanique Pico de Orizaba-Sierra Negra-Cerro Las Cumbres (Sud-Est Mexicain): structure, dynamismes eruptifs et evaluations de areas. Ph.D. dissertation, University Blaise Pascal, Clermond-Ferrand, France, 210 p

Höskuldsson A, Robin C (1993) Late Pleistocene to Holocene eruptive activity of Pico de Orizaba, eastern Mexico. Bull Volcanol 55:571–587

Ibarra-Arzave G, Solleiro-Rebolledo E, Sedov S, Leonard D (2019) The role of pedogenesis in palaeosols of Mexico basin and its implication in the paleoenvironmental reconstruction. Quat Int 502:267–279

Israde-Alcántara I, Garduño-Monroy VH (1999) Lacustrine record in a volcanic intra-arc setting: the evolution of the Late Neogene Cuitzeo basin system (central-western Mexico, Michoacan). Palaeogeogr Palaeoclimatol Palaeoecol 151:209–227

Israde-Alcántara I, Garduño-Monroy VH, Fisher CT, Pollard HP, Rodríguez-Pascua MA (2005) Lake level change, climate, and the impact of natural events: the role of seismic and volcanic events in the formation of the Lake Patzcuaro Basin, Michoacan, Mexico. Quat Int 135:35–46

Israde-Alcántara I, Miller WE, Garduño-Monroy VH, Barron J, Rodriguez-Pascua M (2010) Palaeoenvironmental significance of diatom and vertebrate fossils from Late Cenozoic tectonic basins in west-central México: a review. Quat Int 219:79–94

Israde-Alcántara I, Domínguez-Vázquez G, Gonzalez S, Bischoff J, West A, Huddart D (2018) Five younger Dryas black mats in Mexico and their stratigraphic and paleoenvironmental context. J Paleolimnol 59(1):59–79

Jaimes-Viera MC, Martín del Pozo AL, Layer PW, Benowitz JA, Nieto-Torres A (2018) Timing the evolution of a monogenetic volcanic field: Sierra Chichinautzin, central Mexico. J Volcanol Geotherm Res 356:225–242

Klitgord KD, Mammerickx J (1982) Northern East Pacific rise: magnetic anomaly and bathymetric framework. J Geophys Res 87:6725–6750

Kirianov VY, Koloskov AB, De la Cruz S, Martin del Pozzo AL (1990) The Major Stages of Manifestation of Recent Volcanism in the Chichinautzin Zone. Geological Series 311. USSR Academy of Sciences, pp. 432–434

Komorowski J-C, Navarro C, Cortés A, Siebe C (1994) The repetitive collapsing nature of Colima volcanoes (México). In: Problems related to the distinction of multiple deposits and interpretation of 14C ages with implications for future hazards. Universidad de Colima, Cuarta Reunión Internacional Volcán de Colima: A Decade Volcano Workshop, Colima, 24–28 Jan, pp 12–18

Komorowski JC, Navarro C, Cortés A, Saucedo R, Gavilanes JC (1997) The Colima Volcanic complex: quaternary multiple debris avalanche deposits, historical pyroclastic sequences (pre-1913, 1991 and 1994). In: IAVCEI, Puerto Vallarta, México, 1997, Plenary Assembly, Excursion guidebook: Guadalajara, Jalisco, Gobierno del Estado de Jalisco, Secretaria General, Unidad Editorial, pp 1–38

Kshirsagar P, Siebe C, Guilbaud MN, Salinas S, Layer P (2015) Late Pleistocene Alberca de Guadalupe maar volcano (Zacapu basin, Michoacán): stratigraphy, tectonic setting, and paleo-hydrogeological environment. J Volcanol Geotherm Res 304:214–236

Kshirsagar P, Siebe C, Guilbaud MN, Salinas S (2016) Geological and environmental controls on the change of eruptive style (phreatomagmatic to Strombolian-effusive) of Late Pleistocene el Caracol tuff cone and its comparison with adjacent volcanoes around the Zacapu basin (Michoacán, México). J Volcanol Geotherm Res 318:114–133

Larrea P, Salinas S, Widom E, Siebe C, Abbitt RJ (2017) Compositional and volumetric development of a monogenetic lava flow field: the historical case of Paricutin (Michoacán, México). J Volcanol Geotherm Res 348:36–48

Lermo-Samaniego J, Havskov J, Soto J (2001) Sistema de información sismotelemétrica de México (SISMEX). Veintisiete años de servicio. Memorias Técnicas del XIII Congreso Nacional de Ingeniería Sísmica, 17 p

Limón CG (2011) Estratigrafía y morfología de los flujos de lava y depósitos asociados a la actividad efusiva del volcán Tacaná, México-Guatemala. Master thesis, Posgrado en Ciencias de la Tierra UNAM, México, 125 p

Lipman PW (2000) Calderas. In: Sigurdsson H et al (eds) Encyclopedia of volcanoes. Academic Press, San Francisco, pp 643–662

Lonsdale P (1989) Geology and tectonic history of the Gulf of California. In: Winterer EL, Hussong DM, Decker RW (eds) The eastern Pacific Ocean and Hawaii. Geology of North America. Geological Society of America, Boulder, pp 499–521
Luhr JF, Carmichael ISE (1985) Jorullo volcano, Michoacán, Mexico (1759–1774): the earliest stages of fractionation in calcalkaline magmas. Contrib Mineral Petrol 90:142–161
Luhr JF, Carmichael ISE (1990a) Petrological monitoring of cyclic eruptive activity at Volcán Colima, México. J Volcanol Geotherm Res 42:235–260
Luhr JF, Carmichael ISE (1990b) Geology of Volcán de Colima. Universidad Nacional Autónoma de México, Instituto de Geología. Bol 107. México, D.F., 1–101 + plate
Luhr JF, Simkin T (1993) Paricutin, the volcano born in a cornfield. Geoscience Press, Phoenix, p 427
Luhr JF, Aranda-Gómez JJ, Housh TB (1995) San Quintín volcanic field, Baja California Norte, México: geology, petrology, and geochemistry. J Geophys Res 100:10353–10380
Luhr JF, Navarro-Ochoa C, Savov IP (2010) Tephrochronology, petrology and geochemistry of Late-Holocene pyroclastic deposits from Volcán de Colima, Mexico. J Volcanol Geotherm Res 197:1–32
Lynch DJ, Gutmann JT (1987) Gutmann, Volcanic structures and alkaline rocks in the Pinacate volcanic field of Sonora, Mexico, Spec. Pap. 5, Ariz. Bur. of Geol. and Miner. Technol., Tucson, p. 309–322
Lynch DJ, Musselman TE, Gutmann JT, Patchett PJ (1993) Isotopic evidence for the origin of Cenozoic volcanic rocks in the Pinacate volcanic field, northwestern Mexico. Lithos 29:295–302
Macías JL, Sheridan MF, Espíndola JM, Tilling RI (1993) Refinement of the stratigraphic record of El Chichón Volcano, Chiapas, Mexico. First International Workshop on Environmental Volcanology, Morelia, Mich, Mexico, p12
Macías JL (2007) Geology and eruptive history of some active volcanoes of México. In: Alaniz-Álvarez SA, Nieto-Samaniego ÁF (eds) Geology of México: celebrating the centenary of the Geological Society of México. Geological Society of America special paper 422. Geological Society of America, Boulder, pp 183–232
Macías JL, Siebe C (2005) Popocatépetl's crater filled to the brim: significance for hazard evaluation. J Volcanol Geotherm Res 141:327–330
Macías JL, García-Palomo A, Arce JL, Siebe C, Espíndola JM, Komorowski JC, Scott K (1997a) Late Pleistocene-Holocene cataclysmic eruptions at Nevado de Toluca and Jocotitlan volcanoes, central Mexico. In: Link PK, Kowallis BJ (eds) Proterozoic to recent stratigraphy, tectonics, and volcanology: Utah, Nevada, southern Idaho, and central Mexico, Geology Studies, 42. Brigham Young University, Provo, pp 493–528
Macías JL, Sheridan MF, Espíndola JM (1997b) Reappraisal of the 1982 eruptions of El Chichón volcano, Chiapas, Mexico: new data from proximal deposits. Bull Volcanol 58:459–471
Macías JL, Espíndola JM, García-Palomo A, Scott KM, Hughes S, Mora JC (2000) Late Holocene Peléan style eruption at Tacaná Volcano, Mexico-Guatemala: past, present, and future hazards. Bull Geol Soc Am 112:1234–1249
Macías JL, Saucedo R, Gavilanes-Ruiz JC, Varley N, Velasco-García S, Bursik M, Vargas-Gutiérrez V, Cortés A (2006) Flujos piroclásticos asociados a la actividad explosive del Volcán de Colima y perspectivas futuras. GEOS 25:340–351
Macías JL, Arce JL, García-Palomo A, Mora JC, Layer PW, Espíndola JM (2010) Late-Pleistocene flank collapse triggered by dome growth at Tacaná Volcano, México-Guatemala, and its relationship to the regional stress regime. Bull Volcanol 72:33–53
Macías JL, Arce JL, Layer PW, Saucedo R, Mora JC (2015) Eruptive history of the Tacaná volcanic complex. In: Scholamacchia T, Macías JL (eds) Active volcanoes of Chiapas (Mexico) El Chichón and Tacaná, active volcanoes of the world. Springer Verlag, Berlin, Heidelberg, pp 115–138
Macías JL, Sosa-Ceballos G, Arce JL, Gardner JE, Saucedo R, Valdez-Moreno G (2017) Storage conditions and magma processes triggering the 1818 CE Plinian eruption of Volcán de Colima. J Volcanol Geotherm Res 340:117–129

Macías JL, Arce JL, Capra L, Saucedo R, Sánchez-Núñez JM (2018) Late formative flooding of Izapa after an eruption of Tacaná volcano. Anc Mesoam 29:361–371
Macorps E, Charbonnier SJ, Varley NR, Capra L, Atlas Z, Cabré J (2018) Stratigraphy, sedimentology and inferred flow dynamics from the July 2015 block-and-ash flow deposits at Volcán de Colima, Mexico. J Volcanol Geotherm Res 349:99–116
Mahgoub AN, Böhnel H, Siebe C, Salinas S, Guilbaud MN (2017) Paleomagnetically inferred ages of a cluster of Holocene monogenetic eruptions in the Tacámbaro-Puruarán area (Michoacán, México): implications for volcanic hazards. J Volcanol Geotherm Res 347:360–370
Mahood GA (1977) A preliminary report on the comenditic dome and ash flow complex of Sierra La Primavera, Jalisco. Univ Nal Auton Mexico Inst Geol 1:177–190
Mahood G (1980) Geological evolution of a pleistocene rhyolitic center—Sierra La Primavera, Jalisco, México. J Volcanol Geotherm Res 8:199–230
Mahood G (1981) Chemical evolution of a pleistocene rhyolitic center: Sierra La Primavera, Jalisco, México. Contrib Mineral Petrol 77:129–149
Mahood G, Drake R (1982) K-Ar dating young rhyolitic rocks: a case study of the Sierra La Primavera, Jalisco, Mexico. Geol Soc Am Bull 93:1232–1241
Mahgoub AN, Juárez-Arriaga E, Böhnel H, Siebe C (2019) Late-Quaternary secular variation data from Mexican volcanoes. Earth Plan Sc Letters 519: 28–39
Mammerickx J, Naar D, Tyce R (1988) The mathematician paleoplate. J Geophys Res 93:3025–3040
Martin Del Pozzo AL (1982) Monogenetic volcanism in Sierra Chichinautzin, Mexico. Bull Volcanol 45:9–24
Martín Del Pozzo AL, Romero VH, Ruiz–Kitcher RE (1987) Los flujos piroclásticos del Volcán de Colima, México. Geofis Int 26:291–307
Martin Del Pozzo A, Rodríguez A, Portocarrero J (2016) Reconstructing 800 years of historical eruptive activity at Popocatépetl Volcano, Mexico. Bull Volcanol 78:18. https://doi.org/10.1007/s00445-016-1010-y
Mendoza-Rosas AT, De la Cruz-Reyna S (2008) A statistical method linking geological and historical eruption time series for volcanic hazard estimations: applications to active polygenetic volcanoes. J Volcanol Geotherm Res 176(2):277–290
Mendoza-Rosas AT, De la Cruz-Reyna S (2010) Hazard estimates for El Chichón volcano, Chiapas, México: a statistical approach for complex eruptive histories. Nat Hazards Earth Syst Sci 10(6):1159–1170
Meriggi L, Macías JL, Tommasini S, Capra L, Conticelli S (2008) Heterogeneous magmas of the Quaternary Sierra Chichinautzin volcanic field (central Mexico): the role of an amphibole-bearing mantle and magmatic evolution processes. Rev Mex Cien Geol 25:197–216
Mora JC, Jaimes-Viera MC, Garduño-Monroy VH, Layer PW, Pompa-Mera V, Godinez ML (2007) Geology and geochemistry characteristics of the Chiapanecan volcanic arc (central area), Chiapas, Mexico. J Volcanol Geotherm Res 162:43–72
Mora JC, Layer PW, Jaimes-Viera MC (2012) New 40Ar/39Ar ages from the central part of the Chiapanecan Volcanic Arc, Chiapas, México. Geofis Int 51:39–49
Moziño JM (1870) Informe sobre la erupción del Volcán de San Martín Tuxtla (Veracruz) ocurrida el año de 1793. Bol Soc Mex Geog Estad 2:62–72
Müllerried FKG (1951) La reciente actividad del Volcán de Tacaná, Estado de Chiapas, a fines de 1949 y principios de 1950. Informe del Instituto de Geología de la UNAM, 28 p
Navarro-Ochoa C, Gavilanes-Ruíz JC, Cortés-Cortés A (2002) Movement and emplacement of lava flows at Volcán de Colima, México: November 1998–February 1999. J Volcanol Geotherm Res 117:155–167
Nelson SA (1980) Geology and petrology of Volcán Ceboruco, Nayarit, Mexico. Geol Soc Am Bull 91:2290–2431
Nelson SA, Gonzalez-Caver E (1992) Geology and K–Ar dating of the Tuxtla volcanic field, Veracruz, Mexico. Bull Volcanol 55:85–96
Newhall CG, Self S (1982) The Volcanic Explosivity Index (VEI): an estimate of explosive magnitude for historical volcanism. J Geophys Res 87:1231–1238

Nooren K, Hoek WZ, van der Plicht H, Sigl M, van Bergen MJ, Galop D, Torrescano-Valle N, Islebe G, Huizinga A, Winkels T, Middelkoop H (2017) Explosive eruption of El Chichón volcano (Mexico) disrupted 6th century Maya civilization and contributed to global cooling. Geology 45:175–178

Ort MH, Carrasco-Núñez G (2009) Lateral vent migration during phreatomagmatic and magmatic eruptions at Tecuitlapa Maar, east-central Mexico. J Volcanol Geotherm Res 181:67–77

Ortega-Guerrero B, Vázquez G, Caballero M, Israde I, Lozano-García S, Schaaf P, Torres E (2010) Late Pleistocene: Holocene record of environmental changes in lake Zirahuen, central Mexico. J Paleolimnol 44(3):745–760

Ortega-Guerrero B, Albarrán-Santos MA, Caballero M, Reyes-Corona I, Gutiérrez-Méndez B, Caballero-García L (2018a) Reconstrucción paleoambiental de la subcuenca de Xochimilco, centro de México, entre 18000 y 5000 años antes del presente. Rev Mex Cien Geol 35(3):254–267

Ortega-Guerrero B, García LC, Linares-López C (2018b) Tephrostratigraphy of the late Quaternary record from Lake Chalco, central México. J S Am Earth Sci 81:122–140

Oskin M, Stock J (2003) Pacific-North America plate motion and opening of the upper Delfín basin, northern Gulf of California, Mexico. Geol Soc Am Bull 115:1173–1190

Osorio-Ocampo S, Macías JL, Pola A, Cardona-Melchor S, Sosa-Ceballos G, Garduño-Monroy VH, Layer WP, García-Sánchez L, Perton M, Benowitz J (2018) The eruptive history of the Pátzcuaro Lake area in the Michoacán Guanajuato volcanic field, central México: field mapping, C-14 and 40Ar/39Ar geochronology. J Volcanol Geotherm Res 358:307–328

Ownby S, Delgado-Granados H, Lange RA, Hall C (2007) Volcan Tancítaro, Michoacan, Mexico, 40Ar/39Ar constraints on its history of sector collapse. J Volcanol Geotherm Res 161:1–14

Ownby S, Lange RA, Hall C, Delgado-Granados H (2011) Origin of andesite in the deep crust and eruption rates in the Tancítaro-Nueva Italia region of the central Mexican arc. Geol Soc Am Bull 123:274–294

Panfil M, Gardner TW, Hirth KG (1999) Late Holocene stratigraphy of the Tetimpa archaeological sites, northeast flank of Popocatépetl volcano, central Mexico. Geol Soc Am Bull 111:204–218

Pardo M, Suárez G (1995) Shape of the subducted Rivera and Cocos plates in southern Mexico: seismic and tectonic implications. J Geophys Res 100:12357–12373

Plunket P, Uruñuela G (1998) Preclassic household patterns preserved under volcanic ash at Tetimpa, Puebla, Mexico. Lat Am Antiq 9:287–309

Pola A, Macías JL, Garduño-Monroy VH, Osorio-Ocampo S, Cardona-Melchor S (2014) Successive collapses of El Estribo volcano volcanic complex in the Pátzcuaro Lake, Michoacán, Mexico. J Volcanol Geotherm Res 289:41–50

Pola A, Macías JL, Osorio-Ocampo S, Sosa-Ceballos G, Garduño-Monroy VH, Martínez-Martínez J (2015) El Estribo volcanic complex: evolution from a shield volcano to a cinder cone, Pátzcuaro Lake, Michoacán, México. J Volcanol Geotherm Res 303:130–145

Porter SC (1986) Pattern and forcing of northern hemisphere glacier variations during the last millennium. Quat Res 26:27–48

Rebollar CJ, Espíndola VH, Uribe A, Mendoza A, Pérez-Vertti A (1999) Distribution of stress and geometry of the Wadati-Benioff zone under Chiapas, Mexico. Geofis Int 38:95–106

Reimer PJ, Bard E, Bayliss A, Beck JW et al (2013) IntCal13 and marine13 radiocarbon age calibration curves 0-50,000 years cal BP. Radiocarbon 55:1869–1887

Reinhardt BK (1991) Volcanology of the younger volcanic sequence and volcanic hazards study of the Tuxtla Volcanic Field, Veracruz, Mexico, M.Sc. Thesis, Tulane University, 147 p

Reyes-Guzmán N, Siebe C, Chevrel MO, Guilbaud MN, Salinas S, Layer P (2018) Geology and radiometric dating of Quaternary monogenetic volcanism in the western Zacapu lacustrine basin (Michoacán, México): implications for archeology and future hazard evaluations. Bull Volcanol 80(2):18

Richards AF (1957) Geology of the Islas Revillagigedo, Mexico. Geol Soc Am Bull 68(pt 2):1843

Robin C, Cantagrel JM (1982) Le Pico de Orizaba (Mexique): structure et evolution d'un grand volcán andesitique complexe. Bull Volcanol 45:299–315

Robin C, Mossand P, Camus G, Cantagrel JM, Gourgaud A, Vincent P (1987) Eruptive history of the Colima volcanic complex (México). J Volcanol Geotherm Res 31:99–113

Rodríguez-Elizarrarás S, Siebe C, Komorowski JC, Espíndola JM, Saucedo R (1991) Field observations of pristine block and ash flow deposits emplaced April 16–17, at Volcán de Colima, Mexico. J Volcanol Geotherm Res 48:399–412

Rodríguez-Elizarrarás SR, Morales-Barrera W, Layer PW, González-Mercado E (2010) A quaternary monogenetic volcanic field in the Xalapa region, eastern trans-Mexican volcanic belt: geology, distribution, and morphology of the volcanic vents. J Volcanol Geotherm Res 197:149–166

Rogers G, Saunders AD, Terrell DJ, Verma SP, Marriner GF (1985) Geochemistry of Holocene volcanic rocks associated with ridge subduction in Baja California, Mexico. Nature 315:389–392

Rossotti A, Carrasco-Núñez G (2004) Stratigraphy of the 8.5–9.0 ka B.P. Citlaltépetl pumice fallout sequence. Rev Mex Cien Geol 21:353–370

Santley RS (1992) A consideration of the Olmec phenomenon in the Tuxtlas: early formative settlement pattern, land use, and refuse disposal at Matacapan, Veracruz, Mexico. In: Killion TW (ed) Gardens of prehistory: the archaeology of settlement agriculture in greater Mesoamerica. The University of Alabama Press, Tuscaloosa, pp 150–183

Saucedo R, Macías JL, Bursik MI, Mora JC, Gavilanes JC, Cortés A (2002) Emplacement of pyroclastic flows during the 1998-1999 eruption of Volcán de Colima, México. J Volcanol Geotherm Res 177:129–153

Saucedo GR, Macías JL, Sheridan MF, Bursik I, Komorowski JC (2005) Modeling of pyroclastic flows of Colima Volcano, Mexico: implications for hazard assessment. J Volcanol Geotherm Res 139:103–115

Saucedo R, Macías JL, Gavilanes JC, Arce JL, Komorowski JC, Gardner JE, Valdez-Moreno G (2010) Eyewitness, stratigraphy, chemistry, and eruptive dynamics of the 1913 Plinian eruption of Volcán de Colima, México. J Volcanol Geotherm Res 191:149–166

Sawlan MG, Smith JG (1984) Petrologic characteristics, age and tectonic setting of Neogene volcanic rocks in northern Baja California Sur, Mexico. In: Frizzell AV (ed) Geology of the Baja California Peninsula. Pacific section, vol 39. Society of Economic Paleontologists and Mineralogists, Tulsa, pp 237–251

Schmincke HU (2004) Volcanism. Springer Verlag, Berlin, Heidelberg, 325 p

Scolamacchia T, Capra L (2015) El Chichón volcano: eruptive history. In: Scholamacchia T, Macías JL (eds) Active volcanoes of Chiapas (Mexico) El Chichón and Tacaná, active volcanoes of the world. Springer Verlag, Berlin, Heidelberg, pp 45–76

Scolamacchia T, Macías JL (2005) Distribution and stratigraphy of deposits produced by diluted pyroclastic density currents of the 1982 eruptions of El Chichon volcano, Chiapas, México. Rev Mex Cien Geol 22:159–180

Sedov S, Solleiro-Rebolledo E, Gama-Castro J (2003) Andosol to Luvisol evolution in central Mexico: timing, mechanisms and environmental setting. Catena 54:495–513

Seele E (1973) Restos de milpas y poblaciones prehispánicas cerca de San Buenaventura Nealtican, Puebla. Comunicaciones 7:77–86

Siebe C (2000) Age and archaeological implications of Xitle volcano, southwestern basin of Mexico-City. J Volcanol Geotherm Res 104:45–64

Siebe C, Macías JL (2006) Volcanic hazards in the Mexico City metropolitan area from eruptions at Popocatépetl, Nevado de Toluca, and Jocotitlán stratovolcanoes and monogenetic scoria cones in the Sierra Chichinautzin volcanic field. In: Siebe C, Macías JL, Aguirre G (eds) Neogene-Quaternary continental margin volcanism: a perspective from Mexico. Geological Society of America special paper 402. Geological Society of America, Boulder, pp 253–329

Siebe C, Verma S (1988) Major element geochemistry and tectonic setting of Las Derrumbadas rhyolitic domes, Puebla, Mexico. Chem Erde 48:177–189

Siebe C, Komorowski JC, Sheridan MF (1992) Morphology and emplacement of an unusual debris-avalanche deposit at Jocotitlan volcano, central Mexico. Bull Volcanol 54:573–589

Siebe C, Abrams M, Sheridan M (1993) Holocene block-and-ash flow fan at the W slope of ice-capped Pico de Orizaba Volcano, Mexico: implications for future hazards. J Volcanol Geotherm Res 59:1–31
Siebe CG, Komorowski JC, Navarro C, McHone J, Delgado H, Cortés A (1995a) Submarine eruption near Socorro Island, Mexico: geochemistry and scanning electron microscopy studies of floating scoria and reticulite. J Volcanol Geotherm Res 68:239–272
Siebe C, Macías JL, Abrams M, Rodríguez S, Castro R, Delgado H (1995b) Quaternary explosive volcanism and pyroclastic depositsin east central Mexico: implications for future hazards. Geol Soc Am Field Guide:1–47
Siebe C, Abrams M, Macías JL, Obenholzner J (1996) Repeated volcanic disasters in Prehispanic time at Popocatépetl, central Mexico: past key to the future? Geology 24:399–402
Siebe C, Macías JL, Abrams M, Rodríguez S, Castro R (1997) Catastrophic prehistoric eruptions at Popocatépetl and Quaternary explosive volcanism in the Serdán-oriental basin, East-central México, IAVCEI, General Assembly, 88 p
Siebe C, Rodríguez-Lara V, Schaaf P, Abrams M (2004) Radiocarbon ages of Holocene Pelado, Guespalapa, and Chichinautzin scoria cones, south of Mexico City: implications for archaelogy and future hazards. Bull Volcanol 66:203–225
Siebe C, Arana-Salinas L, Abrams M (2005) Geology and radiocarbon ages of Tláloc, Tlacotenco, Cuauhtzin, Hijo del Cuauhtzin, Teuhtli, and Ocusacayo monogenetic volcanoes in the central part of the Sierra Chchicnautzin, México. J Volcanol Geotherm Res 141:225–243
Siebe C, Guilbaud MN, Salinas S, Kshirsagar P, Chevrel MO, De la Fuente JR, Godínez L (2014) Monogenetic volcanism of the Michoacán-Guanajuato volcanic field: Maar craters of the Zacapu basin and domes, shields, and scoria cones of the Tarascan highlands (Paracho-Paricutin region). In: Field guide for the pre-meeting fieldtrip (13–17 November 2014) of the 5th international Maar conference (5IMC-IAVCEI)
Siebert L, Carrasco-Núñez G (2002) Late-Pleistocene to precolumbian behind-the-arc mafic volcanismin in the eastern Mexican Volcanic Belt; implications for future hazards. J Volcanol Geotherm Res 115:179–205
Sieron K, Siebe C (2008) Revised stratigraphy and eruption rates of Ceboruco stratovolcano and surrounding monogenetic vents (Nayarit, Mexico) from historical documents and new radiocarbon dates. J Volcanol Geotherm Res 176:11–37
Sigurdsson H, Carey SN, Espíndola JM (1984) The 1982 eruptions of El Chichón Volcano, Mexico: stratigraphy of pyroclastic deposits. J Volcanol Geotherm Res 23:11–37
Simkin T, and Siebert L (1994) Volcanoes of the World, 2nd ed. Geoscience Press for the Smithsonian Institution, Tucson, 349 pp. ISBN 0 945005 12 1
Smith IEM, Németh K (2017) Source to surface model of monogenetic volcanism: a critical review. In: Németh K, Carrasco-Núñez G, Aranda-Gómez JJ, Smith IEM (eds) Monogenetic volcanism, Special Publications, 446. Geological Society, London, pp 1–28
Solleiro-Rebolledo E, Macías JL, Gama-Castro J, Sedov S, Sulerzhitsky LD (2004) Quaternary pedostratigraphy of the Nevado de Toluca volcano. Rev Mex Cien Geol 21:101–109
Solleiro-Rebolledo E, Sedov S, Macías JL, Pi T (2007) Late Holocene paleopedological records contained in tephra from El Chichón volcano, Chiapas, Mexico. Catena 71:444–455
Solleiro-Rebolledo E, Sedov S, Cabadas H (2015) Use of soils and paleosols on volcanic materials to establish rates of soil formation at different chronological scales. Quat Int 376:5–18
Solleiro-Rebolledo E, Straubinger M, Terhorst B, Sedov S, Ibarra G, Sánchez-Alaniz JI, Solanes MC, Marmolejo E (2016) Palaeosols beneath a lavaflow in the southern basin of Mexico: the effect of heat on the paleopedological record. Catena 137:622–634
Stock JM, Hodges KV (1989) Pre-Pliocene extension around the Gulf of California and the transfer of Baja California to the Pacific plate. Tectonics 8:99–115
Storey M, Rogers G, Saunders AD, Terrell DJ (1989) San Quintin volcanic field, Baja California, México: "within-plate" magmatism following ridge subduction. Terra Nova 1:195–202
Tilling RI, Rubin M, Sigurdsson H, Carey SN, Duffield WA (1984) Prehistoric eruptive activity of El Chichón volcano, Mexico. Science 224:747–749

Torres-Orozco R, Arce JL, Layer PW, Benowitz JA (2017) The Quaternary history of effusive volcanism of the Nevado de Toluca area, central Mexico. J S Am Earth Sci 79:12–39

Turrin BD, Gutmann JT, Swisher CC III (2008) A 13 ± 3 ka age determination of a tholeiite, Pinacate volcanic field, Mexico, and improved methods for $^{40}Ar/^{39}Ar$ dating of young basaltic rocks. J Volcanol Geotherm Res 177:848–856

Vázquez-Selem L (2000) Late quaternary glacial chronology of Iztaccihuatl volcano, central Mexico. A record of environmentalchange in the border of the tropics. Unpublished Ph.D. dissertation, Arizona State University

Vázquez-Selem L, Heine K (2011) Late Quaternary glaciation in Mexico. In: Ehlers J, Gibbard PL, Hughes PD (eds) Qauternary glaciations—extend and Chronolgy. A closer look. Elsevier, Amsterdam, pp 849–862

Vázquez-Selem L, Lachiniet MS (2017) The deglaciation of the mountains of Mexico and Central America. Geograph Res Lett 43(2):553–570

Walker GPL, Wright JV, Clough BJ, Booth B (1981) Pyroclastic geology of the rhyolitic volcano La Primavera, Mexico. Geol Rundschau 70:1100–1118

Walker M, Johnsen S, Rasmussen SO et al (2008) The global stratotype section and point (GSSP) for the base of the Holocene Series/Epoch (Quaternary System/Period) in the NGRIP ice core. Episodes 31:264–267

Walker M, Johnsen S, Rasmussen SO et al (2009) Formal definition and dating of the GSSP (global stratotype section and point) for the base of the Holocene using the Greenland NGRIP ice core, and selected auxiliary records. J Quat Sci 24:3–17

Walker MJC, Berkelhammer M, Björck S, Cwynar LC, Fisher DA, Long AJ, Lowe JJ, Newnham RM, Rasmussen SO, Weiss H (2012) Formal subdivisión of the Holocene Series/Epoch: a discussion paper by a Working Group of INTIMATE (integration of ice-core, marine and terrestrial records) and the Subcommission on Quaternary Stratigraphy (International Commission on Stratigraphy). J Quat Sci 27:649–659

Williams H (1952) Recent eruptions on San Benedicto Island, Revillagigedo Group, Mexico. Volcanol Lett 517:7

Yáñez-García C, García-Durán S (1982) Exploración de la región geotérmica Los Humeros-Las Derrumbadas, estados de Puebla y Veracruz, México, D.F., Comisión Federal de Electricidad, informe interno, 96 p

Chapter 9
Human Influence Versus Natural Climate Variability

Nuria Torrescano-Valle, Pablo J. Ramírez-Barajas, Gerald A. Islebe, Alejandro A. Vela-Pelaez, and William J. Folan

Abstract This chapter discusses past climate change drivers and ecological responses in southeastern Mexico. Ancient human influence on ecosystems is evident from sediment cores. Past human activities are evident in fossil records but make the interpretation of the past climatic signals more difficult. Conversely the human signal provides important evidence for understanding the level of human impact on the climate system and ecosystems. Four sediment cores at different locations of the Yucatán Peninsula were analyzed for fossil pollen and geochemistry. Human-induced ecological change was evaluated compared to climate-driven environmental change. Fossil pollen gave a clear signal of landscape and precipitation change in the Preclassic and Classic periods. The geochemical ratios provided evidence of local and regional hydrological change. The Chumpich Lake registry reveals that the management of low forests was efficient and indicates good hydrological control in the landscape. However this evidence is different in other sites from Yucatán Peninsula with deficient erosion control. Probably due to the differential climate response in the Chumpich–Uxul region, the drought was not as drastic as in other places.

Keywords Fossil pollen · Geochemistry · Yucatán Peninsula

N. Torrescano-Valle (✉) · G. A. Islebe · A. A. Vela-Pelaez
Departamento Conservación de la Biodiversidad, El Colegio de la Frontera Sur Unidad Chetumal, Chetumal, Quintana Roo, Mexico
e-mail: ntorresca@ecosur.mx; gislebe@ecosur.mx; alejandro.vela@estudianteposgrado.ecosur.mx

P. J. Ramírez-Barajas
Departamento Conservación de la Biodiversidad, El Colegio de la Frontera Sur Unidad Chetumal, Chetumal, Quintana Roo, Mexico

División de Estudios de Posgrado e Investigación, Tecnológico Nacional de México/ I.T. Zona Maya, Othon P. Blanco, Quintana Roo, Mexico

W. J. Folan
Centro de Investigaciones Históricas y Sociales, Universidad Autónoma de Campeche, Campeche, Mexico
e-mail: wijfolan@uacam.mx

N. Torrescano-Valle et al. (eds.), *The Holocene and Anthropocene Environmental History of Mexico*, https://doi.org/10.1007/978-3-030-31719-5_9

Introduction

From a historical environmental point of view, paleoecology is a discipline that reconstructs environmental conditions and their relationships with human activities (Birks 2012; Rull 2010; Delcourt and Delcourt 1998). Disciplines like archaeology and anthropology have provided relevant information for a better understanding of human presences, ecosystems, and climates. The study of cultural collapses during the Late Holocene has contributed to environmental determinism. The role of the environment on human populations has been explained and vice versa. Under this view, ecological aspects like carrying capacity, agricultural productivity, deforestation, and land use change have been discussed (Caseldine and Turney 2010; Coombes and Barber 2005; Erickson 1999). Information on the causes of the collapse of human societies in different parts of the world can be based on these points of view. The evidence of interactions between natural systems and human societies at centennial and millennial scales provides information to understand global change, present and future (Caseldine and Turney 2010).

Climate variability has favored or reduced human existence. The Neolithic revolution is directly linked to these changes (plant and animal domestication some 11,000 years ago). These changes caused radical revisions of human interactions and colonizations in different parts of the world. Day et al. (2004, 2007, 2012) and other authors like Milne et al. (2005), Turney and Brawn (2007), and Clark et al. (2010) have explained how sea-level adjustment around 7000 cal yr BP with a duration of some 500 years has promoted the establishment of civilizations along coastal margins of the world. Changes of monsoon patterns around 6000 cal yr BP promoted cultural development in the Afro-Asian area (Brooks 2006). During the late Holocene ca. 4000 cal yr BP, in Australia and New Guinea significant changes occurred demographically associated with climatic changes, mainly related to the periodicity of ENSO. In the fossil record, stabilization is observed in levels of the coastal and continental lakes, which agree with demographic increases. Decreases in precipitation and increases in the periodicity of ENSO influenced the decline of large human settlements (Anderson et al. 2007).

At various historical moments under certain climatic conditions, the yields in agricultural production have been favored by climate, for example, Angkor in Cambodia reveals that changes in climate favored the development of the Khmer empire between AD 1100 and 1300. In Tikal, Mexico, and a large part of Mesoamerica, the Mayan empire increased its population based on agriculture between AD 700 and 900. In the case of the Inca and other Andean cultures, changes in temperature and precipitation caused by Medieval Warming (AD 800–1100) opened the opportunity to expand the agricultural frontier to higher altitudes and increase deforestation (Diamond 2009; Chepstow-Lusty et al. 2009). On the other hand, adverse effects have also been recorded. Environmental records of Mesopotamia and Egypt show how the cultivation techniques were transformed as the climate changed. Desertification played an important role in human migrations and in the loss of crops that sustained these civilizations (Manzanilla 1997).

During the Middle Holocene (7000–4000 BCE), various climatic adjustments were decisive for human settlement and agricultural expansion (e.g., sea-level stabilization). However, toward the Late Holocene, various regions of the world experienced climatic anomalies in varying patterns (precipitation and temperature). This change was defined as "climate deterioration" (Brooks 2006). These changes promoted the migration of human groups around the world, some groups were favored, which gave way to cultural developments, but mostly large urban developments collapsed. One of the most famous examples in which the influence of climate has been demonstrated is Mayan culture. In the last 60 years, paleoecological studies have obtained sedimentary records in lakes and on the coasts, to explain how climate change influenced the Maya collapse. These records show high climatic variability during the Holocene, fluctuations in sea level, changes in the distribution of vegetation, and the occurrence of several extended droughts. These records identify the changes experienced during the establishment of the Maya civilization after ~ 1000 BC because their presence was an important trigger of changes in landscape dynamics.

The change of land use for agricultural and forestry purposes carried out by the Maya, has been analyzed by different approaches, using various proxies (Islebe et al. 1996; Mueller et al. 2010; Shaw 2003; Scarborough and Valdez 2014; Lentz et al. 2015). The beginning of agriculture in the basin of the Gulf of Mexico coastal areas and the Balsas River watershed in the Pacific has been identified between 8000 and 6200 years before presented by Pohl et al. 2007, Matsouka et al. 2002, Buckler and Stevens 2006, Sluyter and Dominguez 2006, and Buckler and Stevens 2006. For the Yucatán Peninsula region, studies conducted in Belize (Rosenswing et al. 2014) provide as a possible date ~ 6700 years BP for agriculture in the Maya lowlands, while others estimate a more recent date as early as 5400, 4600, and 4000 BP (Pohl et al. 1996; Wahl et al. 2007; Beach et al. 2009; Siemens 1983, 2011). The discussion about early agriculture and its negative effects (deforestation and soil erosion) has been extensively discussed in the last decade, but it is still difficult to get a consensus (Boserup 1965; Harrison 1990; Pohl et al. 1996; Dunning and Beach 2000; Beach et al. 2006; Johnston 2003, 2006; Lohse 2010). There is a strong interest in exploring the interaction between the environment and the human system of the past, especially to establish scenarios and projections of future changes. Climate change policies of mitigation and adaptation have promoted paleoenvironmental research in recent years, recognizing their potential in understanding the climatic and environmental history of the planet. The human signal in the fossil record has been considered as an interference signal and as a signal of change. Distinguishing changes of natural signals constitutes a constant challenge in paleoenvironmental records, providing evidence to separate and identify signals. These signals can provide information for a better understanding of the Maya collapse, actually collapses as these were several during periods of rainfall stress (Gunn et al. 1995). Separating the human signal from the fossil record in the YP is a complex task since there are few sites that do not show signs of occupation during the Middle and Late Holocene. Palynological records can identify the human presence of cultivated taxa (*Zea mays*, Cucurbitaceae, Cheno-Ams) and taxa indicating some ecological disturbance (Poaceae, Asteraceae, Euphorbiaceae). The palynological analysis allows us to understand the different stages of occupation in the region, the beginning of agriculture, the period of highest production, and the agricultural and cultural collapses. Geochemical data provide information about environmental changes that are not clearly identified in palynological records by human interference, and some local changes provide information on changes in the effects of precipitation and erosion (Roy et al. 2013, 2015). Recently, geochemical studies were included in several paleoenvironmental reconstructions in Mexico to identify climate drivers such as ENSO (El Niño), among others (Metcalfe et al. 2010; Sosa-Najera et al. 2010; Lozano-García et al. 2015; Roy et al. 2016). Studies of this nature in YP are still rare (Curtis et al. 1996; Hodell et al. 2005; Carillo-Bastos et al. 2010).

Long-term droughts related to cultural collapses have been a subject widely explored by authors such as deMenocal (2001), who mentions that the adaptation response between cultures has been variable. In addition to explaining how different cultures responded to the climatic variations of the Late Holocene, the role of paleoenvironmental studies, which identify long-term droughts, stands out. In these studies, droughts are seen as a factor of social destabilization, promoters of the abrupt abandonment of numerous human settlements throughout the world (Chew 2007; Gunn et al. 2019). These abandonments coincide with wars, religious upheaval, famine, and deforestation. The magnitude of the effect depends on the duration of droughts, and the effects of a multi-centennial and multi-decadal drought are greater than inter-annual ones. deMenocal et al. (2000) identifies, during the Holocene, synchronic changes in the North Atlantic and Northwest of the USA, associated with the North Atlantic Oscillation, atmospheric couplings, and changes in the Thermohaline circulation. The climate change of the last thousand years (Crowley 2000), associated mainly with changes in solar irradiance (increase) and volcanism (decrease), has left a clear signal of the influence of climatic variability on human populations, being one of the most influential effects of droughts, the Late Antique Little Ice Age (530–660 AD, Büntgen et al. 2016), Medieval Warm Period (ca. 800–1300 AD), and the Little Ice Age (1300–1870 AD).

In the Yucatán Peninsula (YP), various paleoenvironmental studies were carried out through different disciplinary approaches that have used many proxies to formulate their interpretations (Hodell et al. 1995, 2000, 2001, 2005, 2007; Curtis et al. 1996; Leyden et al. 1996, 1998; Brenner et al. 2002; Nooren et al. 2009; Medina-Elizalde et al. 2016; Torrescano-Valle and Islebe 2015; Douglas et al. 2015). The evidence obtained in these previous studies is based on the use of different climatic indicators and presents multiple temporal resolutions, which makes it difficult to establish correspondences between the identified events. However, they reveal the climatic variability of the region, changes in the vegetation distribution, temperature fluctuations, as well as different periods of drought, which have affected heterogeneously the different regions of the Yucatán Peninsula during the Late Holocene (Mayewski et al. 2004; Hodell et al. 2007). The most recent paleoenvironmental studies present high temporal resolution and are based on the records of fossil pollen and other proxies (Aragón-Moreno et al. 2012; Carrillo-Bastos et al. 2013; Gutíerrez-Ayala et al. 2012 in the Northwest of the YP; Torrescano-Valle and Islebe 2015 in the southwest of the YP), which improves many of the interpretations. Each tool has allowed the identification of different signs of environmental change; however discerning between changes of natural and/or anthropogenic origin is always a challenge, since the landscape management carried out by the Maya was conspicuous on spatial and temporal scales. The karstic nature of the sediments of the YP influences the preservation of microfossils. The records must be interpreted considering those effects. Another complication in the region is obtaining sediment columns of the Pleistocene. Most evidence of high-resolution (centennial and decadal) records has been obtained for the Middle and Late Holocene. Given the history of occupation that is interpreted in these records of the region, it is important to consider the last 3500 years. This period allows analyzing the human relationship with the environment and vice versa, the gradual increase of population growth, and the use of natural resources. The northern Peten has been considered as an area of dense population during the Preclassic and Classic (La Roque et al. 2019; Wahl et al. 2006, 2007), which makes it an important region for the identification of this relationship.

This chapter presents an overview of the human-environment relationship of the last 3500 years of the Maya area of the Yucatán Peninsula. Under the motto of Simpson (2012) "The present is the key to understanding the past", we propose the corollary "The past can be used to understand the future". We seek to contribute to the development of emerging disciplines that allow construct more realistic models of climate change for the region. In this case, the "attribution" is an emerging discipline which seeks to quantify how human influence has altered the probability and magnitude of some particular type of climate event (Stott et al. 2016).

Methods

Description of the Study Sites

Ría Lagartos: Site located in the northwest of the Yucatán Peninsula at coordinates 21.5661N and 88.0865W, which is part of the Ría Lagartos biosphere reserve. The geographic position that it occupies allows identification of the influence of diverse climatic patterns, the trade winds, polar air masses, and convection that influences regional precipitation. The site corresponds to the displacement limits of the Intertropical Convergence Zone (ITCZ). The climate is a warm-semiarid type ((Bso (h ') w (x) iw)). Most of the precipitation occurs in summer and the rest in winter with the influence of the polar air masses, low sub- and deciduous forests with *Metopium brownei, Bursera simaruba, Haematoxylum campechianum, Conocarpus erectus, Plumeria sp.*, and *Bravaisia* sp. The core was analyzed and published by Carrillo-Bastos et al. (2013).

Xpuk-Petenes: The core was obtained 3 km from the western coastal margin of the Yucatán Peninsula, at coordinates 20.0752 ″N and 90.2716E. This site is also part of a Biosphere Reserve, Los Petenes. The dominant climates are Aw (warm-subhumid with rain in summer) and BS'H'W (semi dry and dry-warm). The average rainfall varies in the north and center-south, 700–800 mm and 800–1100 mm, respectively. The highest amount of rainfall occurs during June to October. The dominant vegetation types are mangroves, petenes (islands of tree vegetation), low deciduous forests, and deciduous forests composed of *Haematoxylum campechianum*. The data obtained in this core were reported by Gutierrez-Ayala et al. (2012) and Roy et al. (2017).

Silvituc Lake: Lake Silvituc is located in the southwest of the YP in the Mexican state of Campeche at coordinates 18.372N and 90.1741W. The average annual rainfall is 1300 mm (Perez et al. 2011). In the area there are some rivers and flood zones, established on a mature karst with a large accumulation of soils (Marín-Stillman et al. 2004). According to the archaeological studies, the site corresponds to an occupation during the Postclassic period (Ojeda-Mas et al. 1996; Alexander 2000). The dominant medium-statured tropical forest (semi-evergreen tropical forest) and low tropical forest (low semi-evergreen tropical forest) are represented by species such as *Manilkara zapota* (L.) P. Royen, *Brosimum alicastrum* Sw, *Acoeloraphe wrightii* H. Wendl. Ex Mart., Cyperaceae, and Poaceae. The results of the analysis of the column were reported by Torrescano-Valle and Islebe (2015).

Chumpich Lake: Located in the southern part of the YP, the nearest archaeological site corresponds to Uxul, which is ~ 50 km from Calakmul and ~ 30 km from El Mirador. The lake has a length of 1.7 km and 1.3 km; it is surrounded by a low flood forest that connects it with other nearby lakes such as San Felipe and the Mirador basin (Wahl et al. 2007). The karst is ancient and rich in dolines, underground caverns, and poljes (Marin-Stillman et al. 2004). Precipitation ranges from 1300 to 1500 mm per year: most fall between June and December, when the ITCZ migrates north (Hastenrath 1984). The medium forests are characterized by the presence of *Brosimum alicastrum, Manilkara zapota, Talisia olivaeformis,* and *Piper dioica*. The low rainforests have species like *Haematoxylum campechianum, Metopium brownei, Bucida buceras, Cladium jamaicense*, and Cyperaceae. The results of this core are reported for the first time in this chapter.

Chronology

The chronologies of the four sites were standardized; the depth age models (Fig. 9.2) were obtained by means of the Bacon package available in R, which uses Bayesian statistics for the construction of the curves (Blaauw and Christen 2011).

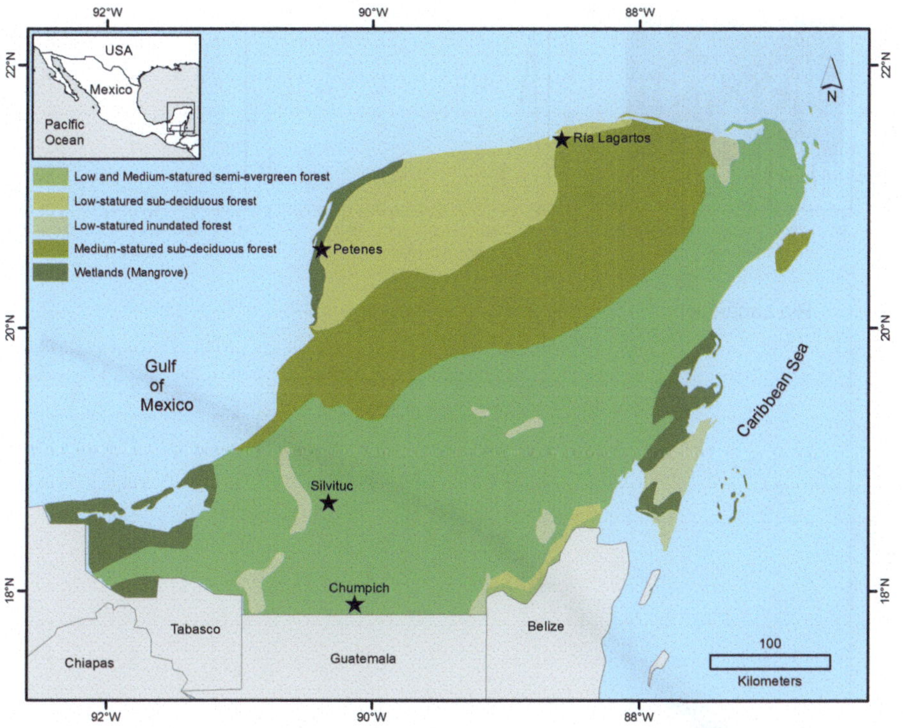

Fig. 9.1 Location of the four coring sites in the Yucatán Peninsula

Fossil Pollen

In the four cores, the same fossil pollen extraction techniques, the physical and chemical methods according to Faegri and Iversen (1989), and the density method of Nakagawa et al. (1998) were used (Fig. 9.1).

Elements Geochemistry

Two of the Xpuk-Petenes (215 cm) and Chumpich (100 cm) cores provide geochemical data for elements, mainly for Ti, K, Fe, and Ca. The columns were cut at 2 cm intervals, dried at 40 °C, homogenized, and ground in an agate mortar. Subsequently they were reviewed by means of a Thermo Scientific Niton XL3T X-ray fluorescence (XRF) analyzer, the data were corrected using a linear equation developed as a calibration curve by Roy et al. (2012).

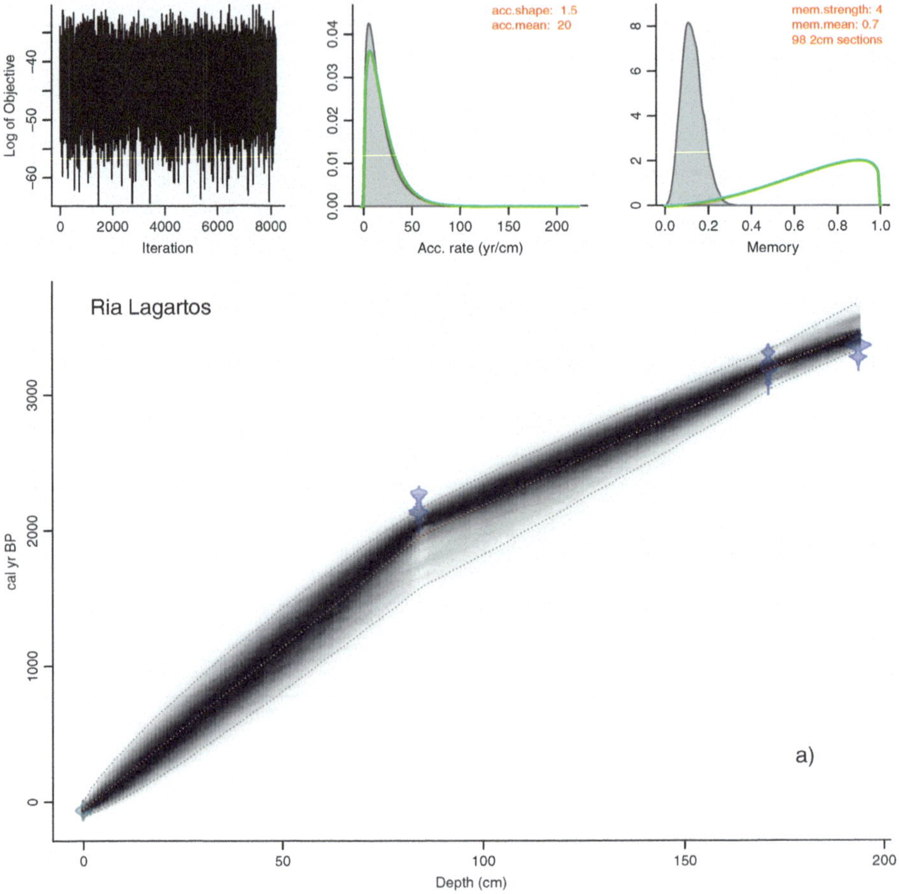

Fig. 9.2 Age–depth models in cal yr BP for the sites (**a**) Ría Lagartos, (**b**) Xpuk–Petenes, (**c**) Silvituc Lake, (**d**) Chumpich Lake

Analysis of Data

The pollen diagrams were constructed using the Tilia and Tilia Graph software, by means of the percentages of each taxon; the pollen zones were obtained by means of the sum of squares of CONISS 1.5.12® (Grimm 2011). The geochemistry data were normalized by standardizing Fisher Z-scores using the standard deviation of the data. For the comparison of data, a linear interpolation was made to calculate continuous values between time (x-axis) and pollen (y-axis) for the four sites, Ría Lagartos, Xpuk–Petenes, Silvituc Lake, and Chumpich Lake, during the last 3750 years.

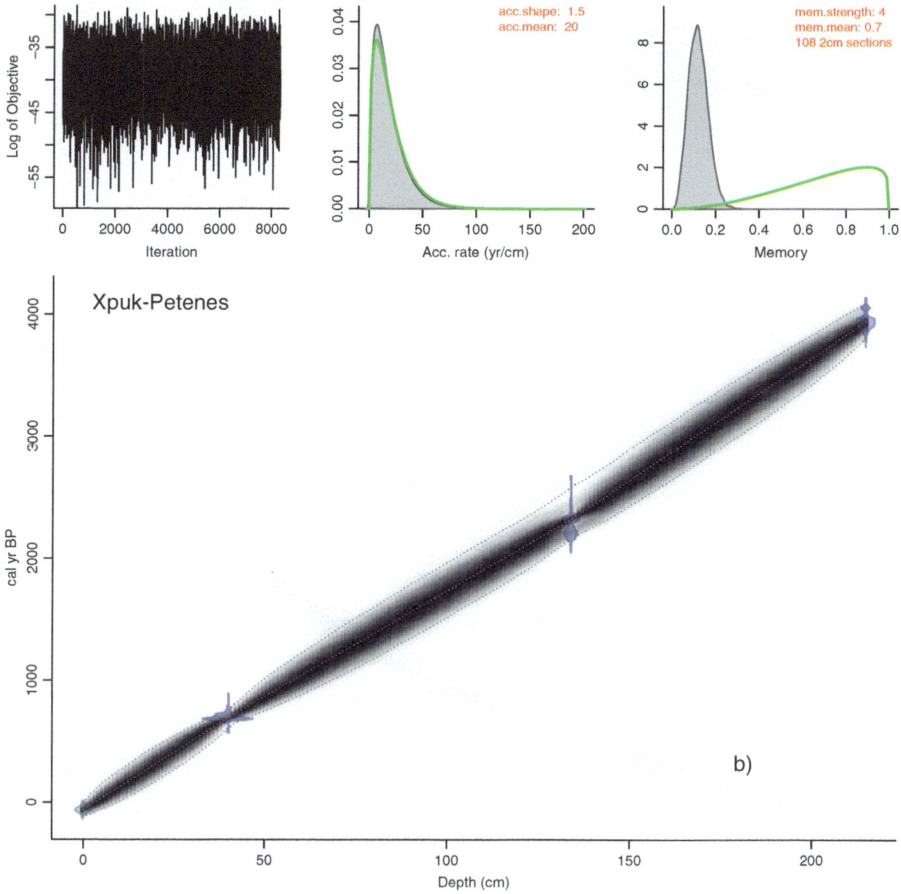

Fig. 9.2 (continued)

Results

Chronology

The age–depth models allowed obtain calibrated dates using the same method (Table 9.1, Fig. 9.2).

Fossil Pollen

Fifty-seven samples were obtained from the Ría Lagartos column, 51 from Xpuk–Petenes, 35 from Silvituc Lake, and 46 from the Chumpich Lake core. In Fig. 9.3, the pollen diagram of Chumpich Lake is shown, and two pollen zones were identified.

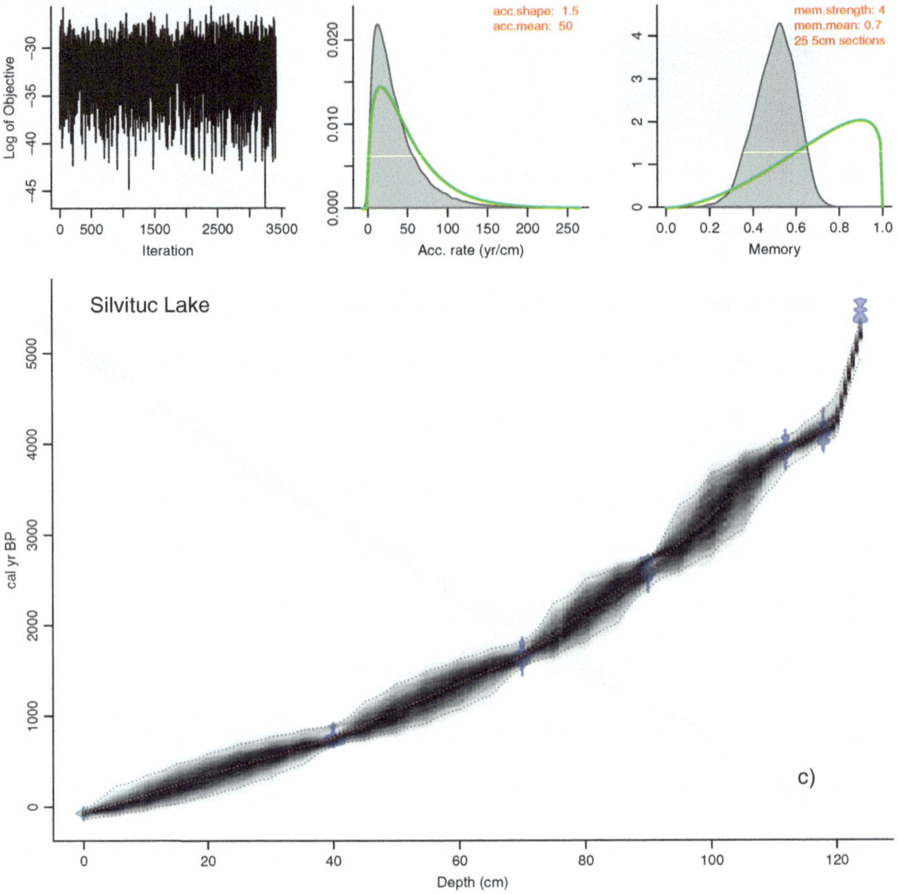

Fig. 9.2 (continued)

Pollen Zone I: The analysis of CONISS suggests two subzones; subzone Ia, the base of the core, shows abundant diatoms and *Botryococcus*. Representative elements of low-flooded forests and tropical forests such as *Bucida buceras*, *Moraceae*, *Fabaceae*, *Euphorbiaceae*, and *Ficus* are abundant (10–30%), Sapotaceae and Euphorbiaceea are present with a percentages lower than 10%. Taxa such as *Zea mays* (15–20%) and Poaceae (20–50%) are present from the base of the nucleus but are more abundant in the subzone Ib. Temperate taxa such as *Pinus, Quercus, Alnus*, and *Podocarpus* are abundant in the base (>10%); in the final part of this subzone there is a change in lithology. Clay sediment rich in CaCO3 covers 20 cm of the column. *Pinus* increases (20%) but *Alnus, Quercus*, and *Podocarpus* decrease (<5%); *Botryococcus* disappears, as well as spores and diatoms; *Nymphaea ampla* increases (20–40%) as does Cyperaceae (5–10%) together with disturbance taxa.

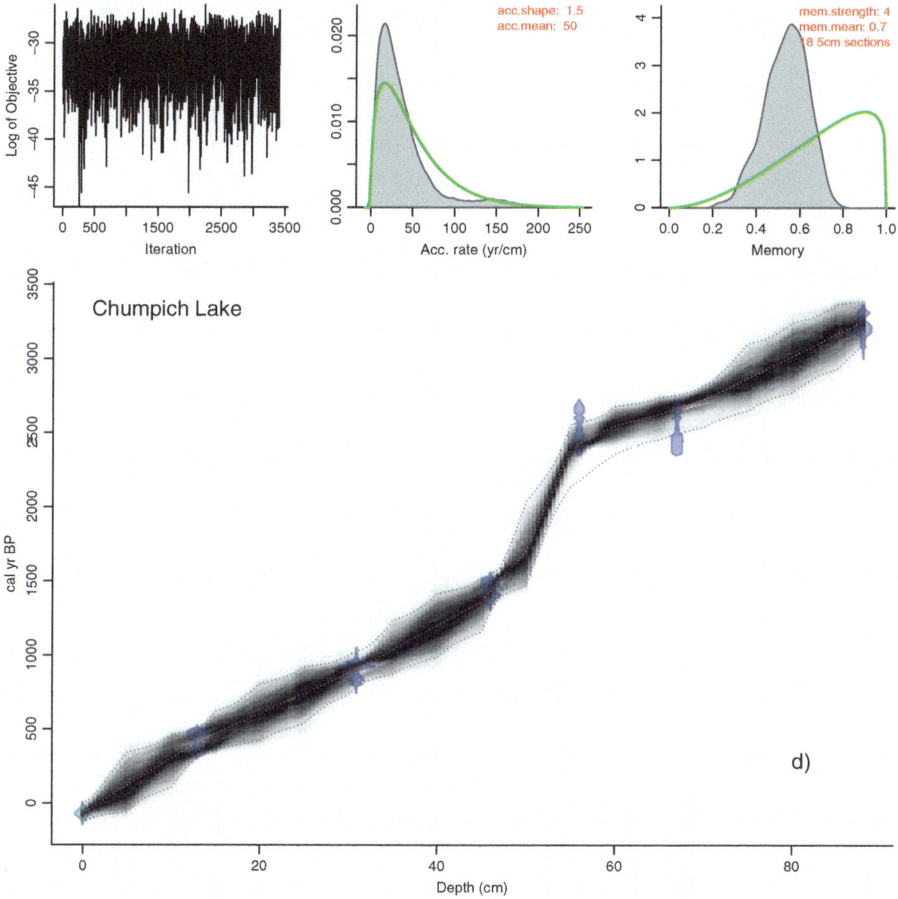

Fig. 9.2 (continued)

Pollen Zone II: Three subzones are identified; IIa shows the lithological change; elements of disturbance and low forest are dominant but at low percentages (<10%); *Bucida buceras*, Fabaceae, Cheno-Ams, *Pinus*, and *N. ampla* reach the highest percentage (10–20%); Moraceae and *Ficus* are present with a percentage <10%; the presence of *Zea mays* decreases until it reaches <5%; and temperate elements increase slightly (10%). In subgroup IIb *Bucida buceras* and Fabaceae dominate the spectrum (20–40%), other elements increase such as Sapotaceae (10–20%), elements of disturbance as Cheno-Ams continue present with 20%, while *Croton, Borreria*, and Solanaceae increase (5–10%). Moraceae and *Ficus* show an increase (15–20%); *N. ampla* reaches 40% representation. In the last subzone IIc, Fabaceae is the most abundant taxon (40%), *Bucida buceras* diminishes (20%), and *Ficus* and

Table 9.1 Radiocarbon and calibrated dates from the four cores

Lab code	Depth (cm)	Radiocarbon age yr BP	Calibrated age BC/AD
IIRLGI	84.1	2160 ± 30	BC 145
IIRLGIII	171.36	3020 ± 30	BC 1237
IIRLGIV	193.8	3130 ± 30	BC 1443
BETA-227732	40	820 ± 40	AD 1190
BETA-227733	70	1740 ± 40	AD 290
BETA-227734	90	2540 ± 40	BC 653
BETA-227735	112	3620 ± 40	BC 1954
BETA-227736	118	3740 ± 40	BC 2215
BETA-227737	124	4710 ± 40	BC 3256
PXPU090540	40	760 ± 30	AD 1255
PXPU0905134	134	2280 ± 50	BC 347
PXPU0905215	215	3650 ± 30	BC 1990
CHUM13132017	13	370 ± 30	AD 1548
CHUM31062014	31	1000 ± 30	AD 1043
CHUM46112013	46	1540 ± 30	AD 521
CHUM56062014	56	2450 ± 30	BC 356
CHUM67112013	67	2430 ± 30	BC 715
CHUM88131217	88	3020 ± 30	BC 1299

IIRLGI = Ría Lagartos, BETA = Silvituc Lake, PXPU = Xpuk-Petenes, CHUM = Chumpich Lake

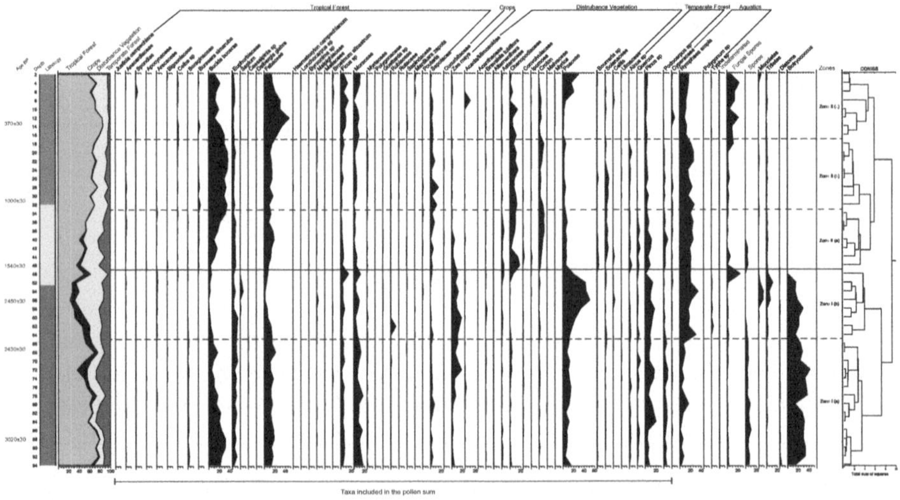

Fig. 9.3 Pollen diagram of Chumpich Lake

Moraceae increase (20%). Euphorbiaceae and Sapotaceae are present with 10%, some taxa that diminish at the top of sub-zone IIb increase slightly until reaching 10% (Poaceae, *Pinus*, Cheno-Ams, *Croton*). Acacia is presented with its highest percentage in the whole column (15%), likewise the fungal spores increase (20%).

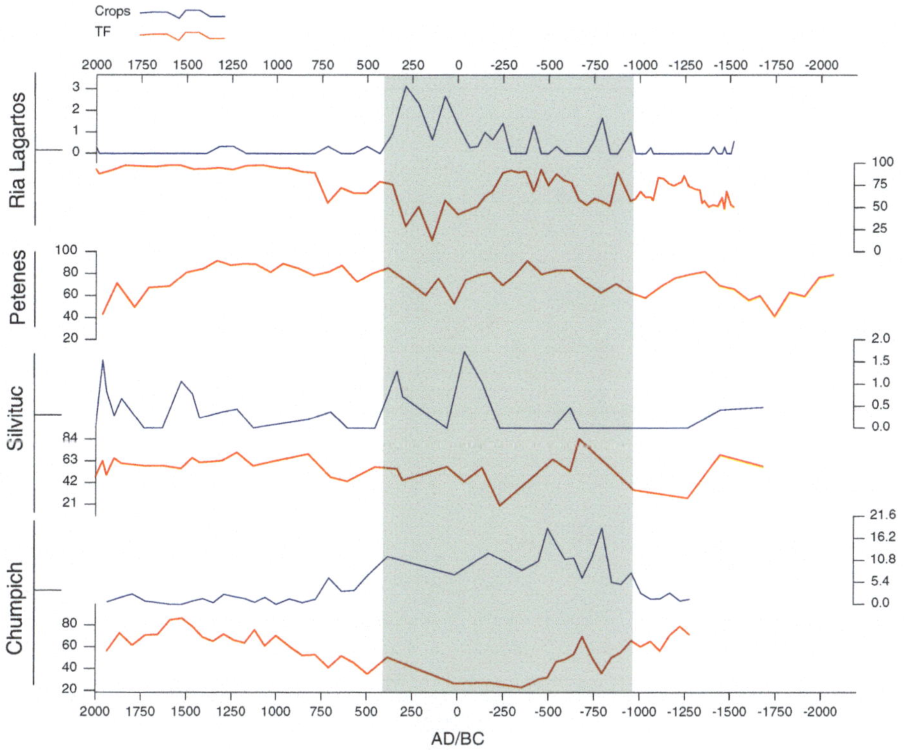

Fig. 9.4 Comparative pollen spectra between tropical forest (TF) and crops, the shaded area shows the period with highest percentages of crops

Figure 9.4 shows the tropical forest vs crop curves of the four fossil columns; the shaded area corresponds to the period recorded with the highest percentages of agriculture from BC ~ 1000 to AD ~ 400. The Silvituc core shows a later agricultural management between AD 1500 and 2000, while Chumpich shows a decrease later than the other cores in AD ~ 750, which corresponds to the abandonment of the nearby city of Uxul under the Kaan dynasty (Grube et al. 2012; Grube and Delvendahl 2016). This comparative shows that the TF decreases up to 70% during the periods of highest agricultural activity.

Geochemistry

The fossil columns Xpuk–Petenes and Chumpich Lake provided geochemical data for environmental interpretation; 109 and 91 samples were measured using a Thermo Scientific Niton XL3t X-ray fluorescence (XRF) analyzer. The elements Ca, Ti, Fe, and K were measured and adjusted by means of the standard curve developed for YP (Roy et al. 2018).

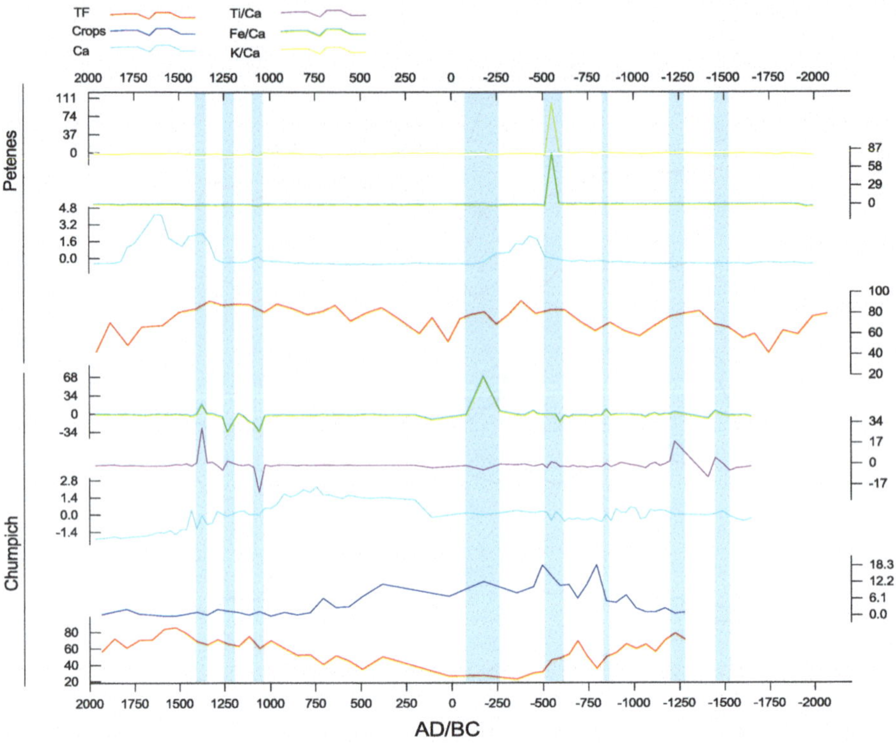

Fig. 9.5 Geochemical elemental and pollen fossil comparison between Chumpich Lake and Xpuk–Petenes core

The adjusted values of Ti, Fe, and K are related to Ca to diminish the effects of dilution. All values were normalized and are shown in Fig. 9.5. The column of Petenes shows uniform behavior over the last 4000 years, significant changes occur at BC ~ 550, and the ratio of Fe / Ca and K / Ca increases significantly. The Ca presents three peaks BC 350–470, another at AD 1330–1400 and AD 1600–1700. Chumpich presents changes in the geochemistry of elements in higher number of dates. The Ti / Ca ratio increases during BC 1200–1400 and in BC 1375, and in AD 1050 it decreases. The Fe / Ca ratio increases considerably at BC 170, decreases to AD 1060 and to AD 1230, and finally increases to AD 1380. The Ca presents an increase in various dates, but the most notable is the period AD 202 to 961.

Discussion and Conclusion

Differentiating the Climate Impact Versus Human Impact

The Xpuk–Petenes core is an example of a record with low human interference; therefore it allows identifying climatic changes independent of human occupation and to establish comparisons with records that do present human interference.

Figure 9.4 shows a pollen comparison of the TF and crop groups in the four fossil cores. The period of highest agricultural activity covers part of the Preclassic period (BC ~ 900) and a large part of the Classic period (AD 400). TF in Xpuk–Petenes shows a lower variation since it does not reveal human occupation. In the other cores, the decrease of TF is evident during the agricultural activity, from 40 to 60% reduction. These two signals are fundamental for the identification of human influence, as well as for identifying the impact on forests.

The normalized data of elemental geochemistry (Fig. 9.5) can be interpreted as an index, according to Roy et al. (2017, 2018) and Gregory et al. (2015), the elemental ratio of Ti / Ca, Fe / Ca, and K / Ca increases when precipitation increases, and these debris enter the basin in higher amounts. Abundant periods of precipitation are observed in Xpuk–Petenes around BC 550, the variation of the index over time is minimal and suggests little soil erosion, as the influx of detritus to the basin is mostly uniform. The increases are mainly related to higher precipitation. On the other hand, the curve of Ca can be interpreted as a sign of drought (Roy et al. 2017). Two periods show an increase in the curve of Ca, BC ~ 400–270 and AD 1300–1700, the latter corresponds to the Little Ice Age (LIA), and a third drought characterized falls within the Medieval Warm Period (AD ~ 1070).

The TF curve shows little variation, but the Mangrove / *Conocarpus* curve (Roy et al. 2017) corresponds to that of Ca. The Chumpich Lake column reveals in the comparison of the tropical forest and crops signal a decrease between 50 and 60% for TF during periods of stronger agricultural activity (BC ~ 1000 to AD ~ 400), which correspond to the Preclassic and Classic period (Dunning et al. 2012; Gill 2008). Agricultural collapse occurs at AD 750 which coincides with the cultural collapse reported by Grube et al. (2012) and Grube and Delvendahl (2016). The geochemical ratios of Fe / Ca and Ti / Ca show an increase in precipitation on the dates BC 1500, BC 1250, BC 200, and AD 1375. Those ratios also reveal low precipitation in the periods around BC 1400, AD 1060, and AD 1200.

The pollen signal and elementary geochemistry allow us to identify events corresponding to the human signal and the climatic signal, especially when there are signals with low interference (Xpuk–Petenes). This leads to the following inferences and questions: 1) In pollen, which signal reveals best the climatic influence? 2) Does the fossil signal allow us to distinguish a climatic gradient of the YP? First, a model was elaborated with the pollen data of the elements of disturbance in the four sites (Figs. 9.6, 9.7, and 9.8); the average value of the abundance was calculated by pollen of disturbance vegetation (DV), crops, and TF taxa, using the software OriginLab (2016) and Minitab 18. Figure 9.7 corresponds to the average of DV and reveals a drought between BC 200 and AD 250. Crops and TF corresponds to a decrease; this drought is recorded by other studies (Conroy et al. 2008; Kennet et al. 2012; Wahl et al. 2014; Akers et al. 2016), which corresponds with increase of ENSO activity and displacement of the ITCZ to the south.

In Fig. 9.9, a summary of the comparative Fig. 9.5 and the results of the average between the pollen curves of TF, crops and DV of the Ría Lagartos, Xpuk–Petenes, Silvituc, and Chumpich sites are presented. This figure allows us to identify the match between the climatic signal interpreted through the geochemical signal and the pollen signal. The droughts based mainly on the average curve of Ca are

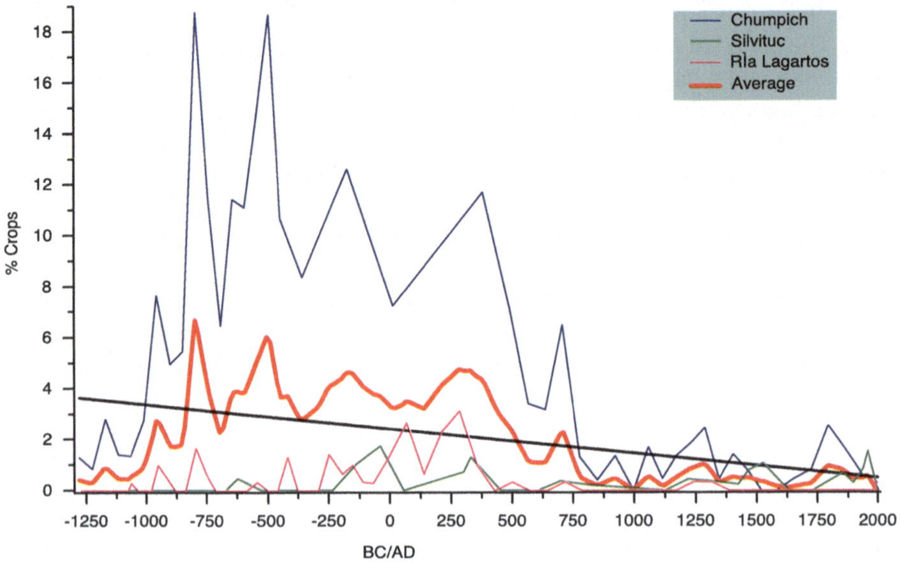

Fig. 9.6 Average abundance of crops in Ría Lagartos, Xpuk–Petenes, Silvituc Lake, and Chumpich Lake

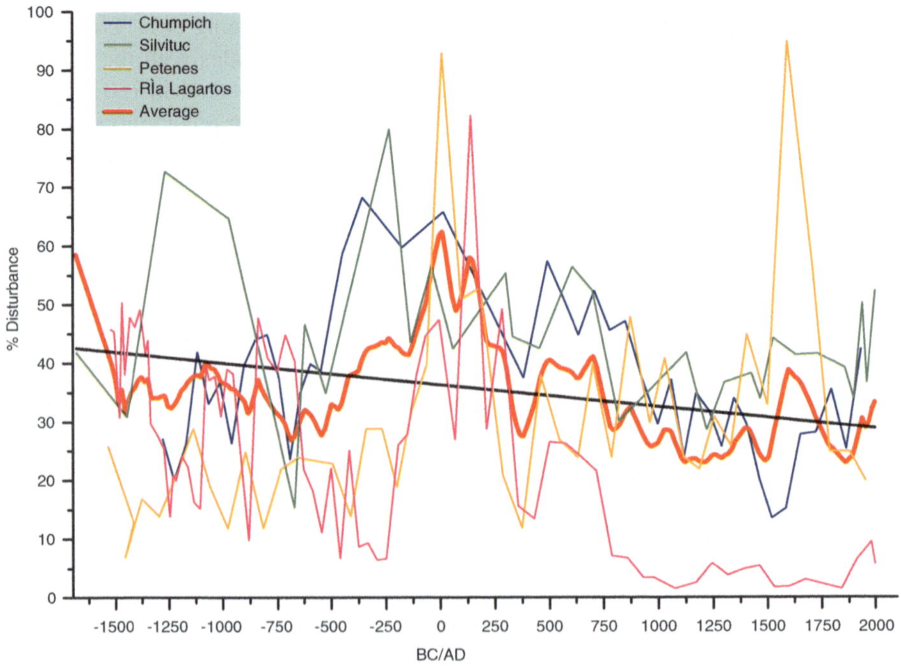

Fig. 9.7 Average abundance of disturbance vegetation in Ría Lagartos, Xpuk–Petenes, Silvituc Lake, and Chumpich Lake

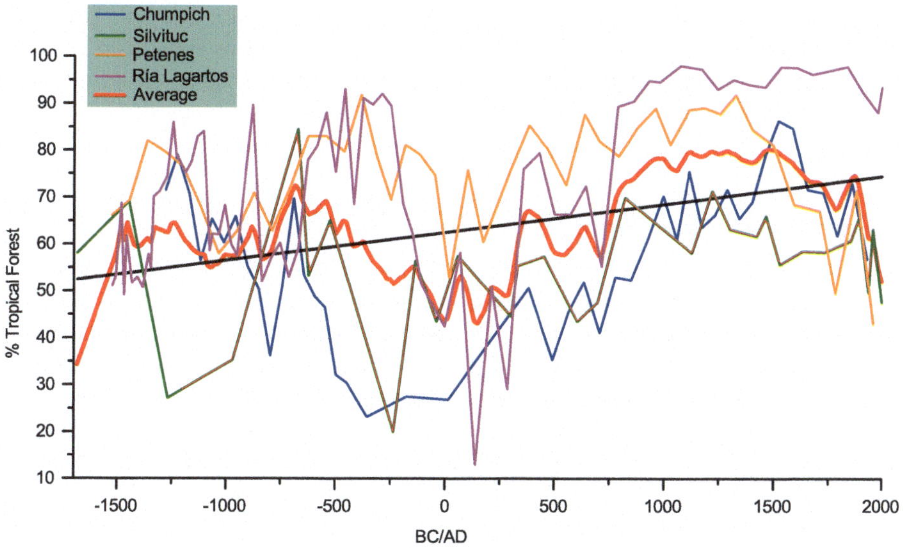

Fig. 9.8 Average abundance of Tropical Forest in Ría Lagartos, Xpuk–Petenes, Silvituc Lake, and Chumpich Lake

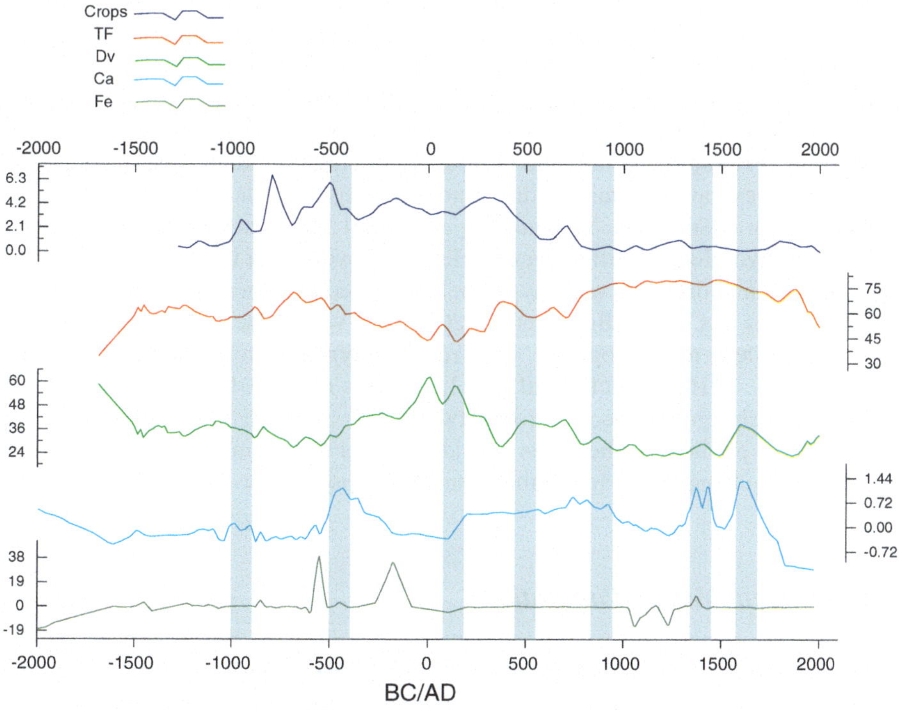

Fig. 9.9 Comparative between averages TF, crops, and DV, with average values of Ca and Fe from cores Chumpich Lake and Xpuk–Petenes

highlighted with blue shading. The vegetation response appears delayed. The geochemical signal corresponds strongly to the DV signal, and this signal can be considered as a climatic signal. The crop signal provides the degree of agricultural development, and the TF reveals the impact of the forest.

Climatic Responses in the YP

Akers et al. (2016) mentioned that some events identified in other sites do not correspond to their data and that the difference responds to regional differences of vegetation and climate expressions of the YP. The comparison between the pollen spectra of the four sites, as well as the comparison of Xpuk–Petenes and Chumpich Lake, shows differences in the responses of vegetation and geochemistry to climatic and anthropogenic changes. These responses have been identified in other studies such as Kenneth et al. (2012), Wahl et al. (2014), and Douglas et al. (2015, 2016).

The most recent comparisons correspond to Douglas et al. (2016), with nine sites in the North (Yucatán–Mexico) and Central Petén (Guatemala–Belize), isotopes of O18, and isotopes of plant waxes. The observed paleoclimate differences between the south and the north reinforce the interpretations of numerous authors about the droughts in the south that favored first the cultural collapse in the south and later in the north. Douglas (2016) summarizes the set of climatic and anthropogenic factors that other authors have identified as promoters of drought and differential climate response: solar variability, variability in the North Atlantic, changes in the seasonal migration of the ITCZ, interoceanic influence between the Pacific and Atlantic, changes in cyclonic frequency, and deforestation. However, there is still a need for analyses that explain in detail the differential response, in relation to the vegetation and climate gradient of the YP. Dunning et al. (2012) argued that the geomorphological and hydrological differences as elements involved in the response capacity and adaptation to droughts will have to be considered in future analyses.

The studies of Carrillo-Bastos et al. (2013) and Vela-Pelaez et al. (2018), based on fossil pollen, reveal together with recent oxygen isotope studies (Medina-Elizalde et al. 2016; Kenneth et al. 2012; Douglas et al. 2014; 2015; Wahl et al. 2013, 2014) changes in precipitation between 20% and 65%. These estimates can change if the gradient of the YP is considered, and the estimated impact for each region would change and explain the cultural collapse in space and time.

Environment Versus Population

Boserup (1965) and later Johnston (2003, 2006) have analyzed the agricultural models in the region, fallow periods and production, to identify and explain the erosion of the soils during the Classic period. They believe that the fundamental change was in the periods of fallow, which were reduced from decades to years, preventing the adequate recovery of the forests. On the other hand, Beach (1998), Dunning et al. (1998), and Dunning et al. (2012) have argued that the regional political structure together with vulnerability and resilience were key in the collapse. Each group or domain established different cultivation techniques according to available resources. This allowed the agricultural success in the appropriate climatic periods; but before the crisis (droughts), the political and social structure linked to a system. Resilience capacities were a key factor to the survival of the large cities.

The techniques used by the inhabitants of Uxul were the result of experimentation and adaptation for centuries, techniques that had even overcome the pre-abandonment drought (AD 200–500). Although probably due to the differential climate response in the region, it was not as drastic as in other places. The Chumpich Lake record reveals that the management of the low forests was efficient, the accumulation of detritus in the basin expressed in the geochemical record, and is related to precipitation rather than the erosion promoted by agriculture, indicating hydrological control. The cultural and agricultural collapse in year AD ~ 750 would be associated then to political and social changes, to a low resilience and high resistance to change, as mentioned by Dunning et al. (2012).

Acknowledgments Conacyt is acknowledged for funding of several projects. We appreciate the technical and financial support of Nicolai Grube through UXUL Project. We acknowledge the review of the present chapter to Dr. Joel Gunn.

References

Akers PD, Brook GA, Railsback LB et al (2016) An extended and higher resolution record of climate and land use from stalagmite MC01 from Macal Chasm, Belize, revealing connections between major dry events, overall climate variability, and Maya sociopolitical changes. Palaeogeogr Palaeoclimatol Palaeoecol 459:268–288

Alexander RT (2000) Patrones de asentamiento agregados en el Sudoeste de Campeche, Una visión desde la Isla Cilvituk. Mesoam 39:359–391

Anderson DG, Maasch KA, Sandweiss DH et al (2007) Climate and culture change: exploring Holocene transitions. In: Anderson DG, Maasch KA, Sandweiss (eds) Climate change and cultural dynamics: a global perspective on mid-Holocene transitions. Elsevier Inc, pp 1–23. San Diego, CA, USA

Aragón-Moreno AA, Islebe GA, Torrescano-Valle N (2012) A ~3800-yr, high-resolution record of vegetation and climate change on the north coast of the Yucatan Peninsula. Rev Palaeobot Palynol 178:35–42

Beach T (1998) Soil constraints on northwest Yucatán: pedo-archaeology and Maya subsistence at Chunchucmil. Geoarchaeology 13:759–791

Beach T, Dunning NP, Luzzadder-Beach S et al (2006) Ancient Maya. Impacts on soils and soil erosion. Catena 65:166–178

Beach T, Luzzadder-Beach S, Dunning N et al (2009) A review of human and natural changes in Maya Lowland wetlands over the Holocene. Quat Sci Rev 28:1710–1724. https://doi.org/10.1016/j.quascirev.2009.02.004

Birks HJB (2012) Ecological palaeoecology and conservation biology: controversies, challenges, and compromises. Int J Biodivers Sci Ecosyst Serv Manag 8(4):292–304. https://doi.org/10.1080/21513732.2012.701667

Blaauw M, Christen JA (2011) Flexible paleoclimate age-depth models using an autoregressive gamma process. Bayesian Anal 6(3):457–474. https://projecteuclid.org/ download/pdf_1/euclid.ba/1339616472
Boserup E (1965) The conditions of agricultural growth: the economics of agrarian change under population pressure. Aldine Publishing, Chicago, p 124
Brenner M, Rosenmeier MF, Hodell DA et al (2002) Paleolimnology of the Maya Lowlands. Anc Mesoam 13:141–157
Brooks N (2006) Cultural responses to aridity in the middle Holocene and increased social complexity. Quat Int 151:29–49
Buckler ES, Stevens NM (2006) Maize origins, domestication, and selection. In: Motley TJ, Zerega N, Cross H (eds) Darwin's harvest: new approaches to the origins, evolution, and conservation of crops. Columbia University Press, New York, pp 67–90
Büntgen U, Myglan VS, Ljungqvist FC et al (2016) Cooling and societal change during the Late Antique Little Ice Age from 536 to around 660 AD. Nat Geosci 9:231
Carrillo-Bastos A, Islebe GA, Torrescano-Valle N et al (2010) Holocene vegetation and climate history of central Quintana Roo, Yucatán Península, Mexico. Rev Palaeobot Palynol 160:189–196
Carrillo-Bastos A, Islebe GA, Torrescano-Valle N (2013) 3800 years of quantitative precipitation reconstruction from the Northwest Yucatan Peninsula. PLoS One 8(12) . Public Library of Science):e84333
Caseldine CJ, Turney C (2010) The bigger picture: towards integrating palaeoclimate and environmental data with a history of societal change. J Quat Sci 25(1):88–93. https://doi.org/10.1002/jqs.1337
Chepstow-Lusty AJ, Frogley MR, Bauer BS et al (2009) Putting the rise of the Inca Empire within a climatic and land management context. Clim Past 5:375–388
Chew SC (2007) The recurring dark ages: ecological stress, climatechanges, and system transformation. Altimira Press, Lanham
Clark JE, Gibson JL, Zeidler JA (2010) First towns in the Americas: searching for agricultura and other enabling conditions. In: Bandy M, Fox J (eds) Becoming villagers: comparing early village societies. Amerind studies in archaeology. The University of Arizona Press, Tucson, pp 205–245
Conroy JL, Overpeck JT, Cole JE et al (2008) Holocene changes in eastern tropical Pacific climate inferred from a Galápagos lake sediment record. Quat Sci Rev 27:1166–1180
Coombes P, Barber K (2005) Environmental determinism in Holocene research: causality or coincidence? Area 37(3):303–311
Crowley TJ (2000) Causes of climate change over the past 1000 years. Science 289:270–277
Curtis JH, Hodell DA, Brenner M (1996) Climate variability on the Yucatan Peninsula (Mexico) during the past 3500 years, and implications for Maya cultural evolution. Quat Res 46:37–47
Day JW, Folan WJ, Gunn JD, Yanez- Aranciabia A (2004) Patrones de productividad costera durante el ascenso del nivel del mar postglacial: Posibles implicacions para la formación del estado pristino. In XIV Encuentro Internacional: Los Investigadores de la Cultura Maya. Campeche
Day JW, Gunn JD, Folan WJ, Yáñez-Arancibia A et al (2007) Emergence of complex societies after sea level stabilized. EOS Trans Am Geophys Union 88(15):170–171
Day JW, Gunn JD, Folan WJ, Yáñez-Arancibia A et al (2012) The influence of enhanced postglacial coastal margin productivity on the emergence of complex societies. J Island Coast Archaeol 7(1):23–52. https://doi.org/10.1080/15564894.2011.650346
Delcourt PA, Delcourt HR (1998) Paleoecological insights on conservation of biodiversity: a focus on species, ecosystems, and landscapes. Ecol Appl 8(4):921–934
deMenocal PB (2001) Cultural responses to climate change during the late holocene. Science 292:667–673. https://doi.org/10.1126/science.1059827
deMenocal P, Ortiz J, Guilderson T et al (2000) Abrupt onset and termination of the African Humid Period: rapid climate responses to gradual insolation forcing. Quat Sci Rev 19:347–361
Diamond J (2009) Maya, Khmer and Inca. Nature 461(24):470–480

Douglas PMJ, Pagani M, Canuto MA et al (2015) Drought, agricultural adaptation, and sociopolitical collapse in the Maya Lowlands. PNAS 112(8):5607–5612

Douglas PMJ, Demarest AA, Brenner M et al (2016) Impacts of climate change on the collapse of Lowland Maya civilization. Annu Rev Earth Planet Sci 44:613–645

Douglas PMJ, Pagani M, Canuto MA, Brenner M, Hodell DA, Eglinton TI, Curtis JH (2014) Drought, agricultural adaptation, and sociopolitical collapse in the Maya Lowlands. PNAS112(18):5607–5612

Dunning NP, Beach T (2000) Stability and instability in prehispanic Maya landscapes. In: Lentz D (ed) An imperfect balance: landscape transformations in the pre-Columbian Americas. Columbia University Press, New York, pp 179–202

Dunning NP, Rue D, Beach T et al (1998) Humaneenvironmental interactions in a tropical watershed: the paleoecology of Laguna Tamarindito, El Petén, Guatemala. J Field Archaeol 25:139–151

Dunning NP, Beach TP, Luzzadder-Beach S (2012) Kax and kol: collapse and resilience in lowland Maya civilization. PNAS 109:3652–3657. https://doi.org/10.1073/pnas.1114838109

Erickson CL (1999) Neo-environmental determinism and agrarian 'collapse' in Andean prehistory. Antiquity 73:634–642

Faegri K, Iversen J (1989) Textbook of pollen analysis, 4th edn. Munksgaard, Copenhagen, p 328

Gill RB (2008) Las grandes Sequías mayas. Agua, Vida y Muerte. Fondo de Cultura Económica, México. p 561

Gregory BRB, Peros M, Reinhardt E et al (2015) Middle-late Holocene Caribbean aridity inferred from foraminifera and elemental data in sediment cores from two Cuban lagoons. Palaeogeogr Palaeoclimatol Palaeoecol 426:229–241

Grimm E (2011) Tilia 1.7.16. Illinois State Museum. Research and Collection Center

Grube N, Delvendahl K (2016) In the wake of the great. Humanit Soc Sci 38(1):16–22. https://doi.org/10.1002/germ.201690012

Grube N, Delvendahl K, Seefeld N et al (2012) Under the rule of the snake kings: Uxul in the 7th and 8th centures. Estudios de la Cultura Maya XL:11–49

Gunn J, Folan WJ, Robichaux HR (1995) A landscape analysis of the Candelaria watershed in Mexico: insights into paleoclimates affecting upland horticulture in the southern Yucatan Peninsula Semi-Karst. Geoarchaeology 10:3–42. https://doi.org/10.1002/gea.3340100103

Gunn JD, JWJr D, Folan W et al (2019) Geo-cultural time: advancing human societal complexity within worldwide constraint bottlenecks—a chronological/helical approach to understanding human–planetary interactions. BioPhys Econ Resour Qual 4:10. https://doi.org/10.1007/s41247-019-0058-7

Gutiérrez-Ayala LV, Torrescano-Valle N, Islebe GA (2012) Reconstrucción paleoambiental del Holoceno tardío de la reserva Los Petenes, Península de Yucatán, México. Revista Mexicana de Ciencias Geológicas 29(3):749–763

Harrison PD (1990) The revolution in ancient Maya subsistence. In: Clancy FS, Harrison PD (eds) Vision and revision in Maya studies. University of New Mexico Press, Albuquerque, pp 99–113

Hastenrath S (1984) Interannual variability and annual cycle: mechanisms of circulation and climate in the tropical Atlantic sector. Am Meteorol Soc 112:1097–1107

Hodell DA, Curtis JH, Brenner M (1995) Possible role of climatic change in the collapse of the Maya civilization. Nature 375:391–394

Hodell DA, Brenner M, Curtis JH (2000) Climate change in the Northern American tropics and subtropics since the last Ice Age: implications for environment and culture. In: Lentz DL (ed) Imperfect balance. Landscape transformations in the Precolumbian Americas. Columbia University Press, New York, pp 13–38

Hodell DA, Brenner M, Curtis JH et al (2001) Frequency in the Maya lowlands. Science 292:1367–1370

Hodell DA, Brenner M, Curtis JH (2005) Terminal classic drought in the northern Maya lowlands inferred from multiple sediment cores in Lake Chichancanab (Mexico). Quat Sci Rev 24:1413–1427

Hodell DA, Brenner M, Curtis JH (2007) Climate and cultural history of the northeastern Yucatan Peninsula, Quintana Roo, Mexico. Clim Chang 83:215–240
Islebe GA, Hooghiemstra H, Brenner M et al (1996) A Holocene vegetation history from lowland Guatemala. The Holocene 6(3):265–271
Johnston KJ (2003) The intensification of pre-industrial cereal agriculture in the tropics: Boserup, cultivation lengthening, and the Classic Maya. J Anthropol Archaeol 22:126–161. https://doi.org/10.1016/S0278-4165(03)00013-8
Johnston KJ (2006) La intensificación de la agricultura Maya Clásica. In: Laporte JP, Arroyo B, Mejía H (eds) XIX Simposio de Investigaciones Arqueológicas en Guatemala 2005. Museo Nacional de Arqueología y Etnología, Guatemala, pp 1090–1100
Kennett DJ, Breitenbach SFM, Aquino VV et al (2012) Development and disintegration of Maya political systems in response to climate change. Science 338(6108):788–791
LaRocque A, Leblon B, Ek J (2019) Detection of potential large Maya settlements in the northern Petén área (State of Campeche, Mexico) using optical and radar remote sensing. J Archaeol Sci 23:80–97
Lentz DL, Magee K, Weaver E et al (2015) Agroforestry and agricultural practices of the ancient Maya at Tikal. In: Lentz DL, Dunning NP, Scarborough V (eds) Tikal, paleoecology of an ancient Maya City. Cambridge University Press, Cambridge, pp 152–186
Leyden B, Brenner M, Whitmore T et al (1996) A record of long-and short-term climatic variation from Northwest Yucatán: Cenote San José Chulchacá. In: Fedick SL (ed) The managed mosaic: ancient Maya agriculture and resource use. University of Utah Press, Salt Lake City, pp 30–50
Leyden BW, Brenner B, Dahlin BH (1998) Cultural and climatic history of Cobá, a lowland Maya city in Quintana Roo, Mexico. Quat Res 49:111–122
Lohse JC (2010) Archaic origins of the Lowland Maya. Lat Am Antiq 21(3):312–352
Lozano-García S, Ortega B, Roy PD et al (2015) Climatic variability in the northern sector of the American tropics since the latest MIS 3. Quat Res 84(2):262–271
Manzanilla L (1997) The impact of climatic change on past civilizations. A revisionist agenda for futher investigation. Quat Int 43/44:153–159
Marín-Stillman LE, Pacheco-Avíla JG, Méndez-Ramos R (2004) Hidrogeología de la peninsula de Yucatán. In: Jímenez B, Marín L (eds) El agua en México vista desde la academia. Academia Mexicana de Ciencias. p 403
Matsouka Y, Vigouroux Y, Goodman MM, Sanchez GS, Buckler E, Doebley J (2002) A single domestication for maize shown by multilocus microsatellite genotyping. PNAS 99(9):6080–6084. https://doi.org/10.1073/pnas.052125199
Mayewski PA, Rohling EE, Curt Stager J et al (2004) Holocene climate variability. Quat Res 62:243–255
Medina-Elizalde M, Burns SJ, Polanco-Martínez JM et al (2016) High-resolution speleothem record of precipitation from the Yucatan Peninsula spanning the Maya Preclassic Period. Glob Planet Chang 138:93–102
Metcalfe S, Jones MD, Davies SJ et al (2010) Climate variability over the last two millennia in the North American monsoon region, recorded in laminated lake sediments from Laguna Juanacatlán México. The Holocene 20:1195–1206
Milne GA, Long AJ, Bassett SE (2005) Modelling Holocene relative sea-level observations from the Caribbean and South America. Quat Sci Rev 24:1183–1202
Mueller AD, Islebe GA, Anselmetti FS et al (2010) Recovery of the forest ecosystem in the tropical lowlands of northern Guatemala after disintegration of Classic Maya polities. Geology 38(6):523–526
Nakagawa T, Brugiapaglia E, Digerfelti G et al (1998) Dense media separation as a more efficient pollen extraction method for use with organic sediment/deposit samples: comparison with the conventional method. Boreas 27(1):15–24
Nooren CAM, Hoek WZ, Tebbens LA et al (2009) Tephrochronological evidence for the late Holocene eruption history of El Chichón Volcano, Mexico. Geofis Int 48(1):97–112
Ojeda-Mas H, Súarez-Aguilar V, Peña-Castillo A (1996) Cilvituk, una economía lacustre: avance de investigación. Investigadores de la cultura Maya. Tomo II. Universidad Autónoma Campeche 450–478

OriginLab (2016) © Corporation. All rights reserved. https://www.originlab.com/2016
Perez L, Bugja R, Lorenschat J, Brenner M et al (2011) Aquatic ecosystems of the Yucatan Peninsula (Mexico), Belize, and Guatemala. Hydrobiologia 661(1):407–433. https://doi.org/10.1007/s10750-010-0552-9
Pohl MED, Pope KO, Jones JG et al (1996) Early agriculture in the Maya Lowlands. Lat Am Antiq 7(4):355–372
Pohl MED, Piperno DR, Pope KO et al (2007) Microfossil evidence for pre-Columbian maize dispersals in the neotropics from San Adnrés, Tabasco, Mexico. PNAS 104(16):6870–6875
Rosenswing RM, Pearsall DM, Masson MA et al (2014) Archaic period settlement and subsistence in the Maya Lowlands: new starch grain and lithic data from Freshwater Creek, Belize. J Archaeol Sci 41:308–321
Roy PD, Jonathan MP, Pérez-Cruz LL et al (2012) A millennial-scale Late Pleistocene–Holocene palaeoclimatic record from the western Chihuahua Desert, Mexico. Boreas 41:707–718
Roy PD, Quiroz-Jiménez JD, Pérez-Cruz LL et al (2013) Late Quaternary paleohydrológical conditions in the dryland of northern Mexico: a summer precipitation proxy record of the last 80 cal ka BP. Quat Sci Rev 78:342–354
Roy PD, Chávez-Lara CM, Beramendi-Orosco LE et al (2015) Pleohydrology of the Santiaguillo Basin (Mexico) since late last glacial and climate variation in southern part of western subtropical North America. Quat Res 84:335–347
Roy PD, Rivero-Navarrete A, Sánchez-Zavala JL et al (2016) Atlantic Ocean modulated hydroclimate of the subtropical northeastern Mexico since the Last Glacial Maximum and comparison with the sou-thern US. Earth Planet Sci Lett 434:141–150
Roy PD, Torrescano-Valle N, Islebe GA et al (2017) Late Holocene hydroclimate of the western Yucatan Peninsula (Mexico). J Quat Sci 32(8):1112–1120
Roy PD, Torrescano-Valle N, Escarraga-Paredes D et al (2018) Comparision of elemental concentration in near-surface late Holocene sediments and precipitation regimes of the Yucatán Peninsula (Mexico): a preliminary study. Bol Geol Min 129:693–706
Rull V (2010) Ecology and Palaecology: two approaches, one objetive. Open Ecol J 3:1–5
Scarborough VL and Valdez F (2014) The alternative economy: resilience in the face of complexity from the Eastern Lowlands. Archeological Papers of the American Anthropological Association 24:124–141
Shaw JM (2003) Climate change and deforestation: implications for the Maya collapse. Anc Mesoam 14:157–167
Siemens AH (1983) Wetland agriculture in pre-hispanic Mesoamerica. Am Geogr Soc 73(2):166–181. http://www.jstor.org/stable/214642
Siemens AH (2011) Subidas y Bajadas: Desafíos en la investigación del factor agua en la ecología humana del mundo Maya. Paper read at the XXI Encuentro Internacional: Los Investigadores de la Cultura Maya. Universidad Autonoma de Campeche, Campeche, México
Simpson GL (2012) Analogue methods in palaeolimnology. Chapter 15. In Birks HJB, Lotter AF, Juggins S, Smol JP (eds) Tracking environmental change using lake sediments. Springer p 496
Sluyter A, Dominguez G (2006) Early maize (Zea mays L.) cultivation in Mexico: dating sedimentary pollen records and its implications. PNAS 103(4):1147–1151
Sosa-Nájera S, Lozano-García S, Roy PD et al (2010) Registro de sequías históricas en el occidente de México con base en el análisis elemental de sedimentos lacustres: El caso del lago de Santa María del Oro. Bol Soc Geol Mex 62(3):437–451
Stott PA, Christidis N, Otto FEL et al (2016) Attribution of extreme wather and climate-related events. Clim Chang 7:23–41
Torrescano-Valle N, Islebe GA (2015) Holocene paleoecology, climate history and human influence in the southwestern Yucatán peninsula. Rev Palaeobot Palynol 217:1–8
Turney CSM, Brown H (2007) Catastrophic early Holocene Sea level rise, human migration and Neolithic transition in Europe. Quat Sci Rev 26:2036–2041
Vela-Pelaez AA, Torrescano-Valle N, Islebe GA et al (2018) Holocene precipitation changes in the Maya forest, Yucatán península, Mexico. Palaeogeogr Palaeoclimatol Palaeoecol 505:42–52

Wahl D, Byrne R, Schreiner T et al (2006) Holocene vegetation change in the northern Peten and its implications for Maya prehistory. Quat Res 65(3):380–389
Wahl D, Byrne R, Schreiner T et al (2007) Palaeolimnological evidence of late-Holocene settlement and abandonment in the Mirador Basin, Peten, Guatemala. The Holocene 17(6):813–820
Wahl D, Estrada-Belli F, Anderson L (2013) A 3400 year paleolimnological record of prehispanic human-environment interations in the Holmul región of the southern Maya lowlands. Palaeogeogr Palaeoclimatol Palaeoecol 379-380:17–31
Wahl D, Byrne R, Anderson L (2014) An 8700 year paleoclimate reconstruction from the southern Maya lowlands. Quat Sci Rev 103:19

Chapter 10
Holocene Paleoecology and Paleoclimatology of South and Southeastern Mexico: A Palynological and Geospatial Approach

Gerald A. Islebe, Alicia Carrillo-Bastos, Alejandro A. Aragón-Moreno, Mirna Valdez-Hernández, Nuria Torrescano-Valle, and Nancy Cabanillas-Terán

Abstract Reconstruction of Holocene paleoecological conditions and paleoclimate of an area with high biological diversity and a variety of climatic conditions like southern and southeastern Mexico is complex. This region is characterized by vegetation types ranging from tropical forest to high mountain vegetation. Additionally, this region was inhabited by the ancient Maya culture, which shaped the landscape for several millennia. Previous paleoecological studies from this region were focused on the Maya culture-environment relationships, to decipher natural and human-induced deforestation. These studies also aimed to understand the effects of climatic regional forcing (El Niño-Southern Oscillation (ENSO), Intertropical Convergence Zone (ITCZ), and North Atlantic Oscillation (NAO) on the natural vegetation. In this chapter we review the paleoecological results and present a new geospatial approach to analyze past precipitation and tropical forest distribution of the Yucatán Peninsula from 1 AD to 1700 AD in 100-year intervals. The geospatial analysis revealed heterogeneity in spatial patterns of precipitation and tropical forest extension during the Late Preclassic, Terminal Classic, and

G. A. Islebe · A. A. Aragón-Moreno · M. Valdez-Hernández · N. Torrescano-Valle (✉)
Departamento Conservación de la Biodiversidad, El Colegio de la Frontera Sur Unidad Chetumal, Chetumal, Quintana Roo, Mexico
e-mail: gislebe@ecosur.mx; aaragon@ecosur.mx; mavaldez@ecosur.mx; ntorresca@ecosur.mx

A. Carrillo-Bastos
División de Estudios de Posgrado e Investigación, Tecnológico Nacional de México/I. T. Chetumal, Chetumal, Quintana Roo, Mexico
e-mail: acarrillo@itchetumal.edu.mx

N. Cabanillas-Terán
Catedras Consejo Nacional de Ciencia y Tecnología, El Colegio de la Frontera Sur Unidad Chetumal, Chetumal, Quintana Roo, Mexico
e-mail: ncabanillas@ecosur.mx

N. Torrescano-Valle et al. (eds.), *The Holocene and Anthropocene Environmental History of Mexico*, https://doi.org/10.1007/978-3-030-31719-5_10

Medieval Warm Period to Little Ice Age transition. The dry periods of the Middle and Late Holocene in the Yucatán Peninsula and southern Mexico can be chronologically placed in the following intervals: 4700–3600 cal year BP, 3400–2500 cal year BP, 2300–2100 cal year BP, 1900–1700 cal year BP, 1400–1300 cal year BP, 730 cal year BP, and 560 cal year BP. We conclude that this region requires additional studies with strong chronological framework due to its heterogeneous environmental conditions.

Keywords Geospatial approach · Tropical forest cover · Holocene · Paleoecology · Ancient Maya

Introduction

Southern and southeastern Mexico covers a vast area with variable climate, biodiversity, and physical conditions. This region includes the states of Chiapas, Tabasco, Campeche, Yucatán, and Quintana Roo. The landscape is diverse and it includes lowland as well as mountain ecosystems. The mountains in Chiapas have elevations up to 4000 m asl and pine-oak highland forests dominate the landscape. Other vegetation types, like mesic mountain forests and wet and dry tropical forests, and mangroves cover different altitudinal ranges of this region (Fig. 10.1). This diverse vegetation and climatic mosaic offers an excellent opportunity to understand ecological variability during the Holocene in an area of high biodiversity, endemism, and with a long history of use by the human societies (Islebe et al. 2015b).

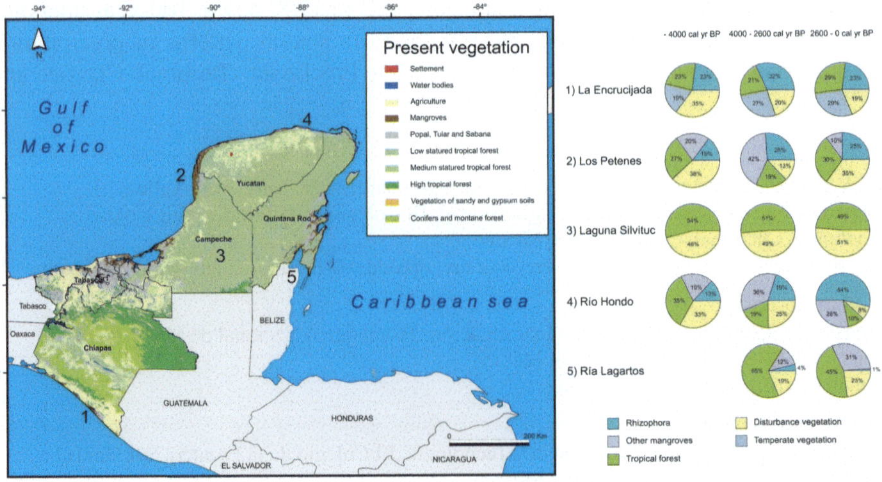

Fig. 10.1 Percentage pie plots of the main sampled vegetation types in different sites across southern and southeastern Mexico during key intervals of the Holocene

From a conservation point of view, this region has major protected areas like the Lacandon rainforest in Chiapas and the Calakmul Biosphere Reserve in Campeche, which represent the northernmost distribution of neotropical forests. The studies of paleoecology of southeastern Mexico, Belize, and lowland Guatemala over the Holocene have a major scientific objective, i.e., to understand the relation of ancient Maya culture with the environment and to decipher their cultural demise (Brenner et al. 2002b; Cowgill et al. 1966). One of the key questions is how the sustainable ancient Maya culture managed their natural resources (Lentz et al. 2014), and these paleoecological studies provided clues to understand the past land use changes.

Palynological research in the wider region was fostered by Leyden (1984) to understand vegetation changes in lowland Guatemala from the Pleistocene to Holocene transition. Later, several relevant palynological studies were published by Crane (1996), Leyden et al. (1998), Islebe et al. (1996), and Dunning et al. (1998), among other studies, to analyze the interactions between vegetation and environment. This growing paleoecological knowledge over the past 50 years in southeastern Mexico allowed a better understanding of the past environmental variability (Islebe et al. 2015a). Paleopalynology in a highly diverse area also depends on accurate identification of fossil pollen grains as more than 8000 plant species can be actually found in southern/southeastern Mexico. Therefore, the accurate pollen identification depends on available bibliography and studies on pollen morphology (Palacios-Chávez et al. 1991) and the correlations between the present plant communities and pollen rain (Correa-Metrio et al. 2011; Escarraga-Paredes et al. 2014; Islebe et al. 2001) and Chiapas (Domínguez-Vázquez et al. 2004). From a paleoclimate view, the regional climate forcings like ENSO (El Niño-Southern Oscillation) and ITCZ (Intertropical Convergence Zone) had conspicuous effect on vegetation (Haug et al. 2001, 2003) and were detected by high-resolution palynology (Aragón-Moreno et al. 2018a). In this chapter, we focus on fossil pollen as a proxy for paleoecological and paleoclimatological reconstruction. We also present a geospatial model of precipitation and vegetation change of the Yucatán Peninsula covering the last 1700 years.

Present Climate

Present climate conditions are complex due to the geography of the region. The ITCZ defines two main seasons of the southern Mexico: a rainy season between May and November and a dry season between December and April (Vela-Peláez et al. 2018). Seasonality is strong and the dry season can extend to May. For the Yucatán Peninsula and areas under the influence of the Gulf region, cold arctic air incursion between December and March is known. Those polar continental air masses can provide an additional amount of precipitation during the winter months (Islebe et al. 2015a). The amounts of precipitation vary enormously in southern and southeastern Mexico. In the Yucatán Peninsula, the annual precipitation ranges from

500 to 1600 mm, along a north-south gradient. An east-to-west gradient of the Yucatán Peninsula shows that precipitation along the coast of the Gulf of Mexico is around 600–800 mm annually, while the precipitation along the Caribbean is higher due to influence of the jet streams (Muñoz et al. 2008). Most of the precipitation falls during summer and autumn. The main sources of humidity are the westerlies and the Caribbean low-level jet (CLLJ; Aragón-Moreno et al. 2018a). Mean annual precipitation in Tabasco is between 1000 and 2000 mm (Gama et al. 2011), and climate of this part is mainly driven by conditions of the Gulf of Mexico. The state of Chiapas presents a complex precipitation regime, as the orographic factors play an important role in the central highlands. Mean annual precipitation ranges from 600 mm along the drier Pacific coast to more than 3000 mm in the high-central mountain ranges and some parts of the Lacandon area (Román-Cuesta et al. 2004). Mean annual temperature in the Yucatán Peninsula varies between 24 and 26 °C, and similar values are recorded for the Tabasco and the low-lying parts of Chiapas. Mean annual temperature of Pacific coastal Chiapas is around 25 °C, while at elevations above 3000 m, mean annual temperature drops below 10 °C. An important climatic factor in south and southeastern Mexico is the El Niño-Southern Oscillation (ENSO). During El Niño periods, the amount of precipitation decreases strongly between June and August in the Yucatán Peninsula and along the Pacific coast of Chiapas (Aragón-Moreno et al. 2018a). It is estimated that the precipitation decreases by 30 and 50% during El Niño years in the region (Mendoza et al. 2007; O'Hara and Metcalfe 1995). A stronger North Atlantic high-pressure center during the winter causes stronger trade winds, and it causes cool sea surface temperatures in the tropical Atlantic. The cooler SSTs in the Caribbean Sea reduce the amounts of spring precipitation. Additionally, south and southeastern Mexico is affected by hurricanes formed in the Pacific Ocean as well as in the Caribbean Sea between June and November (Sánchez-Sánchez et al. 2015).

Holocene Paleoecology

The climate drivers are diverse, and climate of this geographical area is under the influence of the ITCZ (Brenner et al. 2002a), ENSO, North Atlantic high, and change in insolation (Douglas et al. 2015; Hodell et al. 2001). The early postglacial climatic conditions are still not well understood, but the available research indicates establishment of present-day potential vegetation types at around 9000 cal year BP. The late Pleistocene environmental history was reconstructed from several sediment cores of the Lake Petén-Itzá located in the lowland Guatemala (Correa-Metrio et al. 2012; Hodell et al. 2008). Few paleoecological records date back until the Early Holocene, and they are available from the coastal Chiapas and the Yucatán Peninsula. The Middle Holocene was generally characterized by conditions of increased humidity in Mexico and Central America. Evidences came from cores collected from the highland as well as lowland lacustrine cores (Marchant et al. 2009). The Middle Holocene was also relevant as humans started influencing the natural environment and became a major driver of the ecological change (Leyden 2002). The Late Holocene was reconstructed in several studies using different proxies in the southern Mexico (Table 10.1). All these proxy records, generally, have a common trend toward drying conditions and a decrease in precipitation. Humans played a major role in the landscape transformation, and many scholarly discussions analyzed this topic from a variety of disciplines (Gunn et al. 2017; Lentz et al. 2015).

Early Holocene (10–7 ka BP)

The Early Holocene is not well represented in paleoecological records of the southern Mexico. From the Lake Petén-Itzá in the lowland Guatemala, the palynological analysis of Leyden (1984), Hodell et al. (2008), and Islebe et al. (1996) provided some clues to environmental conditions of the Early Holocene. Increased humidity and temperature allowed geographical expansion of the tropical forest species, and

Table 10.1 Paleoecological studies from southeastern Mexico focused on Middle to Late Holocene. The sites marked with an asterisk are included in the geospatial modeling

Site, state	Proxy	Reference
Rio Hondo, Quintana Roo	Pollen, geochemistry	Aragón-Moreno et al. (2018a)
El Palmar, Quintana Roo*	Pollen	Torrescano-Valle and Islebe (2012)
Lake Cobá, Quintana Roo	Pollen	Leyden (2002), Leyden et al. (1998)
Puerto Morelos, Quintana Roo*	Pollen	Islebe and Sánchez (2002)
Lakes, Quintana Roo*	Pollen, modeling	Carrillo-Bastos et al. (2010)
Ría Lagartos, Yucatán*	Pollen	Aragón-Moreno et al. (2018b), Aragón-Moreno et al. (2012)
Ría Lagartos, Yucatán*	Pollen, modeling	Carrillo-Bastos et al. (2013)
Lake Silvituc, Campeche*	Pollen	Torrescano-Valle and Islebe (2015)
Lake Silvituc, Campeche	Pollen, modern analogues	Vela-Peláez et al. (2018)
Los Petenes, Campeche	Pollen, geochemistry	Gutiérrez-Ayala et al. (2012), Roy et al. (2017)
El Triunfo Reserve, Chiapas	Pollen	Joo-Chang et al. (2015)
Lagunas de Montebello, Chiapas	Pollen	Franco-Gaviria et al. (2018)
Lake Najá, Chiapas	Pollen	Domínguez-Vázquez and Islebe (2008)
Yucatán Peninsula	Pollen rain vegetation	Islebe et al. (2001) Correa-Metrio et al. (2011), Escarraga-Paredes et al. (2014)
Chiapas	Pollen rain vegetation	Domínguez-Vázquez et al. (2004)

it resulted in development of the tropical forest. The main pollen taxa indicate development of the Moraceae and *Brosimum* in those tropical forests (Islebe et al. 1996; Leyden 1984). The Early Holocene records from the Quintana Roo (Carrillo-Bastos et al. 2010) do not provide insight about the forest development. Tropical forest taxa in general were higher during the Early Holocene compared to the Middle and Late Holocene.

Middle Holocene and Late Holocene (Last 7 ka)

The Middle Holocene and Late Holocene are characterized by several studies using different approaches and proxies. Response to the 4.2 ka event was identified in the mangrove sediments of the Yucatán Peninsula (Aragón-Moreno et al. 2018a), and it demarcated the initiation of recently defined Meghalayan Stage/Age by Walker et al. (2019). This climate anomaly caused disruption of the westerlies and Asian Summer Monsoon. Disruption of the westerlies had a strong influence on precipitation distribution along the eastern Yucatán Peninsula.

Joo-Chang et al. (2015) presented a palynological study from an extensive mangrove area in the Pacific coast of Chiapas. The area was dominated by *Rhizophora mangle* stands in a mosaic of mangrove vegetation with *Avicennia germinans* and *Laguncularia racemosa*. These mangrove species responded to a decadal-/centennial-scale ITCZ variability, as well as to the coupling of ITCZ and ENSO. Increased ENSO activity from 3500 cal year BP to 2400 cal year BP resulted in higher *Rhizophora mangle* representation, with a sudden drop at 2400 cal year BP (Joo-Chang et al. 2015). Similar environmental changes were observed in different areas of the tropical North Atlantic, e.g., in the Bahamas and Haiti, pointing to a common climatic driving mechanism (Higuera-Gundy et al. 1999; van Hengstum et al. 2018).

A highland pollen record from Chiapas covers the Late Holocene (Franco-Gaviria et al. 2018). Like other records, this pollen record also showed climate conditions changed toward increasing drying trend. Moist oak-dominated montane forests increased after 600 cal year BP. Domínguez-Vázquez and Islebe (2008) presented a pollen record from the Lacandon area in Chiapas. A protracted drought was identified between 1200 and 700 cal year BP, while taxa from moist montane forests increased from 700 cal year BP onward. Pine forests dominated the landscape during the drought intervals, while the montane forest taxa increased their presence during the wetter phases.

Several studies reported the Middle Holocene paleoecology of the Yucatán Peninsula. All those studies observed evidence of drier environmental conditions around 5000 cal year BP, which in turn reflected in the changing tropical forest taxa composition and their diversity (Aragón-Moreno et al. 2018a; Carrillo-Bastos et al. 2010). The tropical forest taxa, mainly Leguminosae and Moraceae, decreased, and pollen of disturbance taxa increased, e. g., Poaceae, Asteraceae, and Malvaceae. Similar conditions of drying trends were found at 5000 cal year BP at Lake Puerto

Arturo of Guatemala (Wahl et al. 2014). A paleoecological study from the Lake Silvituc, southern Yucatán Peninsula, showed the response of vegetation to different drivers like climate and human-induced changes (Torrescano-Valle and Islebe 2015). Vegetation near the watershed changed from a closed to an open vegetation type, with characteristic taxa of disturbed or secondary conditions like Poaceae, Asteraceae, and other herbs. However, the local vegetation reacted to regional variability of precipitation as the near-shore plant communities changed, and specific aquatic taxa appeared, e.g., *Typha* and *Nymphaea*. The sediments of Lake Silvituc is one of the few available paleoecological records showing the influence of volcanic activity in this region. The detected tephra were related to volcanic eruptions of the El Chichón volcano (Nooren et al. 2016).

A Geospatial Approach to Understand Late Holocene Vegetation Change of the Yucatán Peninsula

One of the major points of interest is the understanding of geographical extension of the drought which affected the Late Holocene vegetation distribution. Under this view, we chose several pollen records with adequate chronology to develop a geospatial model (Table 10.1). All the radiocarbon dates were calibrated, and the vegetation maps were built with the ArcGIS 9.3 extension applying ILP (polynomial interpolation technique). The ILP follows a deterministic approach, and it uses values measured at different points inside an area to create a continuous surface (Wadsworth and Treweek 1999). We used the methodology detailed in Carrillo-Bastos et al. (Carrillo-Bastos et al. 2012) and chose local interpolation for polynomials because this does not imply normal distribution of the data and the predictive errors were smaller. The prediction was carried out with a power function of (p) = 2 and under the option of search for the ideal influence for each distance (IDD). These parameters produced surfaces with a smaller prediction error. Extrapolations were calculated to the rest of studied geographical areas and graphed in Fig. 10.2. The thematic layers were added after the paleo-vegetation reconstruction was developed at different chronological sequences. In this case, the geographical distribution of dry tropical forest and present-day mean annual precipitation changes were added to the geospatial modeling. Considering the available chronologies and information about paleo-vegetation changes, the thematic paleo-precipitation maps were reconstructed for the intervals of every 100 years, starting at 1 AD until 1700 AD (Figs. 10.2 and 10.3). The blue-green colors indicate higher reconstructed precipitation levels and the yellow-brownish colors indicate lower precipitation levels. For different time intervals, the information about the change in precipitation compared to the present-day precipitation levels can be drawn. Drier conditions were observed between 200 and 300 AD in the southern and northern Yucatán Peninsula. Drier conditions during this interval were also inferred from a beach ridge, and diatom record was obtained from the Tabasco state (Nooren et al. 2018). This dry interval

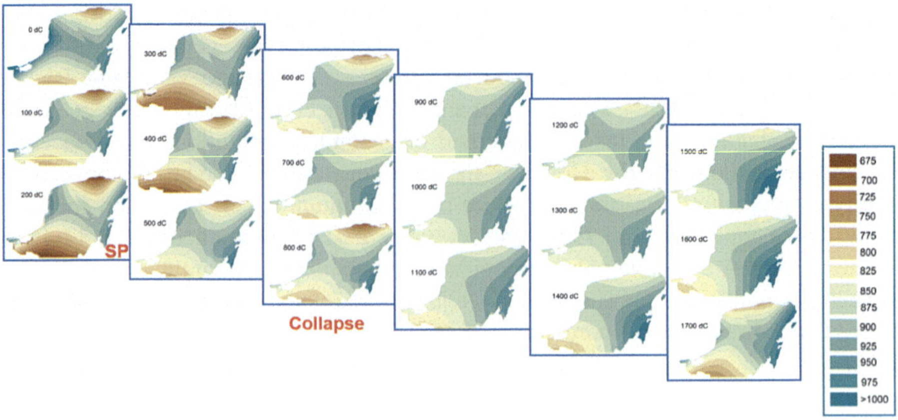

Fig. 10.2 Precipitation reconstruction for the interval 1–1700 AD. Modeled precipitation values are expressed in mm (SP = Late Preclassic drought)

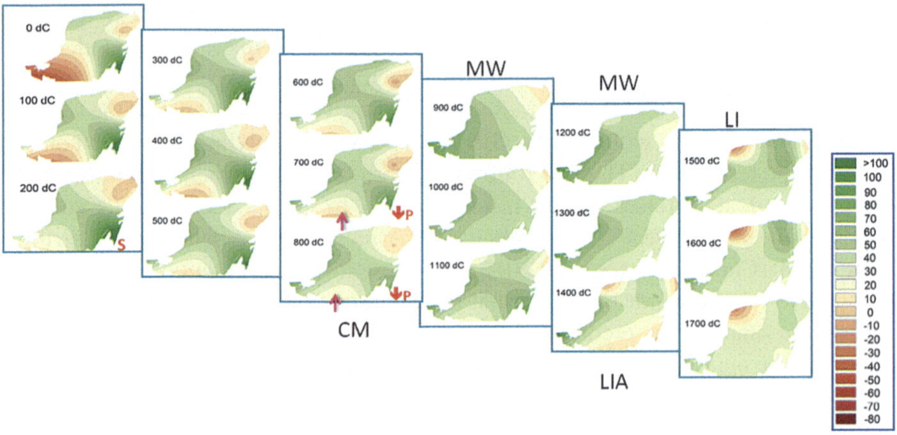

Fig. 10.3 Tropical forest cover reconstruction for the interval 1–1700 AD (CMD Classic Maya drought, MWP Medieval Warm Period, LIA Little Ice Age, P precipitation)

falls within the Late Preclassic drought (Ebert et al. 2017). Many Maya cities were abandoned during this arid period. The reduction in precipitation during the Late Preclassic drought was estimated between 25 and 35% compared to present-day precipitation values. Other recent modeling estimated a reduction of 31% of mean annual precipitation. The regional precipitation values increased between 500 and 700 AD. After 800 AD and within the period of Maya classic droughts, the precipitation reduced, and the southern Yucatán Peninsula received mean annual precipitation between 725 and 850 mm. The general precipitation reduction compared to present precipitation values was at least 20–40% between the Preclassic and Classic periods. The reduction in mean annual precipitation was less pronounced compared to values of the Late Preclassic drought. The following Medieval Warm Period (MWP; 1000–1200 AD) received generally more precipitation compared to the Classic Maya droughts. The Little Ice Age was characterized again by reduced precipitation in the southern, western, and northern Yucatán Peninsula. The eastern part of the Yucatán Peninsula, however, seemed to be less affected by climate conditions of the Little Ice Age.

Figure 10.3 summarizes the paleo-vegetation changes between 1 and 1700 AD. The green to dark red color scales indicate openness of the tropical forest. The past and present tropical forest distributions have been related to mean annual precipitation. Tropical forest conditions were different around 1 AD, presenting a match with a drought period, and it was also detected in the recent modeling of past vegetation (Vela-Peláez et al. 2018). The southwestern and northeastern Yucatán Peninsula vegetations were the most affected, and the open dry forest was the dominant vegetation. During the Late Preclassic drought, the dominant forests in the south and northwest were sub-deciduous forest types with Fabaceae pollen taxa. Vegetation composition in the south, central, and northern peninsula during the Classic Maya droughts was the dry forest types. However, this data modeling did not observe complete regional deforestation in the 100-year time windows. However, the densely populated parts of the southern part of the Yucatán Peninsula and Guatemala, most likely, had some local deforestation. These changes in forest composition were not only driven by climate but also by change in landscape induced by the ancient Maya culture (Islebe et al. 2018). Between 900 and 1100 AD, the forest cover increased on eastern side of the Yucatán Peninsula. Relatively fast forest recovery was also observed at Lake Petén-Itzá during this interval (Mueller et al. 2010). The forests recovered in 80-year period. Projection in the eastern side of the Yucatán Peninsula indicated less affectation in the paleo-vegetation map, suggesting sufficient moisture availability for the tropical forests in this part of the peninsula.

Outlook

The ITCZ, NAO, and ENSO were the main drivers of the Holocene ecological changes in the southern and southeastern Mexico. The southward shift of the ITCZ was related to decreased precipitation in the Yucatán Peninsula, and this was driven by changes in insolation (Haug et al. 2001; Steinhilber et al. 2009). Vegetation responded to those changes at decadal/centennial and millennial scales (Aragón-Moreno et al. 2018a). Response to those changes was evidenced in the pollen spectra recorded on local and regional scales, and it reflected changes in plant communities and exhibited their resilience. Higher precipitation during the Early Holocene was provided by increased jet stream activity and establishment of the CLLJ (Caribbean low-level jet) (Pollock et al. 2016). The geospatial approach of understanding past precipitation and forest distribution also helped to decipher causal changes of landscape, namely, the climate-driven versus human-induced land use changes and the response of vegetation to these changes. This approach needs additional paleoecological data and robust chronological control to define specific areas that probably changed under a variable precipitation scenario. The geospatial approach also helped to understand changes in forest distribution at specific ecotones, e.g., the change from deciduous to sub-deciduous dry forest and wetland to forest transitions, which under past climate conditions responded quickly in species turnover.

Major vegetation change of the Holocene in southern Mexico was recorded during 5500–4000 cal year BP, 3500–2000 cal year BP, and protracted droughts of the Maya Terminal Classic Period (Carrillo-Bastos et al. 2013). Also the intervals of increased precipitation were considered as a driver of vegetation change. The driest period in the southern Yucatán Peninsula was constrained at 2600 cal year BP, and the pollen records indicated a clear decrease in forest type and cover. The forest recovery was observed on a decadal and centennial scale of nearly 80 years. Estimates of precipitation reduction during the Terminal Classic are variable. They were around 50% (Evans et al. 2018) based on the isotope values and ranged up to 35% based on the pollen modern analogues (Vela-Peláez et al. 2018). However, the estimates of precipitation must consider the geographical coverage, as precipitations in southern Mexico and the Caribbean are at present highly variable.

Southern Mexico covers a large biodiverse region, and most of the paleoecological studies were carried out in tropical forests, mangrove ecosystems, and several highland lakes. Additional sediment cores with adequate dating control are required to obtain a complete picture of past vegetation dynamics and climate variability in this region.

Acknowledgments CONACYT is acknowledged for funding several projects.

References

Aragón-Moreno AA, Islebe GA, Torrescano-Valle N (2012) A ~3800-yr, high-resolution record of vegetation and climate change on the north coast of the Yucatan Peninsula. Rev Palaeobot Palynol 178. Elsevier B.V:35–42

Aragón-Moreno AA, Islebe GA, Roy PD et al (2018a) Climate forcings on vegetation of the southeastern Yucatán Peninsula (Mexico) during the middle to late Holocene. Palaeogeogr Palaeoclimatol Palaeoecol 495:214–226

Aragón-Moreno AA, Islebe GA, Torrescano-Valle N et al (2018b) Middle and late Holocene mangrove dynamics of the Yucatan Peninsula, Mexico. J S Am Earth Sci 85:307–311

Brenner M, Rosenmeier MF, Hodell DA et al (2002a) Paleolimnology of the Maya Lowlands. Anc Mesoam 13(01). Cambridge University Press:141–157

Brenner M, Rosenmeier MF, Hodell DA et al (2002b) Paleolimnology of the Maya Lowlands: long-term perspectives on interactions among climate, environment, and humans. Anc Mesoam 13(01):141–157

Carrillo-Bastos A, Islebe GA, Torrescano-Valle N et al (2010) Holocene vegetation and climate history of central Quintana Roo, Yucatán Peninsula, Mexico. Rev Palaeobot Palynol 160(3-4). Elsevier B.V:189–196

Carrillo-Bastos A, Islebe GA, Torrescano-Valle N (2012) Geospatial analysis of pollen records from the Yucatán Peninsula, Mexico. Veg Hist Archaeobotany 21(6) Springer-Verlag:429–437

Carrillo-Bastos A, Islebe GA, Torrescano-Valle N (2013) 3800 Years of quantitative precipitation reconstruction from the Northwest Yucatan Peninsula. PLoS One 8(12). Public Library of Science:e84333

Correa-Metrio A, Bush MB, Pérez L et al (2011) Pollen distribution along climatic and biogeographic gradients in northern Central America. The Holocene 21(4):681–692

Correa-Metrio A, Bush MB, Hodell DA et al (2012) The influence of abrupt climate change on the ice-age vegetation of the Central American lowlands. J Biogeogr 39(3). Wiley Online Library:497–509

Cowgill UM, Hutchinson GE, Racek AA et al (1966) The history of Laguna de Petenxil: a small lake in Northern Guatemala. Connecticut Academy of Arts and Sciences. Connecticut, USA

Crane CJ (1996) Archaeobotanical and palynological research at a late preclassic Maya community, Cerros, Belize. University of Utah Press. Available at: http://agris.fao.org/agris-search/search.do?recordID=US1997059182

Domínguez-Vázquez G, Islebe GA (2008) Protracted drought during the late Holocene in the Lacandon rain forest, Mexico. Veg Hist Archaeobotany 17(3). Springer-Verlag:327–333

Domínguez-vázquez G, Islebe GA, Villanueva-Gutiérrez R (2004) Modern pollen deposition in Lacandon forest, Chiapas, Mexico. Rev Palaeobot Palynol 131(1). Elsevier:105–116

Douglas PMJ, Pagani M, Canuto MA et al (2015) Drought, agricultural adaptation, and sociopolitical collapse in the Maya Lowlands. Proc Natl Acad Sci U S A 112(18). National Acad Sciences:5607–5612

Dunning N, Rue DJ, Beach T et al (1998) Human-environment interactions in a tropical watershed: the paleoecology of Laguna Tamarindito, El Petén, Guatemala. J Field Archaeol 25(2). Maney Publishing:139–151

Ebert CE, Peniche May N, Culleton BJ et al (2017) Regional response to drought during the formation and decline of Preclassic Maya societies. Quat Sci Rev 173:211–235

Escarraga-Paredes D del S, Torrescano-Valle N and Islebe GA (2014) Análisis de la relación vegetación-lluvia de polen actual de las comunidades vegetales en el noroeste de la península de Yucatán, México. Polibotánica (38). Instituto Politécnico Nacional, Escuela Nacional de Ciencias Biológicas: 27–52

Evans NP, Bauska TK, Gázquez-Sánchez F et al (2018) Quantification of drought during the collapse of the classic Maya civilization. Science 361(6401):498–501

Franco-Gaviria F, Correa-Metrio A, Cordero-Oviedo C et al (2018) Effects of late Holocene climate variability and anthropogenic stressors on the vegetation of the Maya highlands. Quat Sci Rev 189:76–90

Gama L, Ortiz-Pérez MA, Moguel-Ordoñez E et al (2011) Flood risk assessment in Tabasco, Mexico. WIT Trans Ecol Environ 145. WIT Press:631–639

Gunn J, Scarborough V, Folan W, et al (2017) A distribution analysis of the central Maya lowlands ecoinformation network: its rises, falls, and changes. Ecology and Society 22(1). The Resilience Alliance. Available at: https://consecol.org/vol22/iss1/art20/

Gutiérrez-Ayala LV, Torrescano-Valle N, Islebe GA (2012) Reconstrucción paleoambiental del Holoceno tardío de la reserva Los Petenes, Península de Yucatán, México. Revista Mexicana de Ciencias Geológicas 29(3):749–763

Haug GH, Hughen KA, Sigman DM et al (2001) Southward migration of the intertropical convergence zone through the Holocene. Science 293(5533). American Association for the Advancement of Science:1304–1308

Haug GH, Günther D, Peterson LC et al (2003) Climate and the collapse of Maya civilization. Science 299(5613):1731–1735

Higuera-Gundy A, Brenner M, Hodell DA et al (1999) A 10,300 14C yr record of climate and vegetation change from Haiti. Quat Res 52(2). Cambridge University Press:159–170

Hodell DA, Brenner M, Curtis JH et al (2001) Solar forcing of drought frequency in the Maya lowlands. Science 292(5520):1367–1370

Hodell DA, Anselmetti FS, Ariztegui D et al (2008) An 85-ka record of climate change in lowland Central America. Quat Sci Rev 27(11–12). Elsevier Ltd:1152–1165

Islebe G, Sánchez O (2002) History of Late Holocene vegetation at Quintana Roo, Caribbean coast of Mexico. Plant Ecol 160(2). Kluwer Academic Publishers:187–192

Islebe GA, Hooghiemstra H, Brenner M et al (1996) A Holocene vegetation history from lowland Guatemala. The Holocene 6(3). SAGE Publications:265–271

Islebe GA, Villanueva-Gutiérrez R, Sánchez-Sánchez O (2001) Relación lluvia de polen-vegetación en selvas de Quintana Roo. Bol Soc Bot Méx 69:31–38

Islebe GA, Sánchez-Sánchez O, Valdéz-Hernández M et al (2015a) Distribution of vegetation types. In: Islebe GA, Calmé S, León-Cortés JL et al (eds) Biodiversity and conservation of the Yucatán Peninsula. Springer International Publishing, Cham, Switzerland, pp 39–53

Islebe GA, Schmook B, Calmé S et al (2015b) Introduction: biodiversity and conservation of the Yucatán Peninsula, Mexico. In: Islebe GA, Calmé S, León-Cortés JL et al (eds) Biodiversity and conservation of the Yucatán Peninsula. Springer International Publishing, Cham, Switzerland, pp 1–5

Islebe GA, Torrescano-Valle N, Aragón-Moreno AA et al (2018) The Paleoanthropocene of the Yucatán Peninsula: palynological evidence of environmental change. Bol Soc Geol Mex 70(1):49

Joo-Chang JC, Islebe GA, Torrescano-Valle N (2015) Mangrove history during middle- and late-Holocene in Pacific south-eastern Mexico. The Holocene 25(4). SAGE Publications Sage UK: London, England:651–662

Lentz DL, Dunning NP, Scarborough VL et al (2014) Forests, fields, and the edge of sustainability at the ancient Maya city of Tikal. Proc Natl Acad Sci U S A 111(52):18513–18518

Lentz DL, Magee K, Weaver E et al (2015) Agroforestry and agricultural practices of the ancient Maya at Tikal. In: Tikal: Paleoecology of an ancient Maya city. Cambridge University Press, Cambridge, pp 152–185

Leyden BW (1984) Guatemalan forest synthesis after Pleistocene aridity. Proc Natl Acad Sci U S A 81(15). National Acad Sciences:4856–4859

Leyden BW (2002) Pollen evidence for climatic variability and cultural disturbance in the Maya Lowlands. Anc Mesoam 13(1). Cambridge University Press:85–101

Leyden BW, Brenner M, Dahlin BH (1998) Cultural and climatic history of Coba, a lowland Maya city in Quintana Roo, Mexico. Quat Res Elsevier:1–12

Marchant R, Harrison SP, Hooghiemstra H et al (2009) Pollen-based biome reconstructions for Latin America at 0, 6000 and 18 000 radiocarbon years. Clim Past Discuss 5(1):369–461

Mendoza B, García-Acosta V, Velasco V et al (2007) Frequency and duration of historical droughts from the 16th to the 19th centuries in the Mexican Maya lands, Yucatan Peninsula. Clim Chang 83:151–168

Mueller AD, Islebe GA, Anselmetti FS et al (2010) Recovery of the forest ecosystem in the tropical lowlands of northern Guatemala after disintegration of Classic Maya polities. Geology 38(6):523–526

Muñoz E, Busalacchi AJ, Nigam S et al (2008) Winter and summer structure of the Caribbean low-level jet. J Clim 21(6). American Meteorological Society:1260–1276

Nooren K, Hoek WZ, van der Plicht H et al (2016) Explosive eruption of El Chichón volcano (Mexico) disrupted 6th century Maya civilization and contributed to global cooling. Geology Geological Society of America: G38739:1

Nooren K, Hoek WZ, Dermody BJ et al (2018) Climate impact on the development of Pre-Classic Maya civilisation. Clim Past 14(8). Copernicus GmbH:1253–1273

O'Hara SL, Metcalfe SE (1995) Reconstructing the climate of Mexico from historical records. The Holocene 5(4). SAGE Publications Ltd:485–490

Palacios-Chávez R, Ludlow-Wiechers B, Villanueva GR (1991) Flora Palinológica de La Reserva de La Biosfera de Sian Ka'an. Centro de Investigaciones de Quintana Roo, Quintana Roo

Pollock AL, van Beynen PE, DeLong KL et al (2016) A mid-Holocene paleoprecipitation record from Belize. Palaeogeogr Palaeoclimatol Palaeoecol 463:103–111

Román-Cuesta RM, Retana J, Gracia M (2004) Fire trends in tropical Mexico: a case study of Chiapas. J For 102(1). Narnia:26–32

Roy PD, Torrescano-Valle N, Islebe GA et al (2017) Late Holocene hydroclimate of the western Yucatan Peninsula (Mexico). J Quat Sci 32(8):1112–1120
Sánchez-Sánchez O, Islebe GA, Ramírez-Barajas PJ et al (2015) Natural and human induced disturbance in vegetation. In: Islebe GA, Calmé S, León-Cortés JL et al (eds) Biodiversity and conservation of the Yucatán Peninsula. Springer International Publishing, Cham, pp 153–167
Steinhilber F, Beer J, Fröhlich C (2009) Total solar irradiance during the Holocene. Geophys Res Lett 36(19):L19704
Torrescano-Valle N, Islebe GA (2012) Mangroves of Southeastern Mexico: palaeoecology and conservation. Open Geogr J 5(1):6–15
Torrescano-Valle N, Islebe GA (2015) Holocene paleoecology, climate history and human influence in the southwestern Yucatán Peninsula. Rev Palaeobot Palynol 217(0). Elsevier:1–8
van Hengstum PJ, Maale G, Donnelly JP et al (2018) Drought in the northern Bahamas from 3300 to 2500 years ago. Quat Sci Rev 186:169–185
Vela-Peláez AA, Torrescano-Valle N, Islebe GA et al (2018) Holocene precipitation changes in the Maya forest, Yucatán peninsula, Mexico. Palaeogeogr Palaeoclimatol Palaeoecol 505. Elsevier:42–52
Wadsworth R, Treweek J (1999) Geographical information systems for ecology. An introduction. Prentice Hall, Addison Wesley Longman, Harlow
Wahl D, Byrne R, Anderson L (2014) An 8700 year paleoclimate reconstruction from the southern Maya lowlands. Quat Sci Rev 103(0):19–25
Walker M, Gibbar, Head MJ et al (2019) Formal subdivision of the Holocene Series/Epoch: a summary. J Geol Soc India 93(2). Springer:135–141

Chapter 11
From Calakmul to the Sea: The Historical Ecology of a Classic Maya City That Controlled the Candelaria/Champoton Watersheds

Joel D. Gunn, William J. Folan, Nuria Torrescano-Valle, Betty B. Faust, Helga Z. Geovannini-Acuña, and Alfred H. Siemens

Abstract The Candelaria River watershed of Campeche, Mexico, and Petén, Guatemala, has shaped millennia of Maya, perhaps from their beginnings and generations of archaeologists. This chapter reviews efforts to understand Candelaria historical ecology over the past four decades, mostly in the northern branch, the Candelaria-El Caribe-Tomatillal system, as it relates to the cities of Calakmul and Uxul. Additional investigations stem from the river system farther north, the Champoton-Desempeño watershed, which also approaches Calakmul at its headwaters. Four other projects (Oxpemul, Yaxnohcah, Naachtun, Uxul) have emerged in the southern branches of the Candelaria and Desempeño in recent years, in addition to a long-standing project at El Mirador. Methods utilized range over global-local climate teleconnections, geology, ethnoecology, soil formation and transformation,

J. D. Gunn (✉)
University of North Carolina at Greensboro, Greensboro, NC, USA
e-mail: jdgunn@uncg.edu

W. J. Folan
Centro de Investigaciones Históricas y Sociales, Universidad Autónoma de Campeche, Campeche, Campeche, Mexico

N. Torrescano-Valle
Departamento Conservación de la Biodiversidad, El Colegio de la Frontera Sur Unidad Chetumal, Chetumal, Quintana Roo, Mexico
e-mail: ntorresca@ecosur.mx

B. B. Faust
Centro de Investigación Científica de Yucatán, Merida, Yucatán, Mexico

H. Z. Geovannini-Acuña
Centre College, Mérida, Yucatán, Mexico

A. H. Siemens
University of British Columbia, Vancouver, BC, Canada
e-mail: asiemens@geog.ubc.ca

N. Torrescano-Valle et al. (eds.), *The Holocene and Anthropocene Environmental History of Mexico*, https://doi.org/10.1007/978-3-030-31719-5_11

geochemistry, pollen, and phytolith sampling. Brief but key statements of results for each method are reported as they contribute to a holistic perspective on the evolution and the death of Calakmul. Key interests that guided the research are the commercial location of the city and watershed in the Maya urban system, the way the watershed shaped the human settlement pattern in the west central Maya Lowlands, the socio-ecological adaptations over time, and its utility as a middle-level unit of study in links between human and Earth system changes. Finally, a nutrient flow model unifies the results into a concept that may yield deeper insights into the narrative and simulation modelling of Maya social evolution.

Keywords Candelaria River · Watershed · Pollen · Sediments · Maya · Campeche · Calakmul · Hydrography

Introduction

As Euro-American history opens on the western Maya Lowlands, Hernan Cortez is reported marching an army from Mexico City to cross the Yucatan Peninsula as a show of force. At Comalcalco, in the delta of the Usumacinta River, he inquired as to the means of crossing the peninsula, to which the men of the city told him that he should push on to Acalan on the Candelaria to be advised by the Chontal who headquartered there and traded with the east coast of the peninsula. At Acalan, Cortez found the thriving city of Itzamkanac/El Tigre (Fig. 11.1). Their king, Paxbolonacha, offered him 600 porters, after which the journey began from the middle reaches of the Candelaria eastward to the vicinity of Lake Misteriósa and then southeast toward Lake Chichen Itza and then on to Honduras (Gunn 2018; Scholes and Roys 1968). On the way out of Acalan territory, Cortez encountered a couple of traders, which he appropriated as guides. Following this route, he would have passed near Calakmul, El Mirador, Nakbe, Tintal, Ceibal, and Caracol, some of the oldest and most important cities in the long Mayan occupation of the region. Most likely he would have been unaware of the rich past of these inland ruins. He only saw the thriving west coastal cities that by then carried the bulk of the international trade of the Yucatan Peninsula's native provinces. Landa in the 1500s, however, reported that the old cites were used as concentration points for goods to be exported to the coasts, a custom that continued into the twentieth century for exporting chicle, lumber, white pepper, textiles, and other products of the interior (Avila Chi 2009).

Three hundred years later (1839), Stephens and Catherwood returned to some of the same trails to catalog and describe the still-mysterious, old interior cities, to be tagged a century later as the "Old Empire" by Morley. Over the twentieth century, the idea of an Old Empire was constructed and deconstructed and, now in the twenty-first century, thanks to gaining the ability to read ancient Maya scripts, reconstructed again. As ancient social relations begin to emerge from Maya written records (Martin and Grube 2008), the Old Empire is being somewhat rebuilt, though as hegemonic blocks. Such "empires" are not uncommon in world history, and in

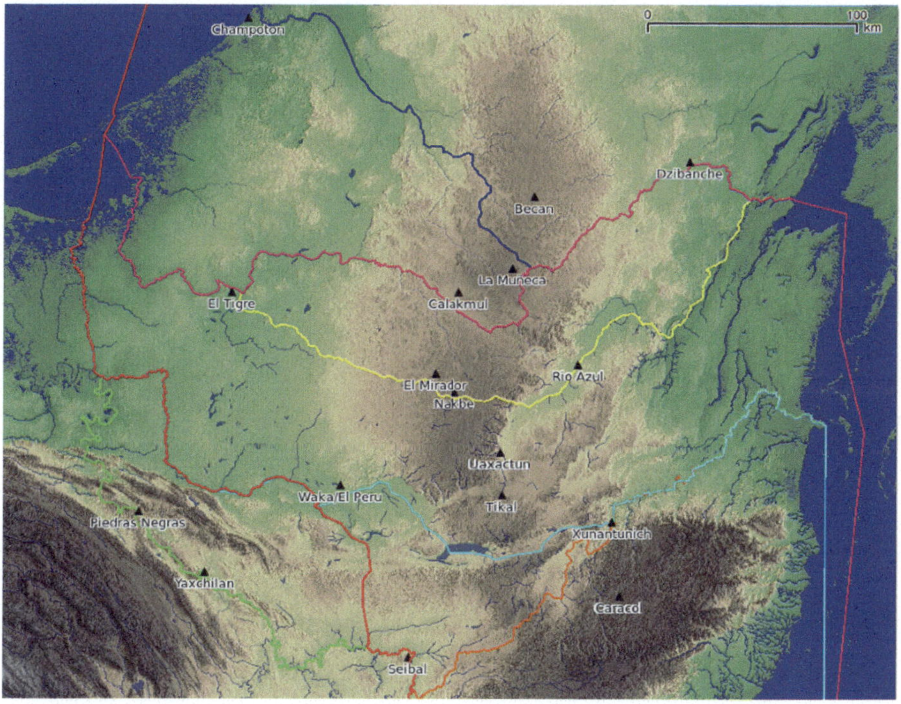

Fig. 11.1 Central Maya Lowlands water routes, Candelaria-El Caribe-Tomatillal (through Calakmul) and La Esperanza (through El Mirador). (Adapted from Volta and Gunn 2012, 2016)

the Maya Classic Period (300–900 CE), an alliance between Calakmul and Caracol (approximately 500–700 CE) came the closest of any to establishing imperial hegemony in the southern Maya Lowlands (Gunn et al. 2014a; Gunn et al. 2017).

Even with the ability to read scripts prepared by scribes of ancient kings, much remains to be reconstructed of those distant lifeways. In this article we review through the methods of historical ecology (Crumley 1994) the surroundings and internal functions of the society that once flourished in the region. To constrain the boundaries of our research, we have focused on the flow of water, nutrients, information, goods, and geological residues down the Candelaria River watershed into the sea, the Gulf of Mexico. Within this landscape, we endeavor to detect the landscape modification that ensued from the evolution of Maya society as it moved from hunter-gatherer bands, to trading villages, to great cities of astonishing scope and reach. Even with reading and reconstruction, we will never know precisely the structure of those ancient societies. We can, however, read the outputs of their daily activities collected under structures, at the bottom of hill slopes, artificial lakes, and the Candelaria River delta.

To examine these outputs, we will first overview the findings of Project Candelaria. Then we will add some details to these findings by offering additional information from cores and test excavations that move us toward a broader understanding of the watershed's social evolution. We will conclude by offering a unified understanding of what has been discovered so far in the form of a flow structure model that can be interpreted from the point of view human nutrients, but in parallel to water, energy, information, ecosystem services, sustainability, or all six simultaneously (Gunn and Folan 2018; Gunn and Members of the ESCHER Workshop 2019). We find that a limited number of geochemical indicators (magnesium, potassium, phosphate) are the critical variables that measure the vitality these flows have for sustaining human life and that the Maya of the Candelaria watershed seem to have had a clever scheme for recycling them to the top of the watershed even though their desalinization process tended to move them down the watershed to the sea.

Overview of Project Candelaria

This is the fourth decade (Table 11.1) of an effort to understand the environment of Calakmul, Campeche, through environmental archaeology and ethnoecology. Calakmul has two apparent access routes to the Gulf Coast, one to the west along

Table 11.1 Candelaria watershed research timeline

1972	Siemens and Puleston characterize semi-karst rivers (Siemens and Puleston 1972)
1980s	Background research, tour, ethnography (Faust 1988, 1998), excavations start (Folan 1983)
1981	GCM study of climate change characteristics (Gunn and Adams 1981)
1983	Historic background of climate and social organization (Folan et al. 1983; Hoggarth et al. 2017)
1989	Siemens published Tierra Configurada (Siemens 1989)
1991	Model mesoscale global climate interactions (Gunn 1991)
1992	Organize biosphere reserve (Folan et al. 1992)
1994	Historical Ecology: Global Climate and Regional Biocultural Diversity (Gunn 1994)
1994	Model past Candelaria discharge from modern 1958 to 1990 data (Gunn et al. 1994)
1995	LIA Calakmul excavations, SW ML hydrological response surface (Joel Gunn et al. 1995)
1996	Tres Rios and Global Change (Gunn and Folan 1996)
2000	Model global climate sensitivity of discharge for adjoining rivers (Usumacinta, Champoton, Grijalva) (Gunn and Folan 2000)
2002	Bajo sediments and Calakmul water system (Gunn et al. 2002)
	Dams on the Candelaria (Siemens et al. 2002)
2008	Panlao core published, Conditions of Interior, Sustainable Urbanism (Gunn et al. 2009a; Torrescano-Valle et al. 2012)
2008	Geovannini Rain Harvesting published (Geovannini-Acuña 2008)
2009	Rosario test pits published (Gunn et al. 2009b)
2009	Candelaria landscape profile (see Fig 11.3) presented to first IHOPE-Maya workshop (Faust et al. 2012) (Gunn 2009 IHOPE)
2009	Siemens reports on pollen cores from Lake Candelaria in lower/middle Candelaria (Siemens 2009)
2013	Nutrient profile published

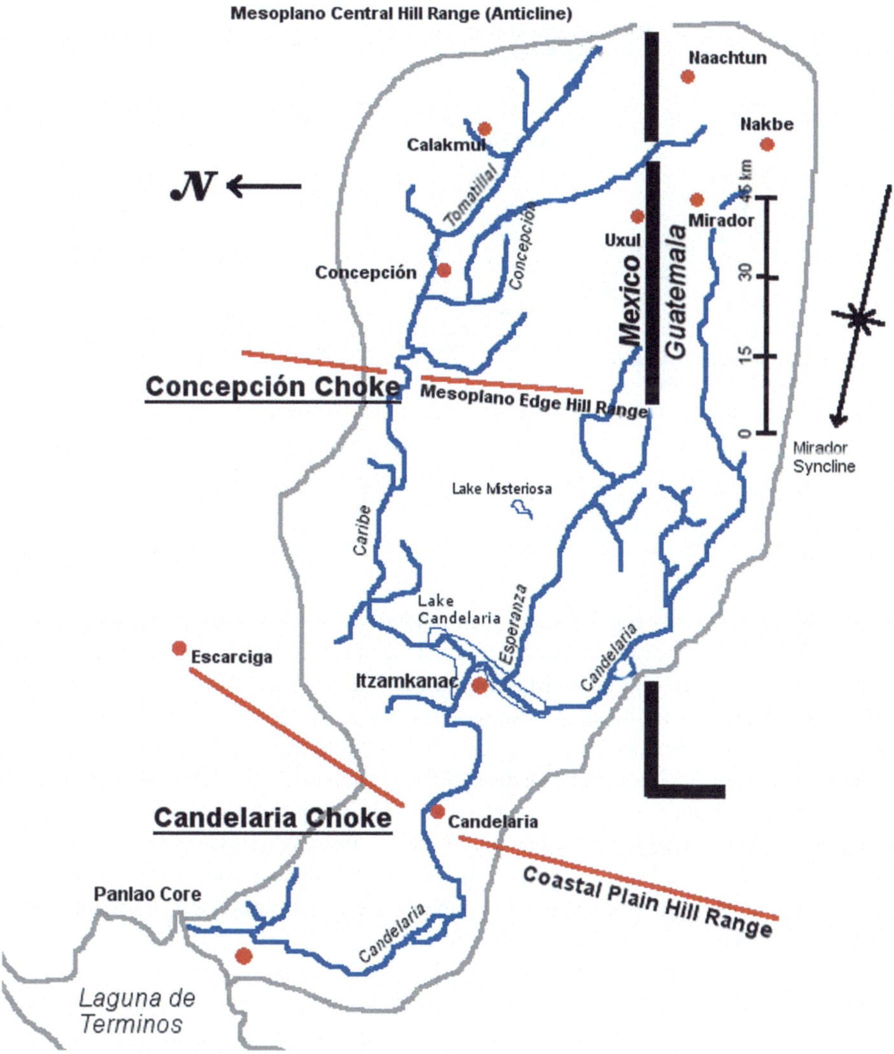

Fig. 11.2 Candelaria River watershed. Note: North is to the left

the Candelaria-El Caribe-Tomatillal system and the other through the Champoton-Desempeño system to the northwest. The study was set initially in the context of the Candelaria River watershed (Gunn et al. 1995) (Fig. 11.2). Aside from the fact that watersheds are excellent units of study for understanding part or whole of human settlement patterns, they also comprise middle-scale (mesoscale) segments of landscapes that can be linked to global environmental change through their water volumes per unit of time (Gunn et al. 2019; Gunn et al. 1995). In addition to water, river systems collect environmental data and deposit it in deltas. To examine these

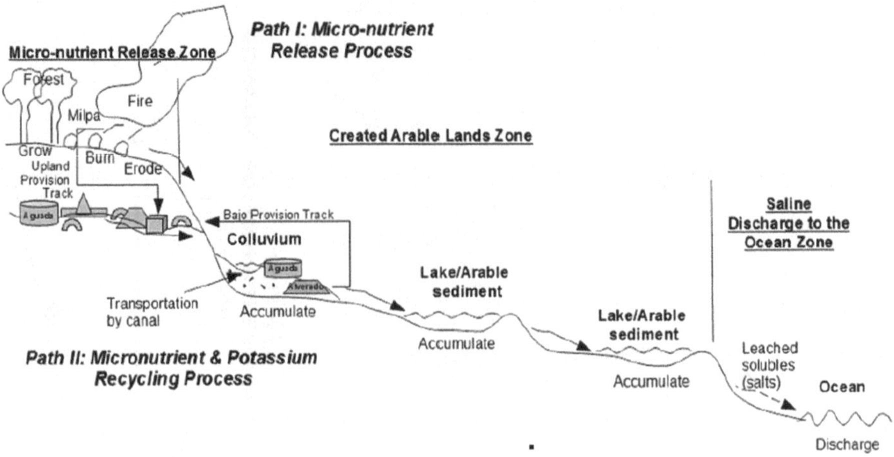

Fig. 11.3 For the Calakmul to the sea project, water, sediment, and nutrient flow profile of the Candelaria-Caribe-Tomatillal River system

accumulated sediments for the watershed narrative we are building, we have cored the Candelaria River delta at Panlao (near Ciudad del Carmen) where we found evidence that the Classic Period Maya were profoundly modifying watershed landscape to manage agricultural production (Gunn et al. 2009a; Gunn et al. 2013; Siemens 2009; Torrescano-Valle et al. 2012). No doubt this accounts for the millennium-long history of Calakmul's sustainability (Folan et al. 2012; Gunn et al. 2013).

Siemens and Puleston (1972) gave considerable impetus to the study of watersheds by bringing semi-karst and rivers of the southern Maya Lowlands to archaeologist's attention. A clear vision of rivers and floodplains unsettled an age-old archaeological myth that the Maya Lowlands were a flat, low-lying, uniform tropical jungle in which climate did not change.

At the headwaters of the Candelaria-El Carib-Tomatillal River system (see Figs. 11.2 and 11.3), Calakmul itself is perched on a promontory and ridgeline between two immense seasonal swamps, El Laberinto Bajo to the south and El Ramonal Bajo to the north. The ridge line approximately halves the Calakmul Basin of the Mesoplano, an elevated interior hill and valley system comprising the central spine of the southern Yucatan Peninsula (Dunning et al. 2012; Folan et al. 2016; Gunn et al. 1995; Volta and Gunn 2016). To address Calakmul's use of their immediate environment, we have cored both seasonal swamps and found that after an initial collapsing of vegetation on the bajo sides in the Preclassic, they began to use the kilometer-wide bajo edges as agricultural lands (Gunn et al. 2002). These capabilities were augmented by construction of reservoirs (or aguadas) in El Laberinto Bajo below the city, one being the largest in the lowlands (Domínguez-Carrasco and Folan 1996; Gunn et al. 2002). Canals and other water distributary elements were also devised (Fletcher and Gann 1994; Geovannini-Acuña 2008).

Within the city the main ceremonial center was constructed overlooking the bajo aguadas and bajo edge. Two large aguadas joined by a stone lined canal were also constructed adjoining the ceremonial center (Domínguez-Carrasco 1991). Water was collected into the city's main reservoirs from the ridge to the southeast by a system of canals (Fletcher and Gann 1994; Geovannini-Acuña 2008; May-Hau et al. 1990). Temples were located along the canals suggesting a means of social control somewhat like Lansing (2003) defines for southeastern Asian communities.

Water was drained from the plaster-lined main ceremonial precinct to the bajo-edge aguadas. By ethnographic analogy we suggest that the great expanse of the city flanking the main ceremonial precinct away from El Laberinto Bajo was organized around aguadas constructed in arroyos, also ultimately draining down to the bajo edges (Faust 1998; Faust et al. 2012; Gunn et al. 2002).

Both Geovannini-Acuña and Gunn have calculated the amount of land available for agricultural production using differing methods. Both found that the city probably could not support the suggested 50,000+ population even by using both dry and irrigated multi-cropped lands available within a day's walk of the main ceremonial center. Three alternate sources of food are suggested. One is to bring food grown along the edge of El Laberinto Bajo to the east from as far as the city of El Laberinto, a distance of about 12 km. Another would be to bring food up the Candelaria-El Caribe from the coastal plain at Itzamkanak where a very substantial agricultural and fishery component was observed in the Historical Period (Scholes and Roys 1968; Siemens 1989; Vargas-Pacheco 2012). The Spanish explorers of the Cortez expedition reported encountering canals and crops as well as organized canal transport of produce. A third option would be to bring food south from Champoton-Desempeño system from Edzna via its Grand Canal. The Champoton-Desempeño watershed has not been as well studied, but it supported a city, El Ramonal, along its banks and might also have been a means of transportation if reengineered.

The Itzamkanac/Acalan slackwater regime (see below) has been cored by Siemens (2002, 2009) who found its rudiments date to 7000 years ago when the shoals of the Candelaria were dammed and nearby uplands were cleared. By the Classic Period sediment accumulation had turned it into vast agricultural fields.

Either of the three solutions would involve extensive canalization and damming, especially of the 10-km-wide El Laberinto Bajo that stretches across most of the Yucatan Peninsula Mesoplano east to west or by the Desempeño drainage. Brown (personal communication, email of October 21, 2011) has made preliminary observation from aerial imagery and found suggestions that such may be the case; LIDAR scans of the bajo are currently ongoing and/or planned that may support these contentions. In any event, El Laberinto Bajo remains the widest, most engineerable passage across the peninsula (Volta and Gunn 2012, 2016). This characteristic may account for the location of Calakmul overlooking El Laberinto Bajo, a situation similar to that of Tikal that overlooks the Holmul watershed (Fialko 2005).

To unify these diverse data, we generated a watershed-long profile from the upper end of El Laberinto and the central spine of the Mesoplano to the Gulf of Mexico and the Laguna de Términos, a distance of some 161 km air (Fig. 11.4) (Faust et al. 2010;

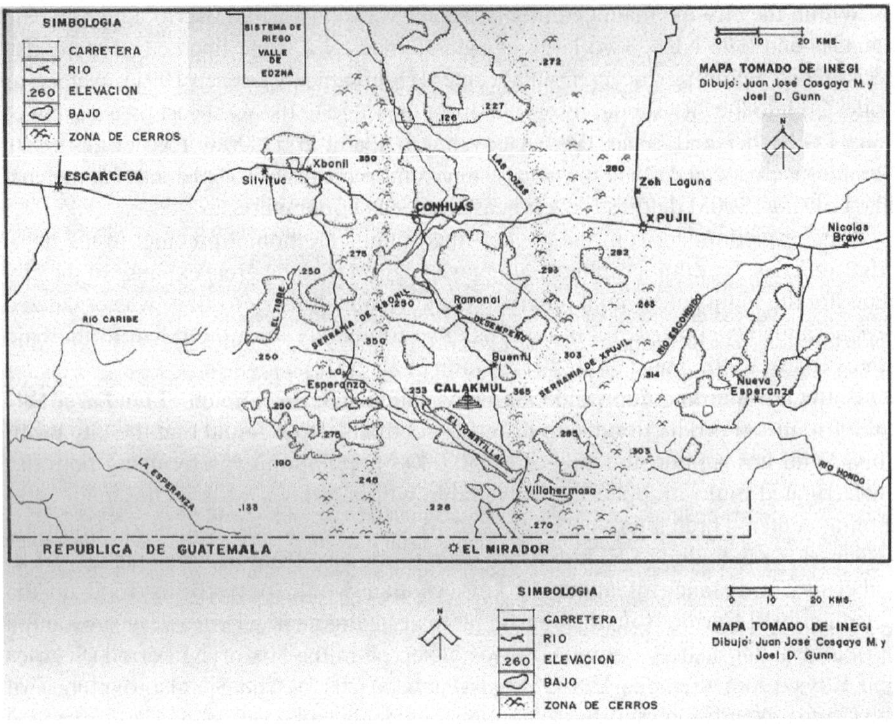

Fig. 11.4 Calakmul Basin and El Laberinto and El Ramonal bajos. (Gunn et al. 2002)

Gunn et al. 2013; Gunn et al. 2009b). The profile could be contemplated in any number of perspectives: nutrition, hydrological, travel, and informational. As a flow of nutrients discussed extensively in a later section on modeling the erosion of phosphorus, potassium, and magnesium that figure as key variables. It might also represent commerce and information flow as river systems and portages were the main routes of communication and commerce for most of Maya history (Gunn 2018; Volta and Gunn 2012). Watersheds could also be the context in which the impact of global climate change on the local watershed is measured (Gunn et al. 1994, 1995).

The northwest Maya Lowlands, basically the Mexican state of Campeche and the NW Guatemalan Petén, has emerged as a grander player in the overall picture of the cultural area than would have been suspected from the previous Tikal-centric perspective. This shift in the center of gravity brings with it certain obvious liabilities in terms of understanding the whole of the lowlands nature-culture system. Calakmul, the capital city of a large part of the region, was only identified as a major player in the 1980s. Thirty square kilometers of the city have been mapped (Folan et al. 1990b). Research on understanding the overall picture of its extent, support network, and social structure have begun by examining sites along the Conhuas-Calakmul road (see Morales-López et al. 2017), sorting out the elite social relations between Calakmul and other cities using ceramics (Domínguez-Carrasco et al. 2000; Gunn et al. 2017; Reents-Budet et al. 2011), and detailed investigations of the regional settlement patterns (Šprajc and Grube 2008). We will, therefore, discuss investigations carried out in the last few years that lay out some of the basic groundwork bringing to light the landscape of the western Classic Maya subsystem.

Calakmul: Top of the Candelaria Watershed

Organizing the City in Its Social and Physical Setting

The Candelaria watershed drains about 13,000 km^2 (Gunn et al. 1995) in the southern part of the Mexican state of Campeche and small portions of northwestern Guatemala (Brant et al. (2010) estimated an area of 5670 km^2 but they did not include El Laberinto Bajo) (Robadue et al. (2004) report 7160 km^2 citing another source, but they do not appear to include Guatemala or El Laberinto, p43). On the coastal plain, the Candelaria River divides into three branches, the Caribe River to the north, La Esperanza (INEGI 1:250,000 Carmen, Las Golondrinas in other maps) in the middle, and the Candelaria main stream to the south that goes into Guatemala (according to Brant et al. 2010). In the headwaters of these branches are the major interior settlements of Calakmul on the Tomatillal, Naachtun (Philippe Nondedeu) and Yaxnohcah (Brewer et al. 2017; Reese-Taylor et al. 2012) on La Esperanza, and El Mirador on the Candelaria branch (Hansen 2004). Calakmul and El Mirador, the two mega-centers of Maya population in the Late Preclassic and Early Classic, appear to have co-evolved in the early period of Maya cultural development (Folan et al. 1995; Gunn et al. 2014b). It is not impossible that the decline in population of El Mirador, though not abandonment, was associated with the Kaan dynasty moving its headquarters to Calakmul (Gunn et al. 2014b), a thought reinforced by the transformation of El Mirador into a pilgrimage site in the Classic Period (Hansen 2004).

In the second and third centuries CE, El Mirador greatly diminished in size and political power (Dahlin et al. 1980; Hansen et al. 2002; Matheny and Matheny n.d.). At least two influences are candidates for contributing to this decline. First, a severe drought (Dahlin et al. 1980, p. 80; Gunn et al. 1994, 1995; Wahl et al. 2006) transpired for which the city was ill-equipped to cope. A cause or co-cause may be that during the Classic Period, it fell within a space thought by R.E.W. Adams to be a buffer zone between Tikal and Calakmul (Adams 2005). The most likely portages across the peninsula (Volta and Gunn 2012, 2016), in fact, suggest that Calakmul and Tikal (Fig. 11.5) may have been struggling for control of waterways, especially in Calakmul's SE sector within which Adam's buffer zone and available river-ways inscribe perpendicular lines across the Mesoplano between the cities (see Fig. 11.1). The buffer zone would have blocked Calakmul's access to the Rio Azul-Hondo but allowed it access to the Escondido River and Dzibanche (Gunn et al. 2014b). Together the two processes suggest that El Mirador's dimming star was due to a realignment of economic powers during a time of inopportune environmental conditions for

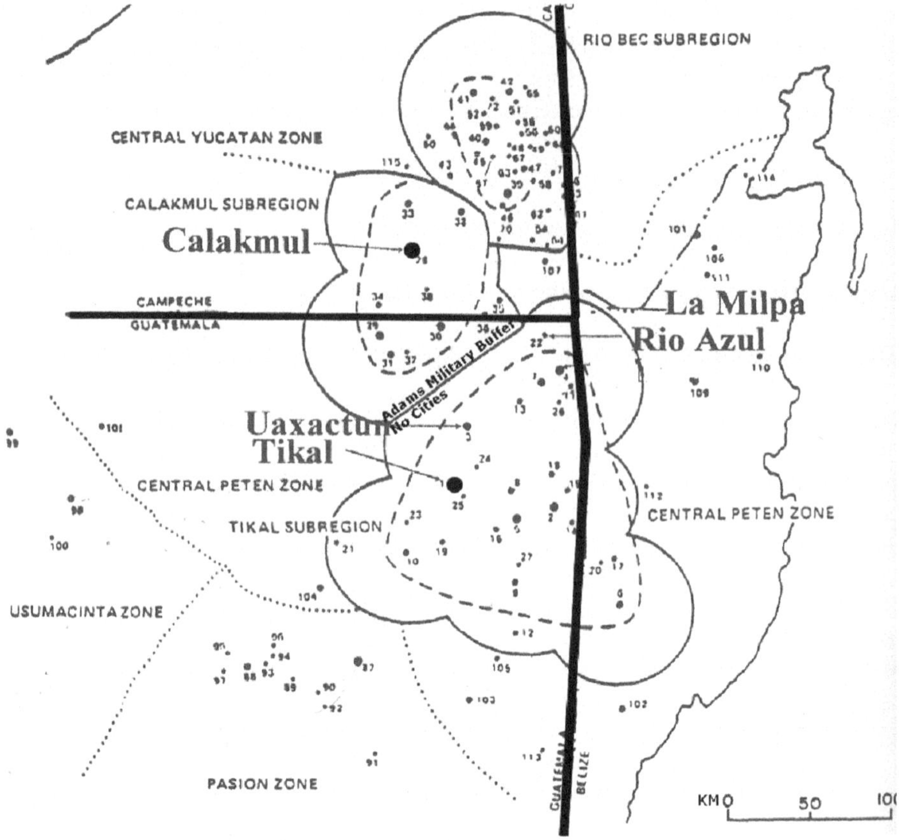

Fig. 11.5 Calakmul-Tikal Buffer. (Adapted from Adams 2005: 174)

maintaining large, interior urban populations, which perhaps relied on trans-peninsular trade. Calakmul recovered from the drought and went on to flourish as one of the great urban centers of the Classic Period in Mesoamerica.

Coring the Bajos Around Calakmul

Calakmul is located in an interior basin of the Mesoplano between two north-south trending ranges of hills with elevations up to 300–400 m (see Fig. 11.4). Within the basin, it is situated on a transverse ridge with large seasonal swamps (bajos) to the north and south. A project was launched with the help of the National Geographic Society to collect and analyze cores from within and on the margins of the bajos to determine the role they played during the lifetime of Calakmul (Gunn et al. 2002).

The results from El Laberinto Bajo south of Calakmul suggest, as do similar findings in other locations such as at Nakbe in the Candelaria branch (Jacob 1995), and on the eastern side of the peninsula (Beach et al. 2008; Dunning and Beach 2010), an initial period of sedimentation following 4000 BCE. In El Laberinto Bajo, this period was marked by heavy metals indicating a well-developed weathering horizon in place for a considerable period of time, probably since the beginning of the Late Holocene (Gunn et al. 2002). Subsequent land clearance released additional sediments into low-lying areas during the Middle Preclassic about 1000 BCE. Such activities may have resulted in unintended consequences, as suggested by Hansen's and Jacob's (Hansen 1998b; Jacob 1995) findings discussed below. Or, they may have been, at least in part, intentional and fruitful as it is possible to see in the bajo investigations around Calakmul where these new lands overlay seasonal swamps poisoned by gypsum salts (Gunn et al. 2002, 2013; Perry et al. 2011). At Naachtun Purdue (2018) found similar uses of soils slumped into aguadas that continued after the Preclassic.

Low-lying walls on these bajo foot slopes below Calakmul indicate raised platforms and a field in one case, in the new bajo space. Masonry-lined reservoirs imply that it was a struggle to keep seasonal rainwater free of underlying gypsum contaminants. The underlying sediments were salty enough to be lethal to plants.

Parallel ethnographic studies in the region imply that Calakmul's monstrous size can be accounted for as a patchwork of aguada-based communities extending along a large part of the transverse ridge (Gunn et al. 2002). C. Brown has found 153 aguadas associated with the El Laberinto Bajo detected in aerial images (Brown, personal communication, email of 2008). Brewer et al. (2017) have developed a LIDAR methodology that could be used to test these observations.

In the Esperanza branch of the Candelaria in and around El Mirador, Hansen has conducted extensive research into farming techniques and architectural design. Jacob's (1995) work at Nakbe near El Mirador suggests that there were also negative consequences to slumping sediments there. They covered sources of carbon-rich mud from the bottoms of lakes making them inaccessible for used to fertilize terrace farms. However, this area featured an aquatic environment, sometimes referred to as the "Lake District." The lakes provide a different dynamic from that of the more elevated and dryer Calakmul 37 km to the north where seasonal swamps replace lakes.

Hansen (1998a) also found that during the Preclassic, in additional to clearing land for agriculture, the inhabitants in NW Guatemala were profligate in the destruction of woodlands for incinerating limestone to make lime for plaster. A key ingredient of which is quick lime, created by heating limestone rocks over burning logs. The early pyramids at Nakbe exhibit thick layers of plaster. Later architecture is less liberally plastered, apparently in an effort to conserve woodlands or large trees that were simply no longer available. This also points out the different dynamics of the moist lowlands and the high interior as at Calakmul. It may have been a dynamic that propelled the rulers of the Calakmul state to move their headquarters to Calakmul in the Early Classic (Gunn 2018; Gunn et al. 2014b). It cautions that very different sourcing and timing may characterize the arms of the Candelaria system, a topic we will return to later.

Geovannini-Acuña (2008) excavated a pattern of test pits in the area of the central ceremonial center and found the site footprint to be confined to that area in the Preclassic but expanded in the Early Classic. Test pit excavations in 1984 in El Laberinto Bajo by Domínguez Carrasco (Domínguez-Carrasco et al. 1998; Gunn et al. 2009b) below the ceremonial center of the city revealed Early Classic sherds but no Late Classic sherds suggesting that the footprint of the city contracted following the volcanic eruptions associated with the AD 536/541 global cooling event (Gunn 2000; see Toohey et al. 2016).

Pollen and Geochemistry of the Upper Candelaria-Champoton Systems

Torrescano-Valle (Torrescano-Valle et al. 2009, 2012; Torrescano-Valle and Islebe 2015) investigated several pollen and/or geochemical sites in the Candelaria and Champoton watersheds. Among them is Lake Silvituc. It is a large, freshwater body at the back edge of the coastal plain (see Fig. 11.4). The water is kept pure by a clay lining formed from volcanic dust (Perry et al. 2011; Tankersley et al. 2011) It was included in a study with Lake Chichen Itza pollen determinations that showed a high correlation with it. Lakes Silvituc and Chichen Itza pollen sequence represent the impact of humans on interior vegetation (Gunn et al. 2017).

Figure 11.6 shows the trends of the four most common trees in the Silvituc core and grasses to represent disturbance vegetation. In the Classic Period, there is a distinct decline in Moraceae, a family of trees related to figures. *Brosimum* is a genus in the Moraceae family that bears an edible nut (ramon). It is of interest that *Brosimum* shows a slight increase during the Late Classic, while the Moraceae family

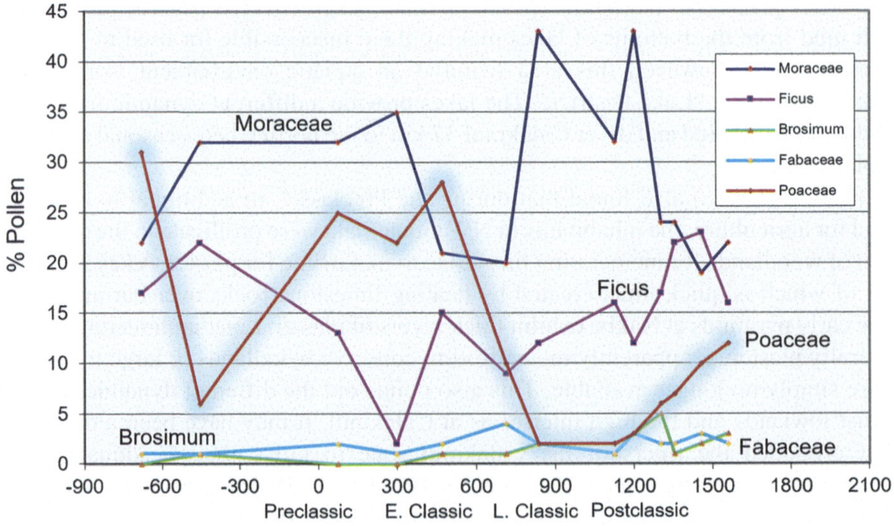

Fig. 11.6 Four most commonly identified tree families/genera/species in the Silvituc pollen core along with grasses (Poaceae). (PolenAnalysis.xls)

shows a sharp decline. This suggests that *Brosimum* was being cultivated for some purpose, quite likely food (Puleston 1978), while Moraceae as a family were being removed en masse. Whether this was for land clearance or whether it served another purpose is worth investigating. It is not currently considered a good wood for construction (O. Silvario Gallegos, personal communication, May 2018).

Grasses (Poaceae, glowing diamonds), disturbance vegetation, move in opposition to Moraceae. Grasses taken to represent the introduction of agriculture indicate that agriculture appeared around 900 BCE, retreated for the mid first millennium BCE, and then surged again through the Late Preclassic and Early Classic. Silvituc is the most northerly of the pollen sequences analyzed by Torrescano-Valle. This pattern of disturbance vegetation is repeated in the most southerly of her cores at Chumpich (see below). If grasses represent the arrival of agriculture, it appears that agriculture appeared across the south-to-north range of the Candelaria and Champoton drainages simultaneously with the appearance of external trade goods such as obsidian and jade around 900 BCE (Lohse 2010) and then fell into retreat until the Late Preclassic. A similar pattern was also observed by Wahl et al. (2006) in the northern Petén at Lago Puerto Arturo. It suggests that after the notable, sudden spread of trade, ceramics, and agriculture across the lowlands at least as far north as Silvituc, there was some sort of calamity that reversed the initial introduction of agriculture.

Ficus (squares), which was valued for paper making, and is valued still today for shade, suffered a devastating decline in the Late Preclassic/Early Classic but was restored to its Preclassic values in the Late Classic. That *Ficus* declines in the interlude between the Preclassic and Classic suggests the drought in that period was involved. It declines a bit in the Late Classic that also ended in extended droughts. Moraceae (diamonds) family, including *Brosimum* (triangles), ramon nut, now considered to be a famine food, was also restored in the Late Classic implying forest management under drought conditions, perhaps replacement by orchards. It might have been before the value of shade to preserve reservoir water a prominent need in the Late Classic, or the introduction of the use of paper, or other reasons. Whichever, it was not being managed at the Preclassic scale until after the late Early Classic.

Fabaceae, which Lentz et al. (2014) catalogs as useful for construction, was only weakly present. It, however, shows a slight increase during the Late Classic suggesting again that it was being managed, as was suggested for Tikal by Lentz, and perhaps managed in the teeth of increasingly frequent, extended droughts.

In another core near Uxul (Grube et al. 2013) from Lake Chumpich, parallel research in the Esperanza branch of the Candelaria (Torrescano-Valle et al. 2016) to the south of Calakmul, tropical vegetation plunges in favor of increases in disturbance vegetation in the early first millennium BCE (Fig. 11.7), the Middle Preclassic around 900 BCE. It recovers briefly in the late first millennium BCE and declines greatly again in the last first millennium (Late Preclassic), recovers modestly in the Classic, and returns to Early Preclassic levels in the Postclassic. This implies that the Uxul/Esperanza watershed area near El Mirador was involved in the Middle Preclassic spread of agriculture across the Lake District around 900 BCE, lost this status for a while, and then joined in the great El Mirador explosion of the Late

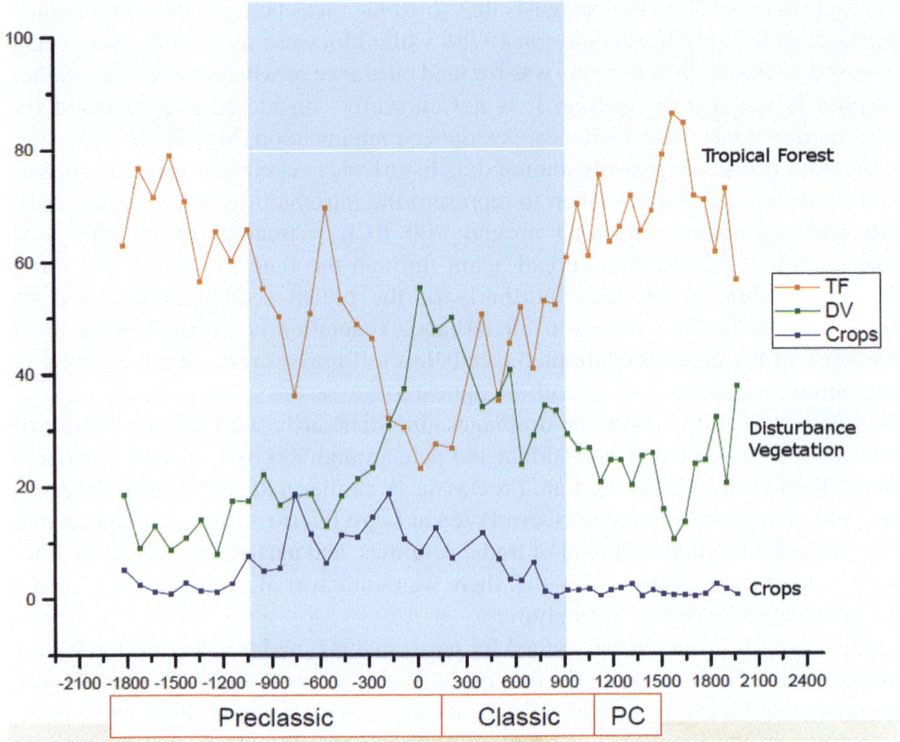

Fig. 11.7 Chumpich vegetation sequence from near Uxul. (Adapted from Torrescano-Valle et al. 2016, slide 13)

Preclassic from around 300 BCE to 200 CE. During the Classic, the landscape was managed to balance tropical forest and disturbance vegetation. This control was lost around 900 CE and the area returned to tropical forest. Comparing the Silvituc and Chumpich findings supports a northward spread of agricultural societies out of the Lake District into the drier bajo district of the Mesoplano in the Late Preclassic after an initial attempt in the early first millennium BCE.

A study of Oxpemul pollen currently underway adds some perspective to the Mesoplano yet further north and inland though still in the Ramonal Bajo (see Fig. 11.4). In the left-hand panel (Fig. 11.8), forest (Selva) and disturbance-cultivation vegetation appear to remain constant through the Classic and then drop off in the early part of the Postclassic. This would be a pattern consistent with the rise of Chichen Itza and Mayapan in the north rather than the Classic of the central Maya Lowlands. Further perspective on this will be available when the bottom 100 cm of the core are finished. It should show when agriculture became important in the north of the Ramonal Bajo.

Turning downstream to Torrescano-Valle's work in the Candelaria delta (Torrescano-Valle et al. 2012; Torrescano-Valle and Islebe 2015), a core was taken

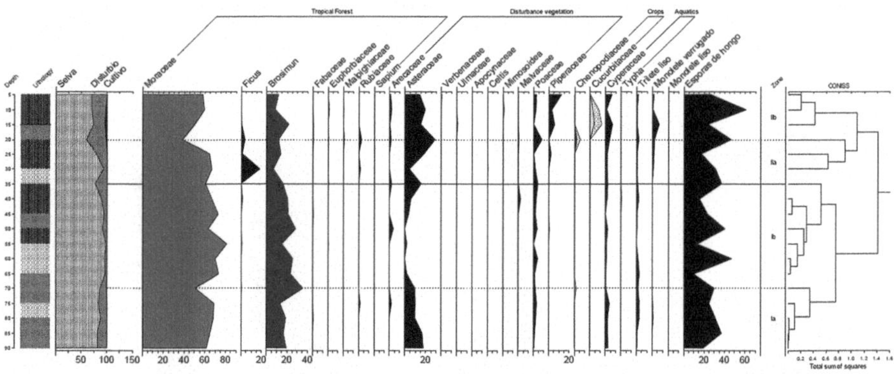

Fig. 11.8 Pollen study from Ramonal Bajo at Oxpemul (Torrescano-Valle in preparation)

for numerous reasons from an island in the Candelaria delta (see Gunn et al. 2012). It was tested for geochemical, grain size, and pollen characteristics (Fig. 11.9). Keeping in mind that these results represent the whole of the Candelaria watershed from Nakbe to Calakmul, Torrescano et al. report that the pollen shows that the Archaic Period experienced high levels of precipitation, a warmer and more humid climate than now. There was a wide distribution of mangroves and tropical forests. The sediments reveal some stability in discharge and low erosion from human activities. This would include the human activity for the previous 3000 years reported by Siemens (2009) in the middle Candelaria (see below).

During the Preclassic, precipitation conditions change substantially. There are periods of high discharge and some periods of erosion. Considering together both the archaeological and pollen data, human occupation is confirmed. This archaeological and agricultural evidence includes dams, raised fields, terraces, and albarradas: the conditions show management and erosion control.

The Classic Period sediments reveal erosion due to deforestation, elements of vegetation of disturbance and increase. The magnitude of the annual cycle decreases, and there is a decline of tropical forest and mangrove vegetation during the ninth century.

The Terminal Classic pollen analysis agrees with the physical and chemical data on the sediments (see below); it is likely that several centers such as Calakmul exercised no further control over erosion during the Terminal Classic Period. The palynological data are not bracketed by high-resolution dating during this period but indicate a drastic decline in all taxa implying a strong drought event.

In the Postclassic tropical forest vegetation recovered, savannahs, moist lowlands, and low deciduous jungles appear. Rainfall levels return but it is not comparable to that of the Archaic Period. After the conquest, repopulation fosters high deforestation and erosion.

Taken as a whole, the Panlao core sediments should present an average of all of the tributary input proportioned according to their average volume of input to the delta. As can be seen in the geochemical and pollen diagram (Fig. 11.10), after about

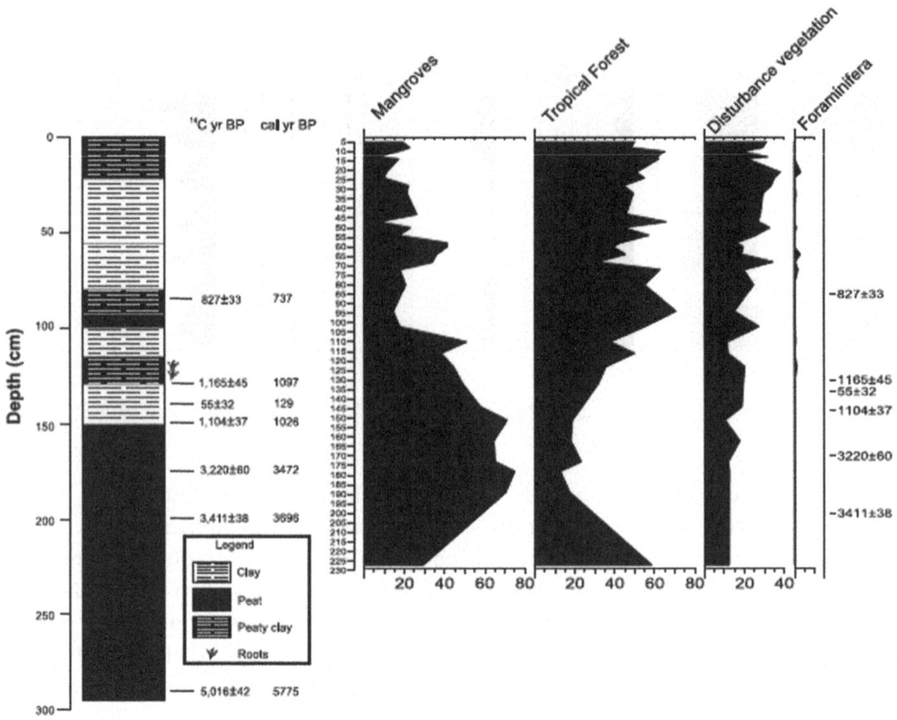

Fig. 11.9 Panlao core pollen and sediments. (Adapted from Torrescano-Valle et al. 2012)

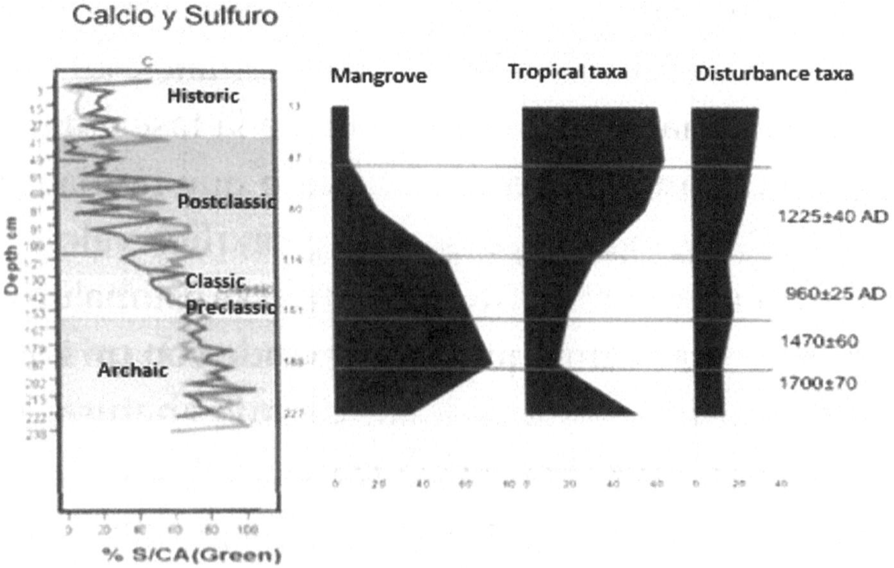

Fig. 11.10 Pollen and geochemical determinations from Panlao (Gunn et al. 2012; Adapted from Torrescano-Valle et al. 2012)

4000 years agov the mangrove vegetation declines and tropical vegetation increases. This is similar to the decline of calcium and sulfur, perhaps suggesting a heathier environment for tropical forest and the reverse for mangrove. Alternatively, it might be more a signal for the core location such as increase distance to the sea or Late Holocene cooling over time. The human signal in vegetation detected at the top of the watershed is obscured for some reason and a topic of forthcoming research. Perhaps the delta records the long-term degradation of the watershed by human activity.

Ethnoecology of the Watersheds

Ethnohistorical/ethnoecological research in the last two decades in nearby towns add some insights into possible habits of residents of the city of Calakmul during the Classic. The research has mostly been carried out along the Champoton-Desempeño watershed and the adjoining Edzna Valley, especially in the communities of Pich and Cauich (Faust 1998, 2010; Faust et al. 2012) Following the patterns of these ethnographic population, residents may have lived in Calakmul part of the year, and each farm family may have spent half or more of the year in a hamlet of associated families near a remote reservoir (aguada) and returned during the dry season to town to do construction work on temples, road building, making pottery, weaving cloth, socializing, etc. (Folan et al. 2000; Lutz et al. 2000; Zetina-Gutiérrez and Faust 2011). Zetina Gutiérrez and Faust report interview data from Pich concerning the pattern of life before the 1970s. While living in the agricultural hamlet, the farmers dispersed to farms, while the women and children stayed closer to the homes where they cared for fruit trees, pigs, chickens, and turkeys and sometimes adopted baby deer whose mother had been shot. The domestic animals were raised to be sold, while the families ate primarily deer, wild pig, and wild turkey for meat.

According to interviews with older people who had lived in these hamlets, every 14 years or so all the families of the hamlet would abandon them and move to another area known to them from previous occupations of their parents or grandparents. The men had continued to visit these sites on hunting expeditions; sometimes women also went and collected medicinal plants, firewood, etc. It has been hypothesized by Zetina and Faust, on the basis of historical reports of persecution by colonial authorities, that in colonial times, these hamlets were intentionally made to look like they were temporary shelters to avoid persecution, a tactic that might well have applied occasionally in the days of the kings as well.

Interview data concerning observations of nearby stone foundations of houses or ruined stone buildings indicate that in Prehispanic times these settlements were more substantial. The hamlet locations probably belonged to lineages, along with the aguada that provided the homes with water. This would indicate that each family during its life cycle maintained a home in town and one in each of two or three hamlet sites. Then each family would have been responsible for three or four houses. Population estimates based on counting remnants of house foundations would therefore need to be divided by three, at least.

Since other crops would rot in the bajo, corn may have been grown there (Silverio, personal communication, August 2018). If so, that was why there was so many dental issues. They could grow a lot of corn because the bajos were extensive and probably reliably moist. The uplands would only produce when the rains were of the correct magnitude and seasonal distribution, 1/2 dry season and 1/2 wet season. However, if the bajos were producing a lot of corn, should there be fewer carries and tartar in the Classic the way there is less porotic hyperostosis?

Camote are grown on slopes near the bajos. The older people formerly took camote into the field to eat; carbs gave them a lot of energy. Beans are grown on slopes, and farther down slope near the bajo the camote, squash, and Macal where they get moisture but the roots do not rot. On the other hand, corn in bajos resists moisture and corn roots do not rot.

Reporting to Lynda Folan, her Comadre (personal communication, 2018) had a traditional lifestyle in her Pueblo, planted and weeded and gathered corn, squash, etc. She remembers the beans were cooked with stones. Her father would bring home fresh ibis (small white beans) from the milpa and her mother would boil them until cooked. She then drained the beans and threw away the water. Also she would cut the stems of small green onions into small pieces and fry them with pepita de calabaza (ground squash seeds). This was added to the cooked white beans and stirred together, and they were put in a pan. Meanwhile her mother found three stones about the size of a fist. She washed and rinsed them well and placed them in the fire until they are red hot. Then she placed them in the pan with the beans, green onion stems, and squash seed powder. This she covered with a cloth for 1 hour. She said the beans are so delicious and they have a flavor of being cooked in the ground. The dish is called Tokzel. She also says she misses eating camote her father always prepared them in the ground.

Hydrography of the Middle Reaches of the Candelaria Basin

A second impediment to sediment travel to the Candelaria delta is being researched by Siemens et al. (2002; Siemens 2009) and Vargas-Pacheco (2012). The middle Candelaria is taken to be the Candelaria system from the first shoals of the river 70 km inland from the Laguna de Términos to the back side of the coastal plain marked by the edge of the elevated interior, the Xbonil Hills (see Fig. 11.4). About 70 km up the Candelaria (Fig. 11.11), the river passes through a ridge of resistant limestone (dashed lines). The elevations on either side of the river rise to over 60 m amsl. The river cuts through this ridge (solid limestone) creating some modestly fast water or shoals in the choke. The ridge line accounts for the convergence of the three arms of the Candelaria system (El Caribe, La Esperanza, Candelaria/San Pedro) that span a much wider area than would be expected if each went straight to the sea. Such geological "chokes" (or "narrows") in river floodplains create sediment traps (slack water deposits) into which sediments accumulated during slow, or

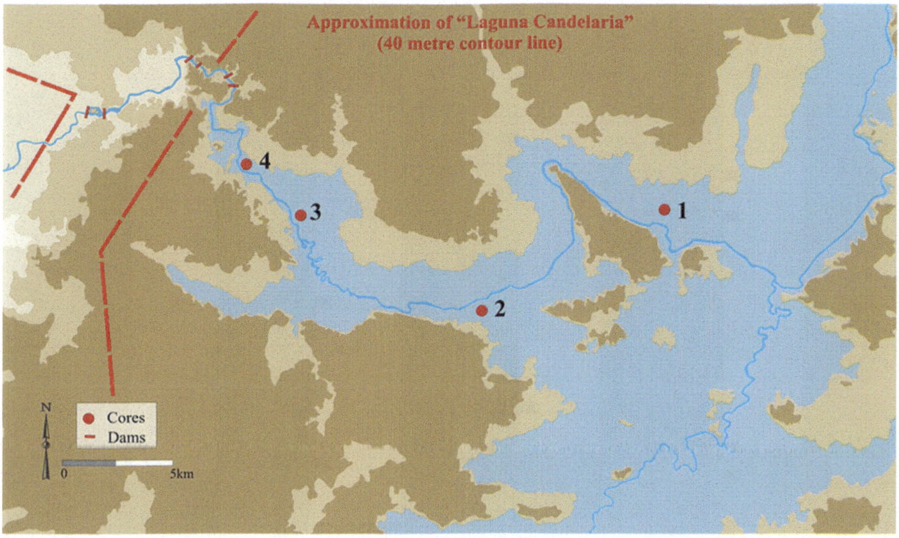

Fig. 11.11 Middle and lower Candelaria in the coastal plain and zone. (Adapted from Siemens 2009)

slack water, episodes during floods as the volume of water is too high to allow for the normal velocities that keep sediments in suspension.

The middle and lower Candelaria in the coastal plain and zone contain a complex lake/laguna topography (Fig. 11.11). The shoals below Candelaria Town funnel the many branches of the Candelaria River through a single choke. This creates a massive slack water deposit upstream that extends as much as 30 km up the streams. The depth of the deposit and Lake Candelaria were extended by damming the choke in ancient times (Siemens et al. 2002; Vargas 2001, pp. 153–155). Seimens working with local resident Soler-Graham (Siemens et al. 2002; Siemens 2009) found six dams in the narrows. The dams appear to have been made originally of logs, since encrusted by carbonates. Most of the dams that Soler and his people reported, confirmed by Seimens snorkeling, were constructed of stones manageable by one or two people rather like stone boundary walls. In the abutments were calcium casts of logs that appeared to have served as reinforcements.

An embayment below the choke indicates a Pleistocene-age, low sea level, well-established channel. The embayment (see Fig. 11.11, triangle) suggests a long-standing and stable relationship between the river and its surrounding uplands extending at least into the Pleistocene when sea level was much lower.

Behind the dams and the ridge on which they reside is the 30-km-long artificial lake. Subsequent to 5000 BCE, the lake was formed accompanied by clearing of surrounding uplands of forest vegetation (Siemens 2009). (This occurred at the same time as occupation of the Tabasco coast (Pope et al. 2001).) Upland clearance was surmised from a pollen analysis of a core (El Chilar; see Fig. 11.12). Major

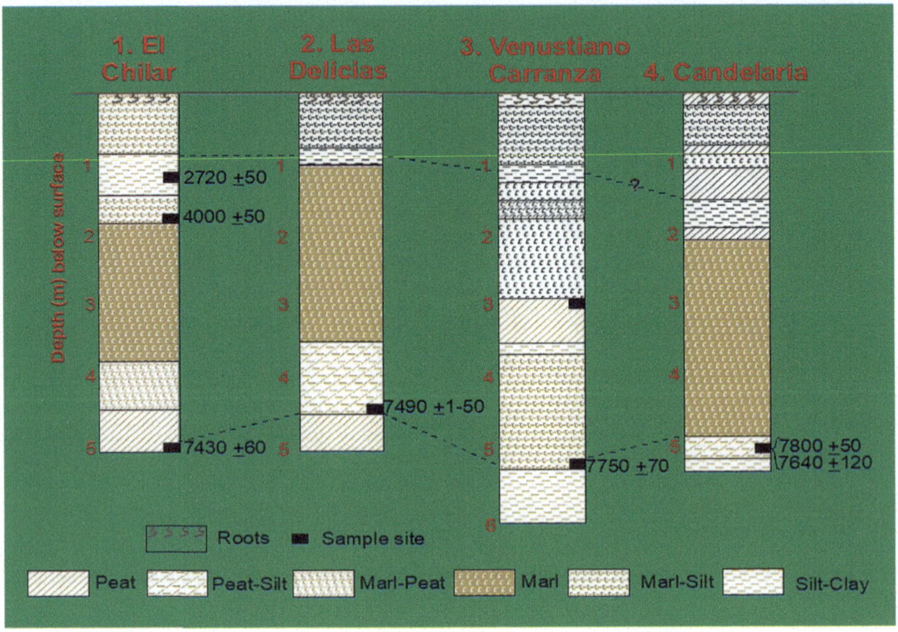

Fig. 11.12 El Chilar core sediment profile. (Adapted from Siemens 2009)

disturbance of the surrounding uplands occurred between 5000 and 3100 BCE. After that, lesser disturbances continued in the uplands until 2100 BCE.

A thick layer of marl within the lake was formed between 4000 and 2000 BCE indicating open water sponsoring a pelagic microfauna (Fig. 11.12). After 2000 BCE the marly regime was replaced by thinner layers of silty marl and clay-silt. The clay-silt sediments indicate that although the lake was file to near capacity, sedimentation continued, apparently from overbank flooding creating a seasonal lake. This was observed by Cortez during his 1524 visit to the area (Scholes and Roys 1968). Thus, the basin above the Candelaria choke would have continued as a sediment trap even without a permanent lake. Water lilies were introduced into the lake at the beginning and continued until 2100 BCE. Lilies are thought to be a Maya source of fertilizer and a water conservation measure in the dry season as their floating leaves reduce evaporation (Matheny et al. 1983). Their root systems also clean impurities from the water.

The clay-silt in the period between 2100 BCE and 1000 CE replaced the lake by agricultural lagoons and raised fields thought to be proto-chinampas (Siemens 2004). The raised fields were identified in zone b of Seimens heuristic model of floodplain annual microvariations. Area-wise, the raised fields are a minor portion of the whole basin, but functionally key. The agricultural system utilized in this period was probably flood recession agriculture as observed in 1524 by Cortez. This occurred in the zone from the 40 m contour down to a variable annual low of the lake during any given year. Exact measurement of the area of the lake was not attempted. The blue area in Fig. 11.11 was defined by a contour level that seemed a likely possible upper boundary. It is a reasonable approximation of high-water levels. Using a total station, attempts were made to deduce successive water levels from the tops of the remains of stone barriers in the river, but no apparent pattern emerged. The various dams may have been adaptations to varying climate regimes. So far no means of dating them has been devised.

This was/is an extremely complex hydrological situation. It was decided to deduce approximate regimes. This particular regime was based on successive plantings following drops in water level referred to as "fugitive" or "flood recessional" agriculture. Raised fields can be thought of as an enhancement of flood recessional agriculture by canalizing land so it is exposed more quickly and reliably for cultivation. Rather than a vast monocrop, lagoons would have had some canalization to facilitate transport by watercraft, webs of connections between settlements. These water bodies would not present one vast mirror surface of a lake, but a scrubby, ambiguous waterscape. Some of that was firm enough to build planting platforms, or "proto-chinampas." Siemens observed such a complicated watery landscape immediately after a flood in Central Veracruz.

The El Chilar sediment sequence is largely reproduced at the three other core sites along the lake (see Fig. 11.11). It appears to imply that inhabitants of the area first established a lake in the Middle Holocene (~5000 BCE), a time suggested by Scarborough for early occupation of coastal lowlands (Scarborough 1998). This would have been part of their water control regime, perhaps to facilitate traffic by watercraft, for farming purposes, for fishing, and for spanning the dry season with adequate water resources. During the Late Preclassic and Classic (~400 BCE–800 CE), the lake bed was turned into a farming area with a complex canal system discussed below.

Additional opportunities to trap sediments exist in the Upper Candelaria in the El Tomatillal, which drains El Laberinto Bajo south of the Calakmul transverse ridge. It passes through a narrow gap in the Xbonil Hills (see Fig. 11.4, the Concepción Choke) near La Concepción settlement into the coastal plain swamp. This is about 90 km up the Caribe branch of the watershed. Damming the Concepción Choke in Fig. 11.2 would likewise have provided an extensive sediment trap within the Calakmul Basin that could have been managed using the strategies suggested in Siemen' s heuristic model.

In Lake Candelaria, Siemens and Puleston (1972) discovered sequences of canals parallel to the stream channels. These canals appear to be repeated incidents of excavating canals, infilling with sediments and excavating a new canal next to an older canal in some instances only a few meters apart (Siemens 1989, Plate IV-7). This could be interpreted to mean either that (1) for some reason it was easier to excavate new canals than dredge old ones or that (2) the canals were more profitably used for other purposes, perhaps agricultural.

1. In the new canals case, in 2015 Faust visited a canal dug by the Bonfil rice project during the 1970s. It was abandoned by 1990, and 15 years later, large trees had taken over the canal, and water was no longer visible except in one area

where a family had cleared the trees. It is possible that if a human population abandoned the area for 15 years due to political or other reasons, it was easier to dig a new canal than remove the roots of large trees fertilized by agricultural activity and open growth areas.

2. In case where the canals were intentionally designed for agriculture use, reflecting on the heuristic model, Siemens proposes that the parallel canals located nearer the river channel may have served to hasten the drainage of water from the upper part of the floodplain recession area early in the growing season (Siemens 2004, p. 349). Later in the growing season when the river level was toward the channel, they would have served to retain water with intervening spaces becoming raised fields. The raised fields would have become the habitat of agriculture during the dry season allowing for multi-cropping. The low water levels would have been relatively reliable because of the Candelaria's semikarstic underground storage of water released slowly through the dry season (Siemens and Puleston 1972). Dry season water levels would also have been maintained by tributary streams emerging from springs. The canals would have additionally provided the routes of a transportation web as observed by Cortez and his recorders at both Acalan and Tenochtitlan and still apparent in Siemens' aerial photographs (Siemens 1989).

In either case, the canals provided another sediment trap explaining the lack of interior sediments during the Classic in the Panlao core. There can be little doubt that the Maya in the Candelaria drainages were busy with canalization for various purposes. These include the canals mapped at Calakmul (Folan et al. 1990a; Folan et al. 2001; Geovannini-Acuña 2008) for potable water storage and watering of crops. Also in the Champoton-Desempeño watershed at Oxpemul (Barnes et al. 2009) and Edzna (Gunn et al. 2002; Matheny et al. 1983), very substantial amounts of water were stored in canals and modified polgars producing by far the largest storage capacity of anywhere in the lowlands (Gunn et al. 2002).

Geochemistry by the Sea: The Laguna de Términos at Panlao

In addition to the pollen, and complementary to coring the bajos, a core was taken from an island in the delta of the Candelaria system in 2004 (Gunn et al. 2009a; Tebbens and van der Borg 2001). It was expected that the Panlao core would lead to an understanding of the impact that settlements in the interior had on sedimentation from the entire Candelaria system as viewed from its terminus (Fig. 11.13). The record of the Panlao core extended back to 4000 BCE. Eight radiocarbon dates were performed to correlate the sediments with the archaeological chronology. Interpreted as they are here, the core seems to indicate relatively rapid sedimentation in the Archaic and Postclassic Periods with lesser sedimentation rates in the Preclassic and Classic. ICP determinations on every third cm down the core (Gunn et al. 2012; Gunn et al. 2009b; Torrescano-Valle et al. 2012) show that the early sediments are

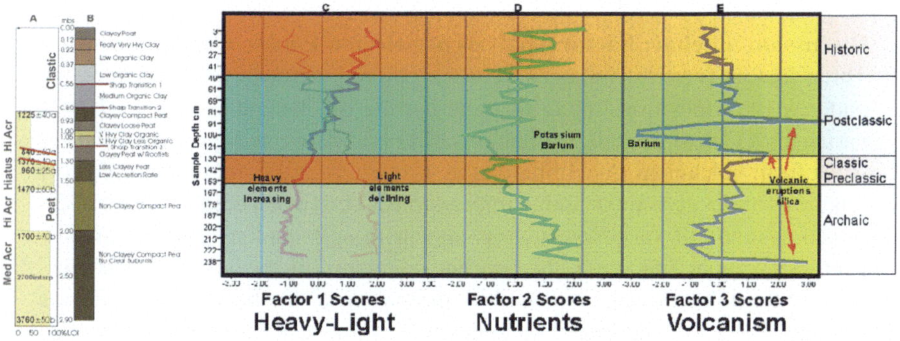

Fig. 11.13 Three sediment factor scores from Panlao core. See text and Table 11.2 for discussion. AD 536 silica spike deserves further investigation. (**A**) Acreation rate, sediment source and Radiocarbon dates, Depth in m below Surface (mbs). (**B**) Sediment grain size organics. (**C**) Factor 1 scores: Heavy and Light elements. (**D**) Factor 2 scores: Nutrients, especially Potassium and Barium. (**E**) Factor 3 scores: Volcanism, Silica, Barium

mostly soluble elements such as carbonates, while those after the Classic contained heavy metals (Fig. 11.13c). We have been studying heavy metals because they would indicate the movement of sediments that have been weathered in place for an extended period as explained above.

In the Panlao core, the heavy metals assume distinctive and opposite patterns relative to lighter sediments such as calcium and sulfur, which in compounds form the bulk of the Yucatan Peninsula platform (Fig. 11.13c). During the Archaic and Preclassic-Classic, there is little evidence of erosion of heavy metals but much evidence of the lighter elements. This pattern reverses after the Classic with much more evidence of erosion of heavy metals. The pattern of light element erosion is parallel to changes in the interior bajos around Calakmul (Gunn et al. 2002). Apparently light elements in uplands were systematically exhausted over the period of study. A group of elements that can be interpreted as nutrients (potassium; barium may not be a nutrient but appears to be sequestered by plants) diminish to a minimum during the Preclassic-Classic but increase both before and after (Fig. 11.13d). A distinctive spike in nutrients appears at the end of the Classic.

A factor analysis of the elements in effect recreates the compounds from which the elements are a part. Compounds are correlated communities of elements. Factor 3 (Table 11.2) contains a strong silica signal indicating volcanic eruptions at the end of the Classic and in the Postclassic (Fig. 11.13e). The mega-spike in the Postclassic is the 1300s CE eruption of El Chichon (Nooren et al. 2009). Beach et al. (2016) point out that there appear to be three episodes of collapse in the Maya chronology but that the correlations of droughts with these changes leave room to question precise cause-and-effect relations. Two of the three silica spikes in the Panlao core strengthen this argument allowing us to step past the cause-and-effect dilemma. Historical observations confirm that the debris of local volcanoes bring troubled times to the lowlands (Dull et al. 2001; Folan et al. 1983; Sheets 1983), especially if they are of sufficient magnitude to bring accompanying droughts rooted in global climate change (Gunn et al. 1995).

Discussion Sidebar: Factor Analysis of Selected Panlao Core Elements
Three factors (vertical linkages between colored elements) show distinctive patterns for deposition of multivariate-correlated elements (= 88% of total variance): (1) heavy and light elements, (2) plant nutrients/biproducts, and (3) silica from volcanism. As Hibbard et al. (2008: 23, 29) point out, simple correlations often give misleading impressions. This is especially so in interlinked systems of variables such as appear in soils and riverine sediments. Factors (vertical linkages) are between elements in the same subsystems. Horizontal linkages (encompassed by double lines) are between subsystems through sharing of individual elements. An example of horizontal or subsystem linkages can be seen in silica. It appears on two of the three subsystems (nutrients, factor 2; volcanism, factor 3). Association with factor 2 indicates some silica is apparently coming down the river with potassium, possibly locked in barium silicates sequestered by plants (phytoliths) and probably a function of human control of sediment exit of the interior. The other silica subsystem is from volcanic events that inject silica without barium. Likewise, potassium is shared between the heavy-light (factor 1) and nutrient subsystems suggesting that potassium comes down the river as a natural matter of course, but human infusions kick part of the potassium allegiance out of the natural orbit in an unrelated pattern with barium. Potassium and silica erosion decline toward the Classic-Preclassic indicating that humans were conserving these elements in the interior (see Fig. 11.13d). A spike in factor 2 deposition at the end of the Classic probably indicates some loss of control of erosion. In terms of number of shared subsystems, barium is the most distributed touching on all three systems. Barium is concentrated in plants so its decline with time in the nature/heavy-light subsystem and decrease in the human/nutrient subsystem during the Preclassic-Classic reflect increasing human control of system-wide vegetation. In factor 3, volcanic injections of silica from volcanoes depress the barium content.

More important than these artifacts of analytical methods are the dynamics of linked systems (Gunn et al. 2017); Gunn et al. 2019) . Some variables are shared by more than one subsystem. They tie the system together in one sense. In another they confuse bivariate correlations by apparently diminishing the variance accounted for in subsystems by variables. In mutualistic systems of variables, the fissure lines between subsystems are the lines of disturbance that reveal partial causations. Factoring these systems sorts out the partial causations, or partial correlations, giving some sense of direction in the more usual human need to see cause and effect in binary relationships. This helps to order one's understanding of events in the perceptual sphere. In this analysis, the variable depth (−0.96), or time, points to factor 1 as harboring much that goes on through time such as the decrease in lighter and more soluble sediments and increases in heavy metals. Humans acting on this system probably control the timing of release of heavy metals and plant nutrients in factor 2. Volcanism in factor 3 is also a centrifical force in the system providing another source of silica, though without barium. This suggests that the barium in factor

2 is sequestered in plants as barium silicates, whereas the silica coming in as volcanic debris is barium-free. This is an hypothesis that was tested using XRD (Faust et al. 2010)

	Factor	1	2	3	
Element	Variable Symbol	Heavy vs Light	Nutrients/Toxins	Volcanism	Atomic Wt
Heavy/light	Depth (time)	–0.96	0.20	0.07	–
Lead	PPb	0.84	–0.13	0.00	207
Barium	PBa	–0.40	0.72	–0.37	137
Copper	PCu	0.84	0.26	–0.10	64
Iron	PFe	0.85	0.12	–0.07	56
Calcium	PCa	–0.83	0.34	–0.06	40
Potassium	PK	0.68	0.60	–0.06	39
Sulfur	PS	–0.93	0.13	–0.08	32
Phosphorus	PP	0.90	0.22	–0.08	31
Silica	PSi	0.09	0.41	0.90	28
Sodium	PNa	–0.86	–0.04	0.04	23

Positive with factor Negative with factor Shared elements

Table 11.2 Factoring of XRF mineral determinations

	Factor	1	2	3	
Element	Variable Symbol	Heavy vs Light	Nutrients/Toxins	Volcanism	Atomic Wt
Heavy/light	Depth (time)	–0.96	0.20	0.07	–
Lead	PPb	0.84	–0.13	0.00	207
Barium	PBa	–0.40	0.72	–0.37	137
Copper	PCu	0.84	0.26	–0.10	64
Iron	PFe	0.85	0.12	–0.07	56
Calcium	PCa	–0.83	0.34	–0.06	40
Potassium	PK	0.68	0.60	–0.06	39
Sulfur	PS	–0.93	0.13	–0.08	32
Phosphorus	PP	0.90	0.22	–0.08	31
Silica	PSi	0.09	0.41	0.90	28
Sodium	PNa	–0.86	–0.04	0.04	23

Positive with factor Negative with factor Shared elements

[a]There are a number of reasons that univariate and bivariate analyses under-analyze multivariate systems. They include masking variables in which relationships are hidden by positive and negative correlations between three variables and reduction in variance and therefore statistical significance because of unaccounted for variables (Draper 1981)

The above-described sequence of events at Panlao might arise for a number of reasons: channel changes, sea level/base level changes, or perhaps another interpretation of the radiocarbon dates. However, taking the interpretation offered above suggests that although there is ample evidence for dissipation of sediments in the interior during the period of Maya occupation dating to the period of Maya civilization as suggested by Deevey et al. (1979), these nutrient and metal-bearing sediments did not reach the ocean until the Postclassic. In other words, the Maya were conserving sediments short of the delta.

There are two identifiable reasons that delayed the sediments arrival. One is, as Gunn et al. (2002) suggest, the sediments were intentionally allowed to slump into the edge of the bajos changing otherwise saline and acid bajo bottoms into arable farmland. Secondly, the Maya would have then constrained these sediments from further erosion and escape to the sea thus accounting for their 1000-year tenure in the Lowlands.

Modeling Nutritional Feedback in the Candelaria Watershed

Continuing the search for key variables, we suggest that in large part, after an initial period of land and sediments modification to create bajo-edge farmland, the Maya gained control of their landscape and retained its fertility by trapping sediments at the base of bajo-edge slopes, canals, lakes, chokes, and other means not discussed here but identified elsewhere: lagoons, terraces, sacbes, and walled fields. Without such measures, the sustained population of the watershed, perhaps the densest in the lowlands, for the millennium or so of Late Postclassic-Classic Maya fluorescence in the watershed, seems unlikely.

An additional ICP analysis on the Panlao core in 2008 revealed four interesting, low-magnitude jiggles in the strontium profile (Fig. 11.14e) during the Preclassic-Postclassic Periods (Gunn et al. 2009a). These might result from incidents like the collapse of El Mirador at the end of the Preclassic presaging the grander Postclassic collapse. In addition to the immediate impact of volcanic ejecta in the region, it would have produced similar droughts that reoccur about every 300 years in the region as a result of the cycling of Earth system variables (Gunn and Folan 2000; Gunn et al. 1995). Thus, the lesser fluxes could be products of disturbance in the Maya sediment world during a time when they were in large part in control of sediments from the interior. They might result from conquest events that temporarily undid the Maya control of their landscape. Since the largest is early during the period, it might prove to correlate with the El Mirador retreat.

Relative to the idea of key variables and the Maya collapse, it is clear that heavy metals inform us that sediments were moving and not just simple rates of sedimentation or other types of sediments such as soluble carbonates (Table 11.2). Lead is especially important because of its great atomic weight. Similar systems appear to be evolving in current conditions on a worldwide basis suggesting to some geologists the naming of a new epoch, the "anthropocene" (Ruddiman 2010; Syvitski et al. 2005). To partially answer our initial question concerning key variables, it would appear that

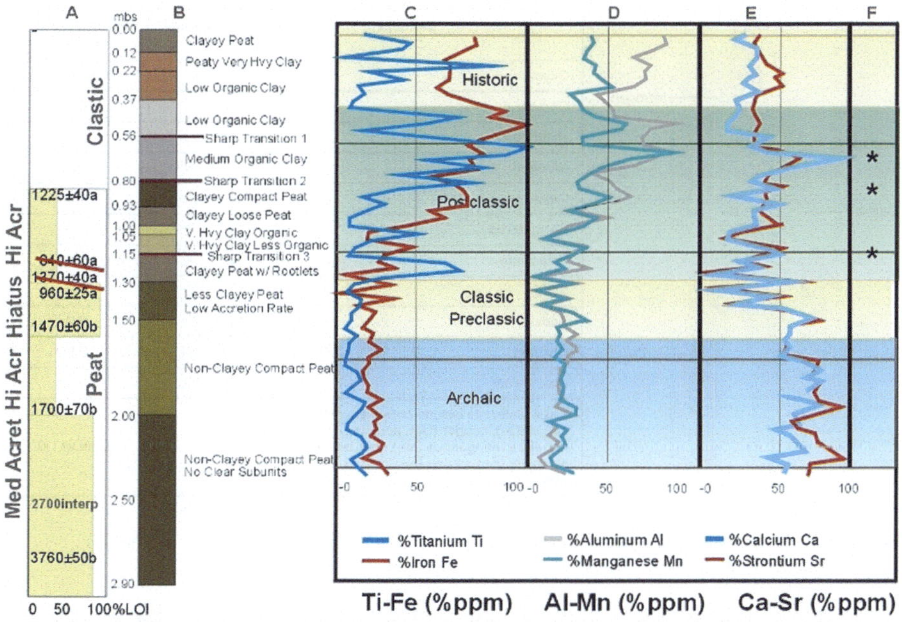

Fig. 11.14 Panlao core 2008 reanalysis for strontium and five other elements. Low-magnitude oscillation during the Preclassic and Classic suggest some disturbances in the Maya control of erosion. * = lithologic volcano beds. (**A**) Acreation rate, sediment source and Radiocarbon dates, Depth in m below Surface (mbs). (**B**) Sediment grain size organics. (**C**) Titanium-Iron. (**D**) Aluminum-Manganese. (**E**) Calcium-Strontium

if critical moments in past and present civilizations can be keyed to the constraint and movement of sediments, there is at least enough comparability to begin thinking about a futurology based on the distant past, an IHOPE design (Hibbard et al. 2008), following during some of this research.

Something that might be relevant to the down-functioning of the interior cities in the Terminal Classic is the depletion of minerals in soils critical to human health, a trend that Chew (2007) identifies with collapses of civilizations, or more frequently transformations like the Maya. It might have been the "push" element along with drought in the migration to the coast situation. In a list of the elements in the human body (Fig. 11.15), the first four (oxygen, carbon, hydrogen, nitrogen) are from the atmosphere and not quantified in sediments by our ICP analysis. The next six (calcium C, phosphorus P, potassium K, sulfur S, sodium Na, chlorine Cl, and magnesium Mg) are well understood to enable body functions such as contraction and relaxation of muscles (K Na P), elimination (Mg), cartilage structure (S), brain function (Mg), etc. Since the Yucatan platform is composed largely of limestone (Tankersley et al. 2011), critical elements other than calcium generally arrive from nearby volcanic eruptions and dust blown from Africa on the tropical easterlies. Depletion of these elements could therefore become critical if there was a long period without eruptions or the easterlies were moved toward the equator by a period of global cooling and probably other situations as well.

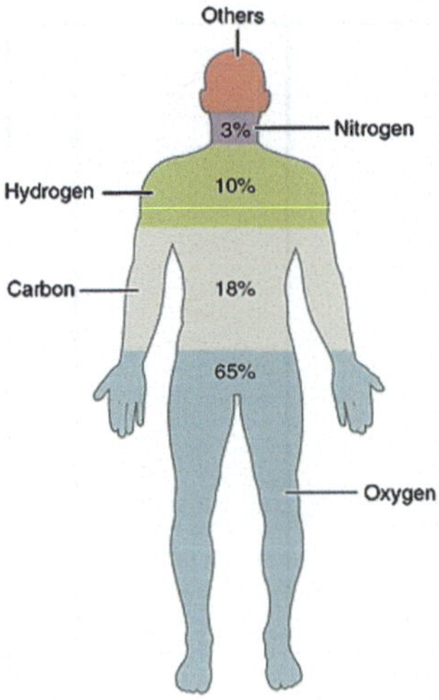

Element	Symbol	Percentage in Body
Oxygen	O	65.0
Carbon	C	18.5
Hydrogen	H	9.5
Nitrogen	N	3.2
Calcium	Ca	1.5
Phosphorus	P	1.0
Potassium	K	0.4
Sulfur	S	0.3
Sodium	Na	0.2
Chlorine	Cl	0.2
Magnesium	Mg	0.1
Trace elements include boron (B), chromium (Cr), cobalt (Co), copper (Cu), fluorine (F), iodine (I), iron (Fe), manganese (Mn), molybdenum (Mo), selenium (Se), silicon (Si), tin (Sn), vanadium (V), and zinc (Zn).		less than 1.0

Fig. 11.15 Percentages of critical elements in the human physiology. (Human Elements)

Since our concern is how well the human population was fairing in the watershed, these critical elements may give us an index of soils and human conditions. In intensively agricultural societies such as the Maya Lowlands, it is not surprising to find that humans can serve as proxies for soils and vice versa. With our ICP analyses, we may have a way to understand how the sediment-human relationships were balanced. Figure 11.16 shows factor 1 and plots of factor 1 scores, which includes the human variables (calcium, phosphorus, potassium, sulfur, sodium, and magnesium; not chlorine) and metals. The calcium-sulfur situation is a little confused because of the veneer of gypsum (calcium sulfate) that overlays the lowlands from the Chicxulub impactor 65 million years ago. There is no shortage of calcium, however, under any circumstances, so it is not very important in terms of sorting out critical variables.

What is notable are the key muscle operators, phosphorus, potassium, and magnesium that are strongly present and separable from the general system interaction network: see the factor loadings in the left panel. In the right panel (factor 1 scores), this complex of elements declines through time from the Archaic to the Early Postclassic and then increases to the present. This implies that the key elements were being used at greater rates than replacement by volcanoes and African dust in Classical times. There are curious spikes in factor 1 inputs at the transitions between the periods. This would be spikes in metals, especially lead, indicating that sediments were on the move, i.e., the soil conservation measures lapsed. The essential conservation measures would be to prevent erosion (indicated by lead) while allowing salts (calcium sulfate and sodium) to pass to the sea to prevent salinization. Factor 1 indicates that in large part the conservationists in the Candelaria drainage were successful at desalinization and erosion prevention. Factor 2 scores (light gray line) shows that during the Archaic very little lead was

Variable	Factor 1
Depth	-0.58
Calcium	-0.29
Sulfur	-0.59
Aluminum	0.80
Magnesium	0.65
Sodium	-0.63
Silicon	0.13
Iron	0.60
Potasium	0.93
Manganese	0.12
Barium	0.12
Nickle	0.48
Boron	0.44
Cobalt	0.12
Lead	0.45
Copper	0.76
Zinc	0.71
Chromium	0.68
Cadmium	0.68
Molybdenum	-0.27
Selenium	0.75
Arsenic	0.37
Phosphorus	0.80
Key:	negative
	human

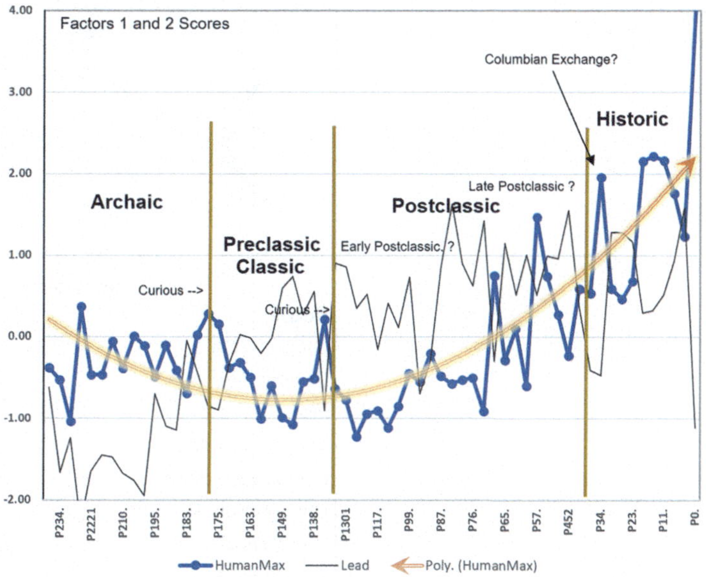

Fig. 11.16 Factor 1 (varimax rotation) and plot of factors 1 and 2 scores from 2008 Panlao core ICP determinations. (FactorAnalyses.xlsx)

coming into the delta but this increased with time and spiked in the Late Preclassic-Classic Period indicating erosion.

That phosphorus, potassium, and magnesium were also coming into the delta indicates the conundrum that the hydrological and soil conservation engineers faced. Desalinization was also removing critical elements for maintenance of the human bodies. Scherer et al. (2007) have documented that fish were largely essential for healthy life in the interior of the peninsula. At Caracol, there is evidence that the upper class was importing whole, live, marine fish for their meals (A. Chase, personal communication, 2016 email). Without fish, porotic hyperostosis develops, a condition of the bones, so easily spotted in mortuary populations. Tiesler et al. (2017, p. 113) (hyperostosis) reports that 54.5% of inland Yucatecan population from the Classic Period have evidence of hyperostosis, while only 29.9% from the coast show symptoms. This analysis suggests that fish were undoubtedly being brought up from the coast either from Itzamkanac and/or Champoton to replace the critical elements being lost to the desalinization process. Faust reports observations and interview data concerning the continuing weekly importation of fish on ice brought by truck to the inland town of Pich from the Campeche coast and grandparents' memories of small, salted fish being brought by oxcart before the highway was built in the 1970s. The importation of fish from Champoton to interior chiclero stations (Classical cities) was reported during the twentieth century by Tomas Arnabar Gunam to WJF (Folan et al. 2003) that "En las épocas de la chiclerias, en La Montaña se llevaba Pámpano fileteado en capas dentro de barriles sin secarlo pero

Fig. 11.17 Pompano nutrients

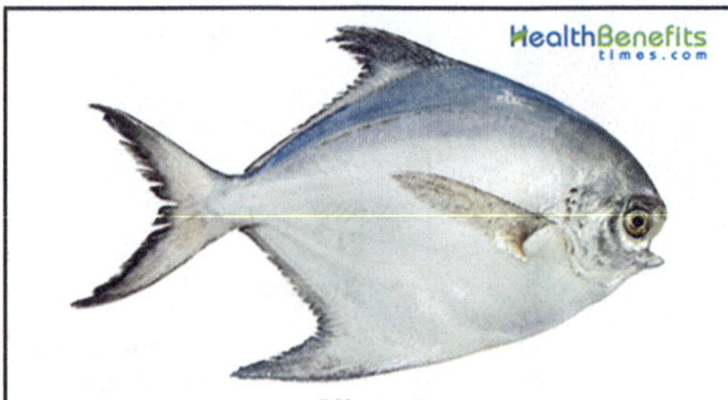

Minerals

Nutrient	Amount	DV
Calcium, Ca	22.00 mg	2%
Copper, Cu	0.038 mg	2%
Iron, Fe	0.60 mg	3%
Magnesium, Mg	27.00 mg	7%
Manganese, Mn	0.013 mg	1%
Phosphorus, P	195.00 mg	20%
Potassium, K	381.00 mg	8%
Selenium, Se	36.5 mcg	52%
Sodium, Na	65.00 mg	3%
Zinc, Zn	0.72 mg	5%

entre una y otra capa se ponían sal; tal vez como podían haberlo hecho durante tiempos prehispánicos utilizando ollas y/o pacas tal vez forradas con hojas grandes." (Translation: "In the times of the chiclerias in La Montaña, pámpano was filleted and placed in layers inside barrels without drying it but between one layer and another they put salt; perhaps as they could have done during pre-Hispanic times using pots and/or bales perhaps lined with large leaves.") As can be seen in Fig. 11.17, pámpano contains the necessary magnesium, potassium, and phosphorus (https://www.nutritionvalue.org/Fish%2C_raw%2C_florida%2C_pompano_nutritional_value.html). They also salted saw fish, bass, stingray, lisa, and charales (small minnows that were dried). Salting would have served the double purpose to preserving the fish and replacing the sodium carried away from the interior by the desalinization process.

A more likely source of fish in Precolumbian times might have been the Campeche Bank red snapper (Pargo) because the pámpano requires sailing further out to sea. Comparing the critical elements of the two species shows that red snapper would have been a better cycling/recycling vehicle than pámpano.

Sodium is also critical: notice that in the recycling processes, it is being captured and purified from calcium sulfate at each step down the watershed landscaping. Figure 11.18 systematizes this process into a flow chart that will enable basic calculations of the parameters of the Candelaria watershed system working from the quantities of critical elements necessary for human construction and maintenance. So, to finally answer the question, what are the critical elements (variables) that will provide insights into sustainability in the elevated interior? Form the point of view of the

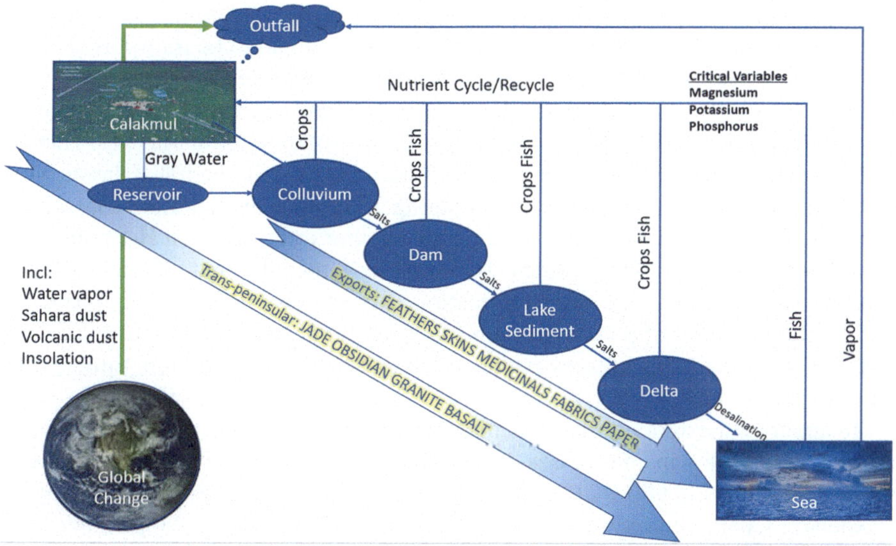

Fig. 11.18 Flowchart of parallel processes in the Candelaria watershed as suggested in Fig. 11.2: agriculture, water/salt/sediment management, nutrient cycling, and recycling by crops and fish

geochemical output analysis, magnesium, potassium, and phosphorus are the indicators of the health of the watershed and would have been cycled and recycled upstream. Calcium, sulfur, and sodium, while clearly important to human well-being, are too entangled in the geochemical web of relationships between the geology of the watershed (the Chicxulub problem) and humans to serve an independent purpose. Metals, especially lead, indicated whether the soil conservation process is healthy or not.

As is seemingly always the case in archaeology, testing alternative models and cross-linking arguments in convincing manners depends on additional information. In this case, we need more refined dating of the critical time periods of the Panlao core to determine cause-and-effect relationships. We cannot tell from the current context, for example, if a lessening of trade or the ninth century drought came first or if they came simultaneously. There are suggestions by indirect evidence that either is a possibility. Also, we need more cores in the Candelaria delta to ensure that we are seeing the integrated behavior of the drainage system rather than local channel and/or base level changes. A project is underway under the directorship of Nuria Torrescano-Valle to obtain better resolved dating of the Candelaria delta sequence.

Discussion

An integrated, holistic Candelaria-Champoton model can be formulated around the following narrative summary of the above. Outputs from the flow structure of the Candelaria provide insights into Maya watershed-level landscapes, ancient and modern. Outputs are collected in foot slopes, dams, and delta deposits.

Calakmul was not agriculturally or hydrologically self-sustaining, so rainwater had to be collected within the city, and additional food had to be transported up to the city. Its social and agricultural network probably extended up the El Laberinto Bajo 12 km to the east and 60 km down to the sea through the Candelaria and/or Champoton watersheds to the west. Among the cities of the central Maya Lowlands, Calakmul and Tikal fell into contentious military postures in the Late Classic with Calakmul, in league with Caracol, dominating Tikal until the late Late Classic. El Mirador, once the central feature of trans-peninsular trade between Honduras and Itzamkanac, fell into the military buffer across the elevated districts of the Mesoplano being the zone of confrontation. It was downgraded to a pilgrimage site, while the center of power was moved into the karst at Calakmul. El Laberinto Bajo becomes the focus of trans-peninsular travel with a parallel route through the San Pedro system to Tikal.

To augment the agricultural capacities of Calakmul, new lands for planting were created on old salt pans. Processes differ in the Lake District to the south and Karst to the north. Around 900 BCE agriculture appears to have spread across the upper Candelaria-Champoton Systems and then retreated shortly not to reappear until the Late Preclassic. This is obviously in strong contrast to what happened in the Nakbe area where the roots of Maya culture flourished. Since this is a period of worldwide cooling (the Greek Dark Age), which amounts to drought in the Maya Lowlands, it seems likely that agriculturists were cast back to the moister conditions of the Lake District. As in the Mediterranean, the late first millennium BCE witnessed a revitalization most strongly expressed at El Mirador. El Mirador represents an early stage of development of hydrological engineering in the form of damming of drainages. This was followed in the Classic by advanced hydrological engineering capable of supporting large cities somewhat sustainably on karstic promontories such as those occupied by Calakmul and Tikal. This process would have required substantial rainfall both to water the system and flush out harmful concentrations of salts. The latter may account for the unusually precipitous dependence of the Central Lowlands on substantial rainfall.

Pollen and geochemistry of the Candelaria delta show that in the Candelaria watershed taken as a whole, tropical forests declined beginning late in the Archaic and remained in a diminished state until the end of the Early Postclassic. Through this time there was a decline in calcium and sulfur deposition indicating soil conservation efforts were evolving upstream. Since disturbance taxa do not decline in the Postclassic, it seems likely that agriculture continued somewhere in the watershed through the whole of the post Archaic occupation. Where it was practiced appears to have shifted from one area of the watershed to others. The primary shift in this period was from the elevated interior to the coastal plain sites such as Itzamkanak and Edzna where it continued on through the Postclassic. There may also have been shifts between branches during the Classic, including but not limited to the shift from El Mirador to Calakmul.

Since Calakmul lacked the soil resources to supply itself with food by rainfed horticulture within a day's walk, it seems likely that some part of the population of the city dispersed to the urban hinterlands during the rainy season to raise milpa crops. Dispersed, seasonal homesteads may account in part for the enormous footprint of Calakmul. A survey along the road north from Calakmul indicates that the whole landscape was occupied, terraced, walled, etc. There were undoubtedly intense house gardens as well, both in the city and in the homesteads.

In the middle Candelaria reach, lillies were introduced into the Lake Candelaria before 4000 years ago suggesting that water quality and evaporation control were being monitored and practiced before that time. By Classical times the deposition of sediments was sufficient to turn the floodplain into agricultural fields and canals into fish nurseries. In the Candelaria delta the Panlao core indicates that in spite of erosion upstream, the sediments input during the Classic Period were limited. Apparently intentionally, and perhaps dating from as early as 7000 years ago, the inhabitants of the Candelaria system were welcoming sediments in Lake Candelaria and ultimately used them to support a substantial agricultural enterprise. The scale of this enterprise may help explain how agriculturally impoverished Calakmul could sustain itself. Such an enterprise would have required reliable transportation of food good up the system. A series of dams, lakes, and canals would have been necessary to catch vital resources such as potassium, phosphorus, magnesium, and sodium in crops and fish and return them to the city. Reserved of these elements would also have been replenished by harvesting fish from the Laguna de Terminus and the Campeche Bank off Champoton.

Conclusions

The geophysical and pollen determinations from the Candelaria watershed provide insights into several issues of interest to the culture history of the western Central Maya Lowlands region. Since the Candelaria watershed spans on the divide between the Lake District and the elevated interior karst, it shows that the ability to cope on the karst in terms of agriculture began in the Middle Preclassic around 900 BCE. Hydrological engineering providing for sustainable urban areas developed in the Preclassic and probably explains the timing of urban migration to the elevated interior for both Calakmul and Tikal. It might also explain in its absence why an earlier attempt to establish agriculture and urbanism in the karst failed. There are a number of reasons the Lowland Maya might have wanted to live in the margins of the karst. Living in elevated circumstances would have avoided the health issues inherent in moist lowlands where tropical biology would have quickly festered among dense urban population into disease. As long as there were sufficient wet season rainfall, the elevated cities were cleansed annually. On the other hand, it also allowed cities both to increase the information densities and sustainabilities important for information-based commerce and to command important trade routes across the peninsula.

Across the panoply of developments, both through time and across the branches of the Candelaria and Champoton watersheds, geochemistry shows that the road to urbanization was not smooth but rather travelled in cyclical patterns that reflected volcanic and global climate disruptions. The disruptive events are marked by spikes in metals appearing in the Candelaria delta indicating that somewhere upstream, some or all of the soil conservationists and hydrologists have lost control of their complex systems of water control and erosion retardation. This system bore its own liabilities. Along with refreshing the landscape by desalinization, it removed some of the elements critical to creation and operation of the human body. These elements (magnesium, potassium, phosphate, and sodium) had to be cycled and recycled up the drainage to maintain the elevated cities in the interior. The overall indications from the delta core are that from the Archaic to Early Postclassic, these elements were being depleted or slightly in balance periodically, thanks to volcanic and dust inputs and/or recycling.

Evidence from the middle Candelaria indicates that the shoals of the river were blocked by log dams about 7000 years ago as the adjacent uplands were cleared. Accumulation of sediments eventually turned a large artificial lake of lagoons into intermittent farmlands and fish farms whose use continued into the historic period. This highly productive nexus was, or eventually became, the headquarters of the Chontal Maya who after 3000 years ago facilitated movement of goods across the peninsula, around it, and between the Mexican Highlands and the Caribbean archipelago. Lake Candelaria would have been a major recycling point for key nutrients coming down the Candelaria through crops and fish. Similar recycling may have been processed through the Champoton-Desempeño system with Edzna being the Itzamkanac-equivalent. Also, fish and salt are known historically to have been cycled inland from the sea. Xicalango/Carmen downstream from Itzamkanac on the Laguna de Términos may have served similar purposes through the Candelaria. Today, coastal fisheries still supply inland towns with fish.

Of special importance to human operation is water, though in a managed environment so as to not foster tropical pathogens and to store it in sufficient purity and quantities to span the dry seasons. Ultimately this key variable proved to be the undoing of the elevated cities when extended droughts in the ninth century brought the horns of the pathogenic-hydrologic dilemma crashing together.

Acknowledgments Gratitude by the authors is greatly deserved by the researchers who participated in the research reported in this article. Of special note are the many employees of the Universidad Autónoma de Campeche, Centro de Investigaciones Históricas y Sociales, who labored tirelessly over many years to collect and prepare the data on which much of this review is based.

References

Adams REW (2005) *Prehistoric mesoamerica*, 3rd edn. University of Oklahoma Press, Norman

Avila Chi R (2009) Andando bajo el monte, picando chicle, cazando lagartos, tumbando palos y haciendo milpa. Una autobiografía. CONACULTA, México

Barnes EB, González Heredia R, Folan WJ et al (2009) Oxpemul, su Corte Real Fortificada y el Patrón de Asentimientodentro de la Cuenca de Calakmul, Campeche, Mexico. Presented at the Los Investigadores de la Cultura Maya XVII, Universidad Autónoma de Campeche

Beach T, Luzzadder-Beach S, Dunning NP et al (2008) Human and natural impacts on fluvial and karst depressions of the Maya Lowlands. Geomorphology 101(1–2):308–331. https://doi.org/10.1016/j.geomorph.2008.05.019

Beach T, Luzzadder-Beach S, Dunning N et al (2016) Climatic changes and collapses in Maya history. PAGES Mag 24(2):66–67

Brant A, Piggott D, Petal R et al (2010) Integrated watershed management plan for Rio Candelaria. Retrieved from https://www.slideshare.net/amberbrant/rio-candelaria-integrated-watershed-management-plan

Brewer JL, Carr C, Dunning NP et al (2017) Employing airborne lidar and archaeological testing to determine the role of small depressions in water management at the ancient Maya site of Yaxnohcah, Campeche, Mexico. J Archaeol Sci Rep 13:291–302. https://doi.org/10.1016/j.jasrep.2017.03.044

Chew SC (2007) The recurring dark ages: ecological stress, climate changes, and system transformation. Altimira Press, Lanham

Crumley CL (1994) Historical ecology: cultural knowledge and changing landscapes, School of American Research advanced seminar series. School of American Research Press, Sante Fe

Dahlin BH, Foss JE, Chambers ME (1980) Project Acalches: reconstructing the natural and cultural history of a seasonal swamp at El Mirador Guatemala; preliminary results. In: Matheny R (ed) El Mirador, Petén Guatemala: an interim report. New World Archaeological Foundation, Brigham Young University, Provo

Deevey ES, Rice DS, Rice PM et al (1979) Mayan urbanism: impact on a tropical karst environment. Science 206:298–305

Domínguez-Carrasco MR (1991) El sistema hidráulico de Calakmul, Campeche. Información 15:51–83

Domínguez-Carrasco MR, Folan WJ (1996) Calakmul, Mexico: Aguadas, Bajos, Precipitatión y Asentamiento en el Petén Campechano. In: Laporte J, Escobedo H (eds) Proceedings of the IX Simposio de Investegaciones Arqueologicas en Guatemala. Museo Nacional de Arqueologia y Etnologia, Guatemala, pp 171–193

Domínguez-Carrasco MR, Gunn JD, Folan WJ (1998) Calakmul, Campeche: Sus Aéreas de Actividades Ceremoniales, Cívicas y Domesticas Observadas de un Análisis de Sus Artefactos de Piedra. In: Los Investigadores de la Cultural Maya 5, vol II, pp 526–540

Domínguez-Carrasco MR, Folan WJ, Lopez M et al (2000) The state of Calakmul, Campeche, and its regional concept. In: Tiesler V, Cobos R, Robertson MG (eds) La organizacion social entre los Mayas. Memoria de la Tercero Mesa Redondo de Palenque, Palenque

Draper NR (1981). Applied Regression Analysis. New York: John Wiley and Sons

Dull RA, Southon JR, Sheets P (2001) Volcanism, ecology and culture: a reassessment of the Volcán Ilopango TBJ eruption in the southern Maya realm. Lat Am Antiq 12:25–44

Dunning NP, Beach T (2010) Farms and Forests: Spatial and Temporal Perspectives on Ancient Maya Landscapes. In: Martini IP, Chesworth W (eds) Landscapes and societies. Springer Science+Business Media B.V., Dordrecht/New York, pp 369–389

Dunning NP, Beach TP, Luzzadder-Beach S (2012) Kax and kol: collapse and resilience in lowland Maya civilization. Proc Natl Acad Sci 109:3652–3657

Faust BB (1988) Cosmology and changing technologies of the Campeche-Maya. Ph.D. dissertation, Syracuse University, Syracuse

Faust BB (1998) Mexican rural development and the plumed serpent: technology and Maya cosmology in the tropical forest of Campeche, Mexico. Bergin & Garvey, Westport

Faust BB (2010) El desarrollo rural en México y la serpiente emplumada: tecnología y cosmología maya en la selva tropical de Campeche. Fondo de Cultura Económica, México

Faust BB, Quintana P, Bautista F, Gunn JD (2010) Effect of Agricultural Burning on Chemicals and Mineralogical soil properties in Maya Milpas: A Key Ingredient in Maya Lowland Sustainability. Presented at the Society for Applied Anthropology

Faust BB, Anaya-Hernández A, Geovannini-Acuña H (2012) Reclaiming the past to respond to climate change: Mayan farmers and the archaeology of ancient agricultural techniques. In: Castro

AP, Taylor D, Brokensha DW (eds) Climate change and threatened communities: vulnerability, capacity, and action. Practical Action Publications, Rugby, Warwickshire, pp 139–151

Faust, B. B., Quintana, P., Bautista, F., & Gunn, J. D. (2010). Effect of Agricultural Burning on Chemicals and Mineralogical soil properties in Maya Milpas: A Key Ingredient in Maya Lowland Sustainability. Presented at the Society for Applied Anthropology

Fialko V (2005) Diez Años de investigaciones arqueológicas en la cuenca del rio Holmul, región noreste de Peten, vol I. Ministerio de Cultura y Deportes, Instituto de Antropología e Historia Asociación Tikal, Guatemala City, pp 253–268

Fletcher LA, Gann JA (1994) Analisis Grafico de Patrones de Asentamiento. El Caso Calakmul. In: Folan Higgins WJ (ed) Campeche Maya colonial. Universidad Autónoma de Campeche, Campeche, México, pp 85–121

Folan WJ (1983) La Prehistoria e Historia de los Mayas Vistas desde los Puntos de Vista de su Paleoclimatologia, Politica y Soceoeconomica. In: Revista Mexicana de Estudios Antropologicos, Tomo XXIX. Sociedad Mexicana de Antropología, Mexico, pp 243–256

Folan WJ, Gunn JD, Eaton JD et al (1983) Paleoclimatological patterning in southern Mesoamerica. J Field Archaeol 10:453–468

Folan WJ, May-Hau J, Couoh-Muñoz R et al (1990a) Calakmul, Campeche, México: Su Mapa. Una introducción. Centro de Investigaciones Históricas y Sociales, Universidad Autónoma de Campeche, Campeche

Folan WJ, May-Hau J, González HR et al (1990b) Mapa de Calakmul. Centro de Investigaciones Históricas y Sociales, Universidad Autonoma de Campeche, Mexico, Campeche

Folan WJ, García-Ortega JM, Sánchez-González MC (1992) Programa de Manejo Reserva de la Biosfera de Calakmul, Campeche, 4 vols. Centro de Investigaciones Históricas y Sociales Universidad Autónoma de Campeche, Campeche, México

Folan WJ, Marcus J, Pincemin S et al (1995) Calakmul: new data from an ancient Maya capital in Campeche, Mexico. Lat Am Antiq 6:310–334

Folan WJ, Faust B, Lutz W et al (2000) Social and environmental factors in the classic Maya collapse. In: Lutz W, Prieto L, Sanderson W (eds) Population, development, and environment on the Yucatan Peninsula: from ancient Maya to 2030. International Institute for Applied Systems Analysis, Laxenburg, pp 3–34

Folan WJ, Fletcher LA, May-Hau J et al (2001) Las Ruinas de Calakmul, Campeche, México: Un Lugar Central y su Paisaje Cultural. Universidad Autónoma de Campeche, Centro De Investigaciones Históricas y Sociales, Campeche

Folan WJ, Morales-López A, Honzalez R et al (2003) Champoton Campeche: su presencia en el desarrollo cultural del Golfo de México y su corredor ecoarqueologico. In: Los Investigadores de la Cultura Maya 11, Tomo II, Universidad Autónoma de Campeche, Campeche, pp 65–71

Folan WJ, Domínguez-Carrasco MR, Gunn JD et al (2012) Calakmul: Poder, Perseverancia y Persistencia. Proceedings. Springer

Folan WJ, Gates G, Gunn JD, Volta B (2016) Capítulo 1, Geología, Hidrología y Clima, En el Peten Campechano. In: Folan WJ, Poot-Franco P (eds) El Peten Campechano. Universidad Autónoma de Campeche, Campeche, México

Geovannini-Acuña H (2008) Rain harvesting in the rainforest: the ancient Maya agricultural landscape of Calakmul, Campeche, Mexico, vol 1879. Archaeo Press, Oxford

Grube N, Delvendahl K, Seefeld N et al (2013) Under the rule of the snake kings: Uxul in the 7th and 8th centuries. Estudios de Cultura Maya 40(0):11–49

Gunn JD (1991) Influences of various forcing variables on global energy balance during the period of intensive instrumental observation (1958–1987) and their implications for paleoclimate. Clim Chang 19:393–420

Gunn J (1994) Global climate and regional biocultural diversity. In: Crumley CL (ed) Historical ecology. School of American Research, Santa Fe, pp 67–97

Gunn JD (2000) A.D. 536 and its 300-year aftermath. In: Gunn J (ed) The years without summer: tracing A.D. 536 and its aftermath, vol 872. Archaeopress, Oxford, pp 5–20

Gunn, J. D. (in preparation). Three Tropical Thoughts. In J. T. Lermon, L. J. Lucero, & F. Valez (Eds.), Path to Sustainability:The Past and Future Role of Water Management. University of Colorado Press

Gunn JD, Adams REW (1981) Climatic change, culture, and civilization in North America. World Archaeol 13:85–100

Gunn JD, Folan WJ (1996) Tres Ríos: Una Superficie de Impacto Climático Global Interregional para las Tierras Bajas de los Mayas del Suroeste (La Cuenca de los Ríos Candelaria, Usumacinta y Champoton). In: Los Investigadores de la Cultura Maya 5. Universidad Autónoma de Campeche, Campeche, pp 57–79

Gunn JD, Folan WJ (2000) Three rivers: subregional variations in earth system impacts in the southwestern Maya lowlands (Candelaria, Usumacinta, and Champoton Watersheds). In: Mcintosh R, Tainter J, Mcintosh S (eds) The way the wind blows: climate, history, and human action. Columbia University Press, New York, pp 223–270

Gunn JD, Folan WJ (2018) LA TRAMPA SEDUCTORA (ATRAYENTE): Las sinergias de simulación y análisis de componentes principales en el estudio de redes sociales de complejos sistemas adaptativos de las Tierras Bajas Mayas. Presented at the Los Investigadores de la Cultura Maya, vol XXVII, Universidad Autonoma de Campeche

Gunn, JD, Day, JW, Folan, WJ, & Moerschbaecher M (2019) Geo-Cultural Time: Advancing Human Societal Complexity within Worldwide Constraint Bottlenecks: A Chronological/ Helical Approach to Understanding Human-Planetary Interactions. BioPhysical Economics and Resource Quality, 4, 1–19

Gunn J, Folan WJ, Robichaux HR (1994) Un Análisis Informativo sobre la Descarga del Sistema del Río Candelaria en Campeche, México: Reflexiones acerca de los Paleoclimas que afectaron a los Antiguos Sistemas Mayas en los Sitios de Calakmul y el Mirador. In: Folan Higgins WJ (ed) Campeche Maya colonial. Universidad Autónoma de Campeche, Campeche, México, pp 174–197

Gunn J, Folan WJ, Robichaux HR (1995) A landscape analysis of the Candelaria watershed in Mexico: insights into paleoclimates affecting upland horticulture in the southern Yucatan Peninsula semi-karst. Geoarchaeology 10:3–42. https://doiorg.libproxy.uncg.edu/10.1002/gea.3340100103

Gunn JD, Foss JE, Folan WJ et al (2002) Bajo sediments and the hydraulic system of Calakmul, Campeche, Mexico. Anc Mesoam 13:297–315

Gunn JD, Folan WJ, Day JW et al (2009a) Los Mayas y el Mar: Un Estudio Sobre la Variación de Estroncio in la Laguna de Panlao, Campeche. Presented at the Los Investigadores de la Cultura Maya, vol XVIII, Universidad Autonoma de Campeche

Gunn JD, Folan WJ, Domínguez-Carrasco MR et al (2009b) Explicando la Sustentabilidad de Calakmul, Campeche: Eslabones Interiores en el Sistema de Energía del Estado Regional de Calakmul. In: Encuentro Internacional de Los Investigadores de la Cultura Maya, vol XVIII, Tomo 1. Universidad Autónoma de Campeche, Campeche, pp 13–40

Gunn JD, Folan WJ, Day JW et al (2012) Laguna de Terminos/Rio Candelaria delta core: conditions of sustainable urban occupation in the interior of the Yucatan Peninsula. Estudios de Cultura Maya, vol XXXIX, Universidad Nacional Autónoma de México (UNAM), pp 67–97

Gunn JD, Folan WJ, Domínguez-Carrasco MR et al (2013) Calakmul y el agua: La resiliencia y vulnerabilidad en las tierras bajas mayas occidentales. In: Los Investigadores de la Cultura Maya 21, November 2012, vol 21. Universidad Autónoma de Campeche, Campeche, Mexico, pp 307–323

Gunn JD, Day JW, Yáñez-Arancibia A et al (2014a) The Maya in global perspective: the dawn of complex societies, the beginning of the anthropocene, and the future of the earth system. Austin, TX, p 103

Gunn JD, Folan WJ, Isendahl C et al (2014b) Calakmul: agent risk and sustainability in the western Maya lowlands. In: Chase AF, Scarborough VL (eds) The resilience and vulnerability of ancient landscapes: transforming Maya archaeology through IHOPE, vol 24. American Anthropological Association, Toronto, pp 101–123

Gunn JD, Scarborough VL, Folan WJ (2017) A distribution analysis of the central Maya lowlands ecoinformation network: its rises, falls, and changes. Ecol Soc 22:20. https://doi.org/10.5751/ES-08931-220120

Hansen, R. D. (1998a). Continuity and Disjunction: The Pre-Classic Antecedents of Classic Maya Architecture. In S. D. Houston (Ed.), Function and Meaning in Classic Maya Architecture (pp. 49–122). Washington, D.C.: Dumbarton Oaks

Hansen RD (1998b) Incipient Maya wetland agriculture: definition of ancient systems and sustainable application in contemporary rainforest populations. MS on file, Institute of Geophysical and Planetary Physics, University of California, Los Angeles, and Foundation for Anthropological Research and Environmental Studies (FARES)

Hansen RD (2004) El Mirador, Guatemala. El Apogeo del Preclásico en el Area Maya. Arqueología Mexicana 11:28–33

Hansen RD, Bozarth S, Jacob J et al (2002) Climatic and environmental variability in the rise of Maya civilization: a preliminary perspective from the northern Peten. Anc Mesoam 13:273–295

Hibbard KA, Costanza R, Crumley C et al (2008) Developing an integrated history and future of people on Earth (IHOPE): research prospectus. IGBP Secretariat, Stockholm

Hoggarth JA, Restall M, Wood JW et al (2017) Drought and its demographic effects in the Maya lowlands. Curr Anthropol 58(1):82–113

Jacob JS (1995) Archaeological pedology in the Maya lowlands. In: Collins ME et al (eds) Pedological perspectives in archaeological research. Special Publication 44. Soil Science Society of America, Madison

Lansing JS (2003) Complex adaptive systems. Annu Rev Anthropol 32:183–204

Lentz DL, Dunning N, Scarborough VL et al (2014) Forests, fields, and the edge of sustainability at the ancient Maya city of Tikal. Proc Natl Acad Sci 111:18513–18518

Lohse JC (2010) Archaic origins of the lowland Maya. Lat Am Antiq 21:312–352

Lutz W, Prieto L, Sanderson W (2000) Population, development, and environment on the Yucatán Peninsula: from ancient Maya to 2030. International Institute for Applied Systems Analysis (IIASA), Laxenburg

Martin S, Grube N (2008) Chronicle of the Maya Kings and Queens: deciphering the dynasties of the ancient Maya, 2nd edn. Thames & Hudson, London

Matheny RT, Matheny DG (n.d.) Introduction to investigations at El Mirador, Petén, Guatemala. Salt Lake City: Papers of the New World Archaeological Foundation: Mirador series, Brigham Young University

Matheny RT, Gurr DL, Forsyth DW, Hauck R (1983) Investigations at Edzna, Campeche, Mexico: the hydraulic system. Papers of the New World Archaeology Foundation, No. 46, Brigham Young University Press, Provo, UT

May-Hau J, Couoh-Muñoz R, González-Heredia R et al (1990) Mapa de Calakmul, Campeche, México. CIHS, Universidad Autónoma de Campeche, Campeche

Morales-López A, González-Heredia R et al (2017) La Arqueología, Geología, Hidrología y Florística del Petén Campechano. Universidad Autónoma de Campeche, Campeche

Nooren CAM, Hoek WZ, Tebbens LA, Del Pozzo ALM (2009) Tephrochronological evidence for the late Holocene eruption history of El Chichón Volcano, México. Geofísica Internacional 48:97–112

Perry E, Velazquez-Oliman G, Wagner N (2011) Preliminary investigation of groundwater and surface water geochemistry in Campeche and Southern Quintana Roo. In: Oswald Spring U (ed) Water resources in Mexico. Springer, Heidelberg, pp 87–97

Pope KO, Pohl MED, Jones JG et al (2001) Origin and environmental setting of ancient agriculture in the lowlands of Mesoamerica. Science 292:1370–1373

Puleston DE (1978) Terracing, raised fields, and tree cropping in the Maya lowlands: a new perspective on the geography of power. In: Harrison PD, Turner BL II (eds) Pre-Hispanic Maya agriculture. University of New Mexico Press, Albuquerque, pp 225–245

Purdue L (2018) Operación Iv.7a: Agricultura Y Subsistencia. In: Nondédéo P, Michelet D, Begely J, Garrido L (eds) Proyecto Petén-Norte Naachtun 2015–2018: Informe De La Octava

Temporada De Campo 2017. Instituto de Antropología e Historia de Guatemala, Guatemala, pp 348–349
Reents-Budet D, Boucher Le Landais S, Palomo Carrillo Y et al (2011) Cerámica del Estilo Códice: nuevos datos de producción y patrones de distribución. In: Arroyo B, Paiz-Aragón L, Linares-Palma A, Arroyave AL (eds) XXIV Simposio de Investigaciones Arqueológicas en Guatemala. Ministerio de Cultura y Deportes, Instituto de Antropología e Historia, y Asociación Tikal, Guatemala, pp 841–856
Reese-Taylor K, Anaya-Hernandez A, Walker D et al (2012) Resultados y Interpretaciones Preliminaries de la Primera Temporada de investigaciones en Yaxnohcah, Campeche, Mexico. In: XXI Encuentro Los Investigadores De La Cultura Maya. Universidad Autónoma de Campeche, Campeche
Robadue D, Calderon R, Oczkowski A, et al (2004) Characterization of the Region of the Laguna de Términos Campeche, Mexico: Level One Profile, Fresh Water Inflow to Estuaries Project. The Nature Conservancy and the University of Rhode Island
Ruddiman WF (2010) Plows, plagues, and petroleum: how humans took control of climate, 2nd edn. Princeton University Press, Princeton
Scarborough VL (1998) Ecology and ritual: water management and the Maya. Lat Am Antiq 9:135–159
Scherer AK, Wright LE, Yoder CJ (2007) Bioarchaeological evidence for social and temporal differences in diet at Piedras Negras, Guatemala. Lat Am Antiq *18*(*1*):85–104
Scholes FV, Roys RL (1968) The Maya Chontal Indians of Acalan-Texchell: a contribution to the history and ethnography of the Yucatan Peninsula. University of Oklahoma Press, Norman
Sheets P (1983) Archaeology and volcanism in Central America: the Zapotitan Valley of El Salvador. University of Texas Press, Austin
Siemens AH (1989) Tierra Configurada: Investigaciones de los Vestigios de Agricultura Precolumbina en Tierras Inundables Costeras desde el Norte de Veracruz Hasta Belice. Consejo Nacional para la Cultura y las Artes, Mexico D.F.
Siemens AH (2004) Modelling the tropical wetland landscape and adaptations. Agric Human Values *21*:243–254
Siemens AH (2009) Un Río en Tierra Maya (Vol. XXVIII). Presented at the Los Investigadores de la Cultura Maya, Universidad Autónoma de Campeche
Siemens AH, Puleston DE (1972) Ridged fields and associated features in southern Campeche: new perspectives on the lowland Maya. Am Antiq *37*:228–239
Siemens AH, Soler-Graham JA, Hebda R et al (2002) "Dams" on the Candelaria. Anc Mesoam *13*:115–123
Šprajc I, Grube N (2008) Arqueología del Sureste de Campeche: Una Síntesis. In: Sprajc I (ed) Reconocimiento arqueológico en el sureste del estado de Campeche, México: 1996–2005, BAR International Series, vol I, 1742. Archaeopress, Oxford, pp 263–275
Syvitski JPM, Vorosmarty CJ, Kettner AJ et al (2005) Impact of humans on the flux of terrestrial sediment to the global coastal ocean. Science *308*:376–380. https://doi.org/10.1126/science.1109454
Tankersley KB, Scarborough VL, Dunning N et al (2011) Evidence for volcanic ash fall in the Maya lowlands from a reservoir at Tikal, Guatemala. J Archaeol Sci *38*:2925–2938. https://doi.org/10.1016/j.jas.2011.05.025
Tebbens LA, van der Borg K (2001) Middle-late Holocene floodbasin record of the Usumacinta-Grijalva Delta (SE Mexico): deltaic evolution and El Chichón eruptive history. (ms. in possession of the authors)
Tiesler V, Cucina A, Stanton TW et al (2017) Before Kukulkán: bioarchaeology of Maya life, death, and identity at classic period Yaxuná. University of Arizona Press, Tucson
Toohey M, Krüger K, Sigl M et al (2016) Climatic and societal impacts of a volcanic double event at the dawn of the Middle Ages. Clim Change *136*:401–412. https://doi.org/10.1007/s10584-016-1648-7
Torrescano-Valle N, Islebe GA (2015) Holocene paleoecology, climate history and human influence in the southwestern Yucatan Peninsula. Rev Palaeobot Palynol *217*:1–8

Torrescano-Valle N, Islebe GA, Gunn JD, Folan WJ (2009) Palynological and historical evidence of environmental change from Lake Silvituc (Yucatan Peninsula) during the Late Holocene. Presented at the IIth International Paleolimnology Symposium, p 146

Torrescano-Valle N, Gunn JD, Folan WJ (2012) Historia Climática y Ambiental del Río Candelaria, Laguna Panlao 4000 Años Antes de Cristo. In: Cobos R (ed) Arqueología de La Costa de Campeche: La Época Prehispánica. Ediciones de la Universidad Autónoma de Yucatán, Mérida, Yucatán, pp 49–63

Torrescano-Valle N, Mas JF, Folan-Higgins WJ et al (2016) Agricultural and climatic history from Calakmul empire: a 3600-year record. Presented at The XIV International Palynological Congress and X International Organisation of Palaeobotany Conference, 23–28 October 2016

Vargas PE (2001) Itzamkanac y Acalan. Tiempos de Crisis Anticipando el Futuro. Instituto de Investigaciones Antropológicas, UNAM, Mexico, México

Vargas-Pacheco E (2012) El Rio Candelaria, Rio de Encuentros: Ofrendas, Diques, Canales, Campos Levantados. In: XXI Encuentro Los Investigadores De La Cultura Maya. Universidad Autónoma de Campeche, Campeche

Volta B, Gunn JD (2012) Análisis de costo mínimo de posibles rutas de intercambio transpeninsulares en el Petén campechano. Paper Presented at the 54th International Congress of Americanists, Vienna, Austria, 15–20 July 2012

Volta B, Gunn JD (2016) The political geography of long-distance trade in the Maya lowlands: comparing proxies for power structure and exchange networks. Paper Presented at the Abstracts of the Society for American Archaeology 81st Annual Meeting, Orlando, p 464

Wahl D, Byrne R, Schreiner T et al (2006) Holocene vegetation change in the northern Peten and its implications for Maya prehistory. Quat Res *65*:380–389

Zetina-Gutiérrez M, Faust BB (2011) Cuántas casas cada familia? Un reto para la arqueología maya de la agricultura nómada. In: Estudios de La Cultura Maya XXXIII. Instituto de Investigaciones Filológicas UNAM, Mexico, pp 97–120

Chapter 12
Lidar at El Pilar: Understanding Vegetation Above and Discovering the Ground Features Below in the Maya Forest

Anabel Ford and Sherman W. Horn III

Abstract Lidar data from El Pilar shows great potential for understanding the ancient and contemporary Maya forest landscape. Exploring these rich three-dimensional data with ground visualization strategies using Geographic Information Systems (GIS), our field validation strategy integrates the twenty-first-century tools Lidar, Global Positioning Systems (GPS), and GIS with time-tested methods of field observation and assessment of surface features and vegetation. While there is no doubt Lidar is a stimulating addition to the geographical and archaeological tool kit, we recognize it is essential to understand the sources of features our visualizations reveal. Our survey protocol evaluates human impacts on the forest environment by identifying and mapping ancient cultural features, recording basic characteristics of vegetation, and deriving information to extrapolate to the expanding database of Lidar coverage in the Maya Lowlands. Based on emerging results supporting the viability of the milpa-forest garden land-use cycle at the regional and local scales, we hypothesize the Maya created land-use strategies that can be modeled and tested at the site scale at El Pilar.

Keywords Lidar · Maya settlement · Maya forest

A. Ford (✉)
ISBER/MesoAmerican Research Center, University of California, Santa Barbara, CA, USA
e-mail: anabel.ford@ucsb.edu

S. W. Horn III
Grand Valley State University, Allendale, MI, USA
e-mail: hornsher@gvsu.edu

N. Torrescano-Valle et al. (eds.), *The Holocene and Anthropocene Environmental History of Mexico*, https://doi.org/10.1007/978-3-030-31719-5_12

Introduction

The Maya Forest Landscape

Interpretations of preindustrial land use depend on archaeological survey to locate and document settlements. Yet in the case of the Maya area, where the landscape itself was domesticated (Ford and Nigh 2015; Ford in press), land-use studies must also include the forest to capture human impacts on the environment through time. Archaeologists view the tropical forest as a challenge to understanding the ancient Maya, impeding progress toward mapping and identifying ancient settlement remains, when it should be seen as an historical archive of human adaptation and ingenuity. Most major Maya sites have been identified (Fig. 12.1), yet El Pilar, the largest site in the Belize River area, was not recorded until 1983. Today, the El Pilar Archaeological Reserve for Maya Flora and Fauna is defined by contiguous boundaries incorporating 20 km^2 in Belize and Guatemala (Fig. 12.2; http://marc-ucsb.opendata.arcgis.com/). With new Lidar technology expanding our views of the Maya forest (e.g., Canuto et al. 2018; Chase et al. 2011; Ford et al. 2013; Ford 2014; Magnoni et al. 2016; Reese-Taylor et al. 2016), we have the opportunity to see above and below the canopy and perceive the Maya forest, along with archaeological remains, as part of the cultural landscape.

Explorers began to report lost cities hidden beneath the Maya forest canopy in the mid-nineteenth century (Stephens 1969), and these reports gained considerable traction in the Western imagination when Sylvanus[1] Morley led the Carnegie Institute of Washington's expeditions to the Maya Lowlands in the early twentieth century (Adams 1969). This work set the stage for scientific efforts to understand Maya settlement, such as Bullard's Northeastern Petén surveys (Bullard 1960) and Puleston's Tikal surveys (Puleston 1973, 1983), which created standards for subsequent settlement pattern research (e.g., Rice 1976, 1978; Ford 1981, 1986, 1990, 1991; Ford and Fedick 1992; Ford and Horn 2017). An understanding of settlement patterns from these pioneering studies provides insight into ancient Maya land use, but the importance of forest cover has largely been ignored.

Technological developments in the decades since these first studies have enhanced survey tools and allowed an unprecedented expansion of settlement pattern studies in the Maya Lowlands, but the need for field survey under the forest canopy persists. Today, Lidar has brought a new perspective on the Maya forest (Canuto et al. 2018; Chase et al. 2011, 2014, 2017; Ebert et al. 2016; Ford and Horn 2018; Hutson et al. 2016; Prufer et al. 2015; Reese-Taylor et al. 2016; Rosenswig et al. 2013; Yaeger et al. 2016, among others in the region), and coverage has expanded to include more than 3000 km^2 across the Maya heartland of the greater Petén (Fig. 12.3; Ford et al. 2018). We now have substantial geospatial datasets, stretching from the ground surface to the top of the canopy, embedded in Lidar point clouds that await continuing efforts in remote visualization and field validation for reliable extrapolations. Diverse projects are working with this regional coverage,

[1] Interestingly the Roman god of forests

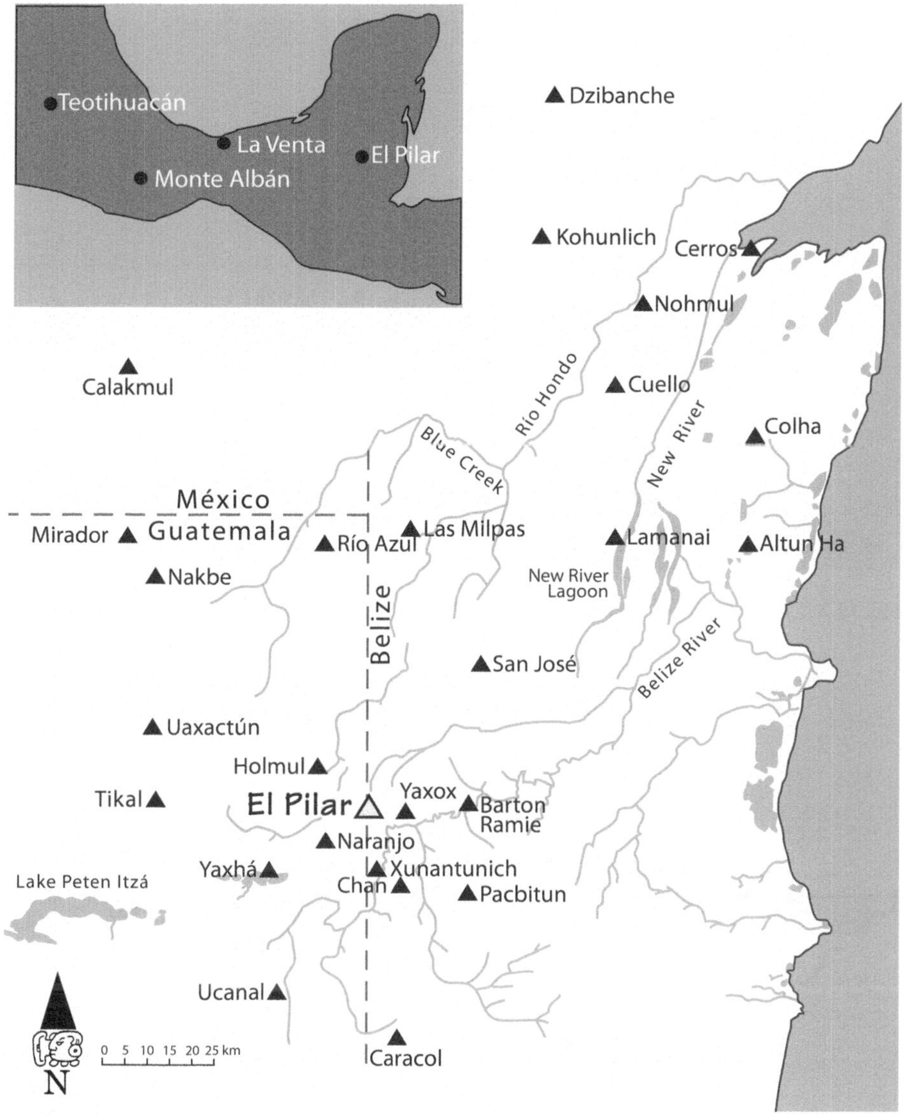

Fig. 12.1 Regional Maya area with El Pilar and major sites noted

and we see real value in this phenomenal resource being shared among all investigators of ancient settlement and land use. Recognition of features in visualizations and assessment of these features on the ground has remarkable potential to expand interpretation to the regional scale (Stular et al. 2012). Increasing data validation efforts (Ford and Horn 2018) point to an intricate landscape that is best understood by including studies of vegetation and investigating impacts, past and present, on the forested areas (Ford 2008; Prufer et al. 2015).

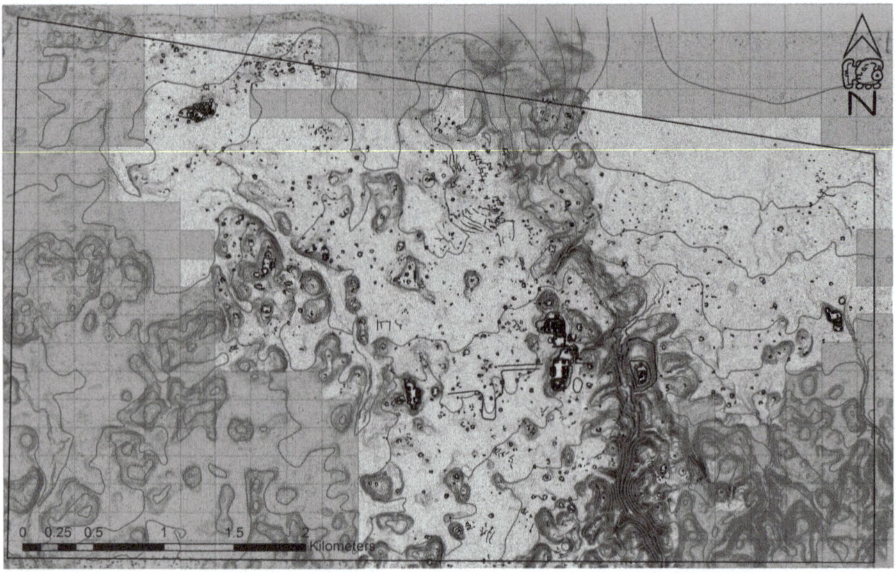

Fig. 12.2 Lidar coverage of the El Pilar Archaeological Reserve for Maya Flora and Fauna

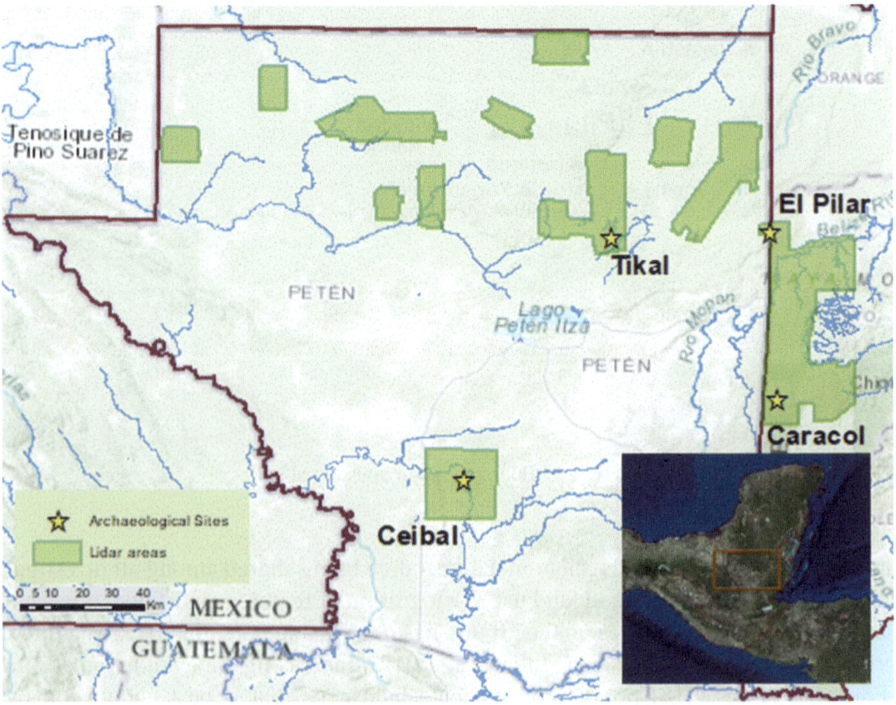

Fig. 12.3 Lidar coverage of the central Maya Lowlands of Guatemala and Belize

In this chapter, we consider Lidar data from El Pilar (Ford et al. 2013, 2015; Ford 2014; Ford and Horn 2018; Horn and Ford in press; Fedick et al. 2016), including the visualization strategy (Pingel et al. 2015) and survey protocol (Horn et al. 2019) we developed based on past field research. Our field validation strategy integrates the twenty-first-century tools Lidar, Global Positioning Systems (GPS), and Geographic Information Systems (GIS), with time-tested methods of field observation and assessment of surface features and vegetation. We find Lidar an exciting addition to the geographical and archaeological tool kit while recognizing the need to understand the source of features observed in visualizations. In the field, we identify and map cultural features and itemize vegetation characteristics that relate to human impacts on the forest environment. We hypothesize that the Maya created viable land-use strategies testable at the regional and local scales (Canuto et al. 2018; Ford and Clarke 2019), and our aim is to test a model of the milpa-forest garden cycle at the site scale at El Pilar. This begins with Lidar data and culminates with field validations.

Lidar Potentials and Pitfalls

The National Oceanic and Atmospheric Administration (NOAA) describes Lidar (Light Detection and Ranging) as a remote sensing method using aircraft fitted with a laser scanner and specialized, survey-grade GPS receivers. The scanners acquire pulsed laser returns that measure ranges or variable distances and pinpoint them in three-dimensional space (NOAA 2019). Originally developed with US government funding for NASA in the 1960s and brought to world attention when Apollo 15 transmitted images of the moon in 1971, this technology has continued to improve over five decades, with first efforts primarily concerned with aerospace development and research applications from the private sector and academy following. Dependent on the quality and capability of the laser scanners, by the 1990s, applied geospatial applications were developing results that promised feature extraction and forest characterizations. Resolution is the next frontier of development, and terrestrial and areal coverages have improved in the last decades (Britannica 2019). Lidar applications now provide dense 3D point clouds reflecting tops of trees, foliage and branches, trunks and roots, and ground surfaces with precision and resolution over large areas, and costs have decreased with more widespread use of the technology (Gaurav 2018).

In the last decade, Lidar has made a splash on the archaeological scene. Broad Lidar coverage in Europe at the beginning of the twenty-first century has been used to great advantage by archaeologists (e.g., Devereux et al. 2005). A little more than a decade later, the technology was applied in the Maya world (Chase et al. 2011), and it has been adopted for research in forested areas across the globe (e.g., Evans et al. 2013; Johnson and Ouimet 2014; Parton et al. 2018). Returning laser pulses that ultimately reach the land surface produce visualizations with astonishing topographic detail when processed with GIS software, as can be seen in the work of

Canuto et al. (2018). These applications demonstrate the importance of Lidar for archaeological prospection, yet more information is encoded in the point cloud than simply the ground surface. Vegetation density, height, and the shape of trees can be detected. Attention to the vegetation can provide clues to ancient as well as modern influences on the forest (Hightower et al. 2014).

Broad swaths of Lidar coverage in Guatemala and Belize (Fig. 12.3) provide a view of the landscape essential to understanding regional settlement and environmental patterns. Analyses of these coverages produce reasonable estimates of 80–120 people/km^2, which extrapolate to population estimates of 7–11 million people across the 95,000 km^2 of the central Maya Lowlands (Canuto et al. 2018). Regional population estimates based on Lidar visualizations are developed remotely without on-the-ground validation or the incorporation of the topographic character and vegetation influence recognized by traditional ground surveys (Bullard 1960; Ford 1986). At this point in the application of Lidar, we are still in the experimental stage, and field validation is essential for landscape interpretations. Detection of small features and vegetation character, environmental impact analyses of settlement, and geographic variables such as slope, soil, drainage, and vegetation must be included in settlement studies to understand land-use and human-environment interactions. Local areas provide insight into variability by combining spatial variables for deciphering environmental influences (cf. Ford et al. 2009), but understanding these influences at the detailed site scale is most challenging.

Our examination of settlement patterns has demonstrated that the Maya, from the regional and local perspectives, managed a landscape with the logic of living in well-drained areas with access to lands amenable to hand cultivation (Ford et al. 2009; Ford and Nigh 2015; Ford and Clarke 2019). The local-scale model for land use with the milpa-forest garden proposes a cyclical procession from forest to field and back again, which creates a diverse landscape over time, consisting of short-term annuals and long-term perennials that meet the daily requirements of the populace. Trees and plants selected for the long term persist as the forest we see today (Campbell et al. 2006; Ross 2011; Ross and Rangel 2011; see Dove 1983). Data on topography, settlement, vegetation, and soil at the site-specific scale are needed to test our model of cyclical field-to-forest cultivation strategies by the ancient Maya, and we are working to accumulate these data on the El Pilar project.

Site-scale Lidar coverage at El Pilar (Fig. 12.2) can contribute to considerations of ancient land use, vegetation cover, and sustainability in the tropics, where contemporary land use is expanding at the expense of the forest. The standard narrative views deforestation as the only outcome of living in the Maya forest (e.g., Turner and Sabloff 2012). This is difficult to contemplate when considering certain facts. For instance, how could the Maya forest be among the most diverse on Earth (Mittermeier et al. 2000) and be composed of dominant plants that are economically useful (Campbell et al. 2006; Dussol et al. 2017a, b; Ford 2008; Thompson et al. 2015) if the area were denuded of forest cover in the past? The seed bank harbored in the soil reflects millennia of directed human impact, and forest gardens flourish today among traditional Maya farmers as they did in the past, providing a source of seeds that sprout and recommend themselves in their appropriate habitat throughout the region (Ford and Nigh 2015). The archaeological sites and forest vegetation are both part of ancient Maya heritage.

Breaking Down the Point Cloud

Examining Lidar point clouds for vegetation cover and terrain features is essential to assessing ancient and contemporary land use and land cover in the Maya forest and beyond. The fidelity between what can be identified remotely at large scales, such as the regional scale of 1:250,000 offered by Canuto's team (2018) and even the local scale of 1:50:000 presented by Ford et al. (2009), may not be as easily matched in the Maya area at the site scale of 1:10,000 or less. Large features are readily identified, yet the complexity of small features, natural and archaeological, can be bewildering. With the data coverage at hand, we are in a position to appreciate the concordance of features identified remotely and targeted for field validation. As concordance studies increase in number and sophistication, we will be able to extrapolate on a firmer basis from field validations to remotely identified features at the local and regional scales. We also advocate the examination of topographic variability and vegetation qualities, which are clearly an influence on feature identification in Lidar imagery (see Prufer et al. 2015).

Appreciating vegetation influences, past and present, involves unpacking the point cloud and understanding the field conditions. The point cloud is usually divided into zones dependent upon the first laser return, reflecting the height of the biomass or forest canopy, the last laser return representing the land surface (frequently visualized as a "bare-earth" hillshade image), and intermediate zones corresponding to different levels of the understory. Canopy and understory density constrain how many pulses make it to the ground surface and thus the resolution provided by Lidar imagery (Fernandez et al. 2014; Fig. 12.4).

Ground Truth

Understanding the settlement patterns of ancient societies is among the highest priorities for archaeologists, and Lidar visualizations provide a direct means of deriving an image of these patterns. At the regional and local scales, general patterns of

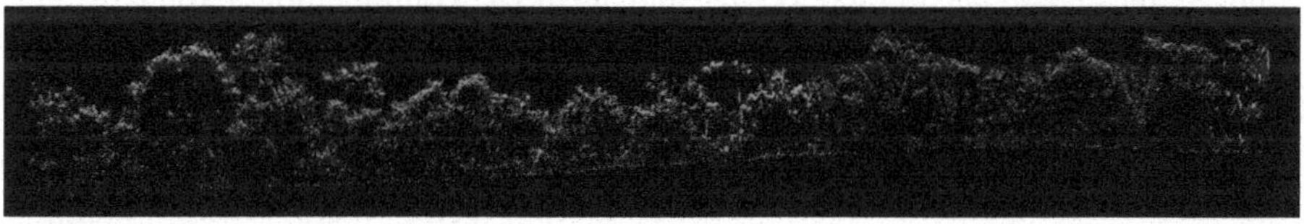

Fig. 12.4 Cross section of terraces at El Pilar

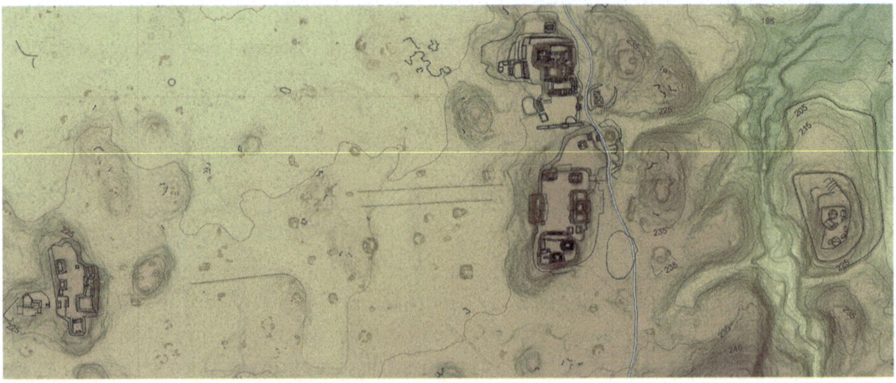

Fig. 12.5 El Pilar core area features with Lidar visualization

Maya settlement located in well-drained uplands have been recognized since Bullard's work 60 years ago (1960, 1964). These general settlement and environmental relationships continue to prove valid (Fedick and Ford 1990), as can be seen in the identified features of Canuto and colleagues (2018; Fig. 12.6) and in our work at El Pilar (Fig. 12.5). Yet there are many more useful data harbored in the Lidar point cloud. It is common that 1%–5% of the total returns penetrate to the surface. Thus 95–99% of the point cloud relates to the vegetation biomass.

At the El Pilar Archaeological Reserve for Maya Flora and Fauna, we have mapped 1647 unique locations in an area of 12 km^2 of which 1360, or 83%, were field-validated as cultural features and 287, or 17%, were rejected. Confirmed features include structural remains, depressions and aguadas, chultuns (underground storage pits), quarries, terraces, and linear features or berms. Rejected elements identified in Lidar imagery fall into several categories of natural features (tree falls, palm debris, ant mounds, or slumped earth) and errors generated in the Lidar visualization process.

Mapped features in our survey include those targeted (763, 56%) and those discovered (597, 44%) in our current coverage of 12 km^2. Features identified in the field, and not a priori detected on the visualizations, provide insight into the inherent challenges of interpreting Lidar visualizations. Considering the domestic architecture, we have mapped 529 residential units, comprising solitary or grouped small structures. Such domestic architecture constitutes the cultural remains most likely to impact our interpretation of population and land use. We discovered 54 of these residential units, or 10%, in the field. Other features, such as chultuns, quarries, depressions and aguadas, and berms and terraces, combine for a total of 831 recorded features, of which 543, or 65%, were discovered in the field. On the positive side, since the majority of residential units (90%) were features identified in Lidar imagery and targeted for validation, these features can contribute to land-use discussions. Residential units, however, make up only 39% of all mapped features, with the remaining 61% comprising the features mentioned above. These features make a significant impact on the landscape and will be missed in the absence field

Fig. 12.6 Linear features confirmed as terraces at El Pilar

survey. Landscape features represent much variety, and all features are not residential. This must be considered when inferring human interactions with the landscape, considering settlement density, and developing population estimates.

Terrain features remotely sensed and attributed to terraces and channels need field validation. Confirmation of these features is essential, as not all linear features will be validated (Figs. 12.6 and 12.7). When confirmed, agro-engineering features have been equated with land-use intensification, and while this may be so, it actually implies land *limitations*, where water must be slowed down (terraces) or drained (channels) to increase land productivity. Indications of agricultural intensification not tied to infrastructure, such as increases in labor inputs and changes in crop scheduling, will not be reflected in visual surface features. Infield forest garden management is more intense than that of distant outfields, whether on terraces, in drained fields, or on unmodified terrain. Only field validation will provide the appropriate means to assess the validity of agro-engineering interpretations and the relative importance of agricultural intensification.

The archaeological survey methods developed in the Maya forest over the past 60 years are essential to the Lidar validation process. Putting archaeological "boots on the ground" is the only way to determine if features identified in Lidar images are cultural remains or one of several potential "false positives" listed above. Equally challenging are linear or curvilinear features, which are presumed to be signatures of land modifications but are often artifacts of the Lidar visualization process

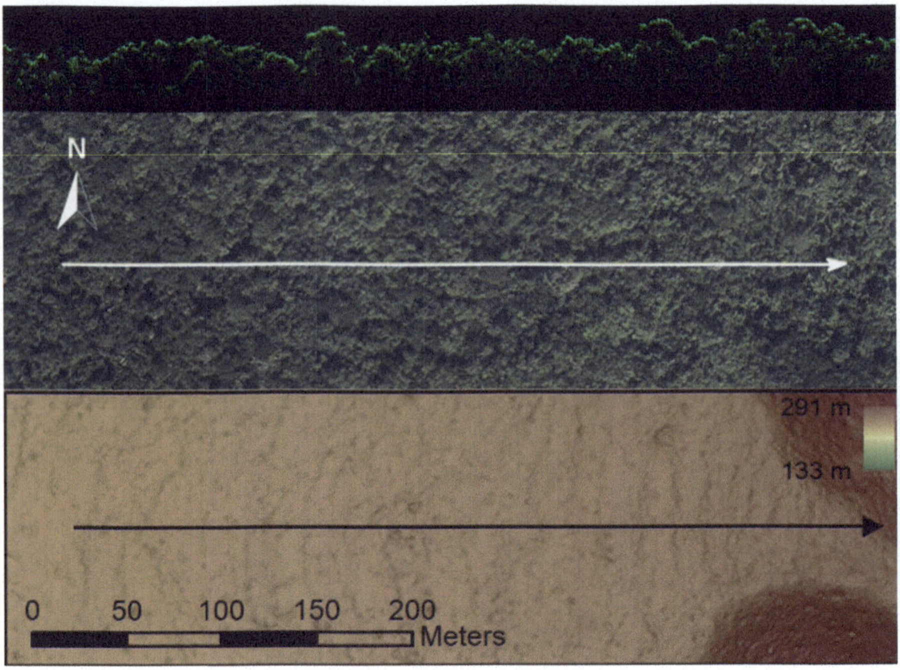

Fig. 12.7 Linear features interpreted in the Lidar visualization

(Fig. 12.7). We are in the early stages of testing the terrestrial patterns revealed by Lidar in the Maya forest, and there is no doubt of its importance in visualizing the topography, drainage, slopes, and vegetation across wide areas. With standard GPS navigation units, we can locate and validate surface features identified remotely within c. 7–14 m (Ford et al. 2013), which is sufficient resolution for mapping at the site scale. Structure counts and locations, and the relationships between these remains and other features, are better understood with field inspection (Ford 2014). As stimulating as the imagery is, we need to understand where small, and presumed domestic, structures may or may not be present and how other features impact human-environment interactions. Strides are being made with Lidar technology and significantly in methods of visualization (Pingel et al. 2015; Canuto et al. 2018), where we think the greatest potential lies.

Vegetation and the Canopy

Relatively unexplored by archaeologists employing Lidar are investigations of vegetation biomass data derived from point clouds (but see Hightower et al. 2014 for an example). Vegetation character, density, and height are part of the three-dimensional data points gathered by Lidar sensors (Fig. 12.8). The Lidar point cloud incorporates the entire forest biomass, from the surface (last laser pulse return) to the top of

Fig. 12.8 Oblique view Lidar

the canopy (first return), and includes shape characteristics of the vegetation that allow identification of particular trees (Fig. 12.4).

We have only begun our exploratory forays into the potential of Lidar and the concordance of visualizations to field identification. How can the Lidar help in the remote assessment of vegetation composition and height? What can the field identification of forest characteristics and documentation of the dominant plants reveal about human impacts on the forest? These are questions we are examining at El Pilar.

Our field protocol calls for the validation of Lidar targets and on-site visual assessment of terrain characteristics, vegetation density, and the identification of the dominant trees by master forest gardeners (Horn and Ford In Press; Horn et al. 2019). Evaluations of the Maya forest have determined that the dominant plants in the region show distinct influences of human selection based on alfa and beta diversity (Campbell et al. 2006; Ross 2011). Our data on the 20 dominant plants of the Maya forest (Table 12.1; based on Campbell et al. 2006) provide a basis for understanding human influence, past and present, on forest composition.

A total of 7395 dominant plants have been identified at 907 mapped sites in 12 km^2 of survey at El Pilar. This makes for an average of 8 dominant trees per site and 638 per km^2. We recorded the number of trees (Fig. 12.9), providing a relative impression of the distribution of the dominant plants across the site. The trees are not uniformly distributed and while some are ubiquitous, others are infrequent. While most mapped sites have the dominant trees present (Fig. 12.10), those that lacked trees were found in disturbed areas primarily covered by the bracken fern, *Pteridium*. Observed distributions of the dominant plants suggest variations that reflect environmental factors across the surveyed areas.

Given our interest in variability, we examined vegetation density and height using the Lidar point cloud. Density and height reflect modern impacts on the landscape. The binational space of 20 km^2 includes a western half surrounded by the Reserva de la Biosfera Maya in northern Guatemala and an eastern half surrounded by private land holdings in Belize. The difference in land use and land tenure is

Table 12.1 Dominant plants of the Maya forest, their pollinators and uses

Common name	Scientific name	Pollinator syndrome	Primary uses
Bay leaf	*Sabal morrisian**	Insects	Food, production
Breadnut	*Brosimum alicastrum**	Wind	Food, fodder
Cabbage bark	*Lonchocarpus castilloi*	Insects	Construction
Chicle	*Manilkara zapota**	Bats	Food, latex
Cohune	*Attalea cohune**	Insects	Food, construction
Drunken baymen	*Zuelania guidonia*	Bees	Medicine
Fiddlewood	*Vitex gaumeri*	Bats	Construction
Give-and-take	*Cryosophila stauracantha*	Beetles	Production
Guaya	*Talisia oliviformis**	Bees	Food
Gumbo-limbo	*Bursera simaruba**	Bees	Medicine
Hog plum	*Spondias radlkoferi*	Insects	Food
John Crow redwood	*Simira salvadorensis**	Moths	Construction
Mahogany	*Swietenia macrophylla*	Insects	Construction
Mamee cirila	*Pouteria campechiana*	Insects	Food
Mayflower	*Tabebuia rosea*	Bees	Construction
Monkey apple	*Licania platypus*	Moths	Food
Mylady	*Aspidosperma cruentum*	Insects	Construction
Wild mamey	*Alseis yucatanensis*	Moths	Construction
Wormwood	*Piscidia piscipula*	Bees	Poison
Zapotillo	*Pouteria reticulata*	Insects	Food, latex

*Plants found in local contemporary home gardens

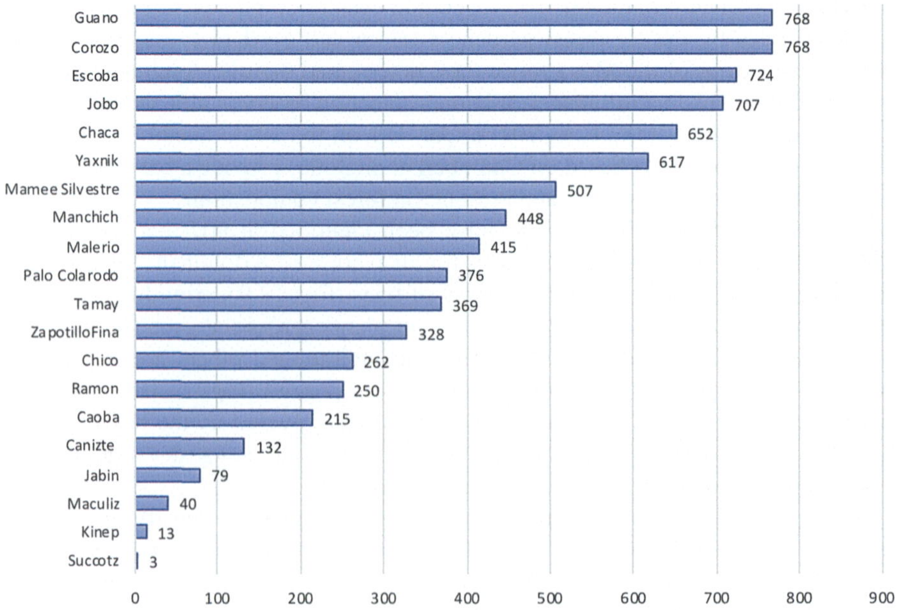

Fig. 12.9 Distribution of dominant plants at El Pilar

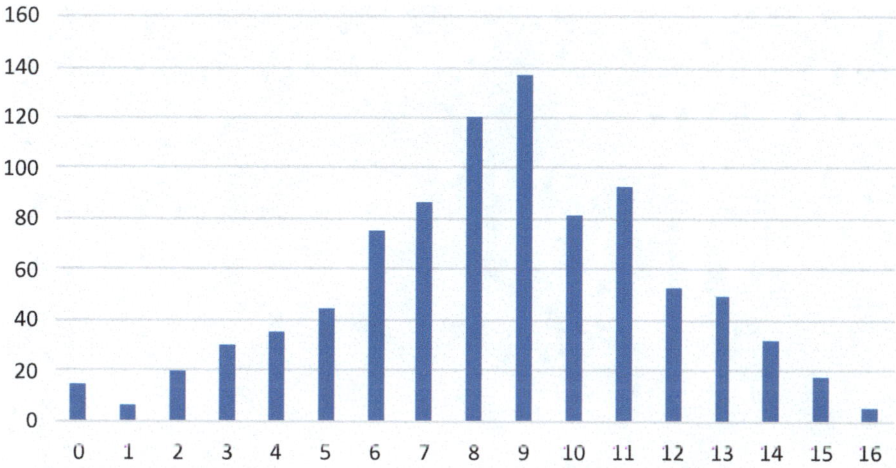

Fig. 12.10 Frequency of dominant plants at mapped sites

evident at the local scale with Landsat imagery (Fig. 12.11) and is apparent at the site scale provided by Lidar visualizations (Fig. 12.12). While the El Pilar Archaeological Reserve for Maya Flora and Fauna was established in 1998, the imprint of recent farming activity is still evident. Today, the expansion of forest clearance and conventional farming north and east of the reserve, and the presence of pastures and traditional farming in the south, continues to impact the local area. There are notable areas where the land cover has been damaged, illegal lumber extraction has reduced the number of large hardwoods, and low canopy areas consist of exclusionary flora such as the bracken fern.

We isolated three areas within the reserve for comparison of land use and land cover indications (Fig. 12.12; note red boxes). We include the transitional wetlands of the NE (Corozal), the monumental core area (El Pilar), and the NW uplands (Amatal). These areas represent examples of topography and slope influencing vegetation (Fig. 12.13) and surrounding land use impacting forest cover. From these three areas, we can see how land use and management have influenced the forest at El Pilar.

The first area to consider is the NE Corozal transitional wetlands. Drainage gradually slopes from the east and south toward the north and west, descending more than 25 meters across 1.5–2 km. Elevations change from above 165 m to below 140 m. Settlement is sparse and concentrated in the slightly higher southern sector, with an average of 51 residential units per km^2. This area is located adjacent to contemporary farms that are experiencing heavy use and impact by fires. Fires in the surrounding area have penetrated the reserve over the past 10 years, impacting canopy height and density (Fig. 12.14). Heavy land clearing is obvious at the NE corner of the reserve, where vegetation has been reduced to just above the surface. Canopy height ranges from 0 to 25 m (Fig. 12.15), with an average around 10 m. This is an

Fig. 12.11 Local El Pilar area as viewed from Landsat 2014

area with few archaeological remains, and where vegetation is dominated by *Attalea cohune;* this the corozo palm from which the area takes its name.

The second area is the core of El Pilar, where the site's impressive monuments are concentrated. This is an area of variable drainage, with a flow gradient running from north to south separating the two ridges supporting monumental complexes. There is considerable relief in the southeastern and southwestern sections, where

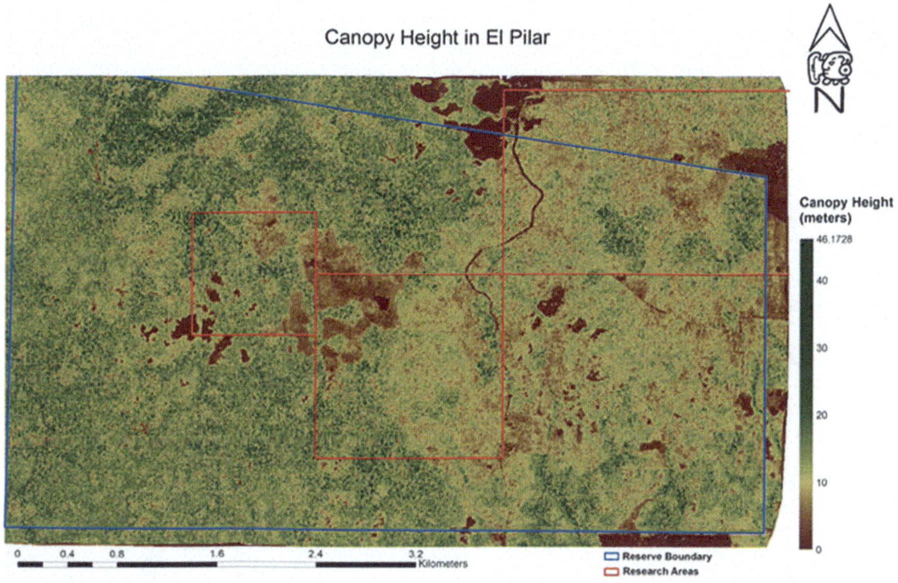

Fig. 12.12 Lidar visualization of the canopy height based on the first laser return

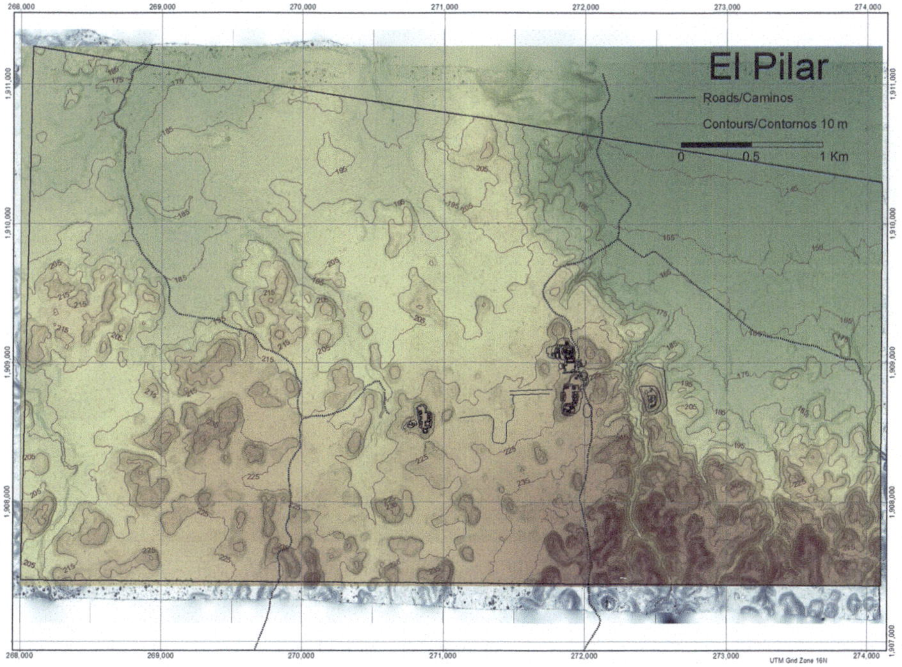

Fig. 12.13 El Pilar Archaeological Reserve for Maya Flora and Fauna topography

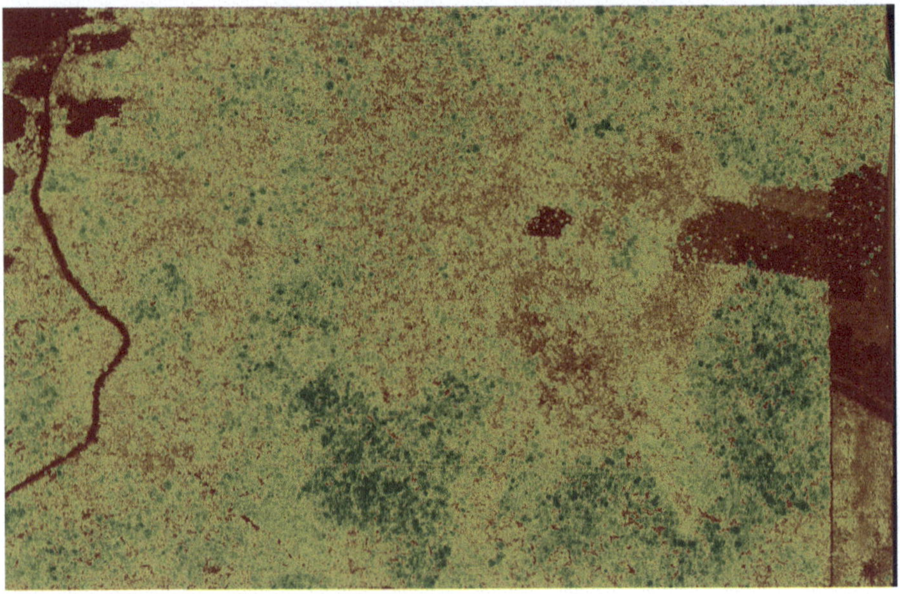

Fig. 12.14 NW Corozal wetland forest canopy cover variation

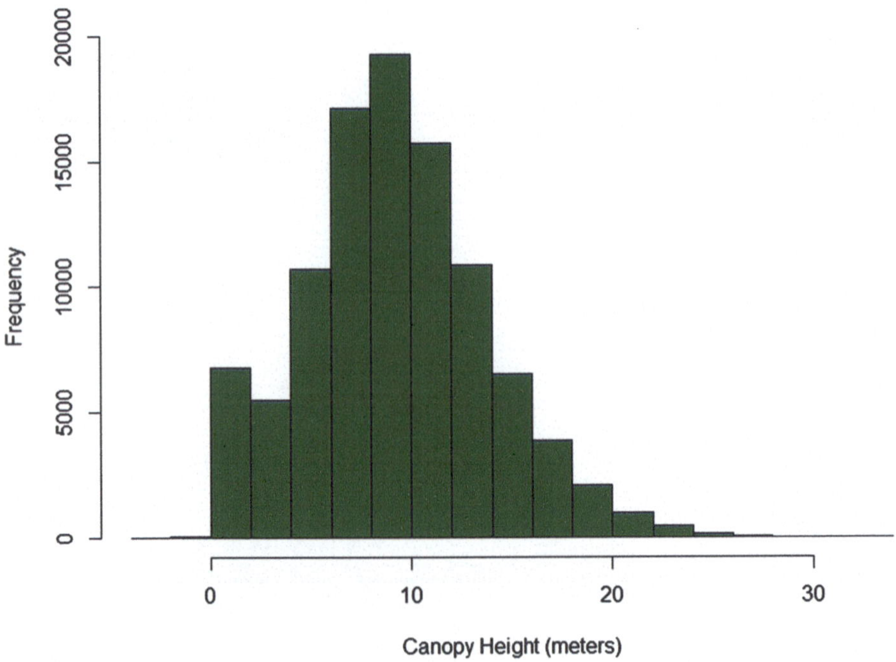

Fig. 12.15 Distribution of canopy height of NW Corozal wetlands

elevations vary from 260 to 270 m on the ridges to less than 210 m in the south-north running drainage. Settlement in this area is dispersed around the monuments, averaging 74 residential units per km^2. A modern road intersects the site core on the east, impacting the monuments on that side. In historic times, this area has been one of significant human activity. The road access is recorded before the 1950s and was mapped by the Central American Geodetic Survey in the early 1960s. There was an established camp for lumber and chicle on the site at that time, and smallholder farming and plantings were noted from the 1970s. Archaeological research began in the 1980s, followed by the delimitation of the archaeological reserve in the 1990s and recognition of government management plans in the 2000s. The road is clearly visible (Fig. 12.16), and the "brizantha" grass of the irregular square in the NW segment is also apparent. Overall the canopy range is similar to the NE Corozal, averaging around 10 m, but the highest trees reach to 30 m (Fig. 12.17). Much of the core area is dominated by corozo palms, but in higher areas the trees are more diverse and include taller hardwoods – at least those that have not fallen to the saws of illegal loggers. Archaeological remains of the major civic center El Pilar span this area and extend into Belize and Guatemala.

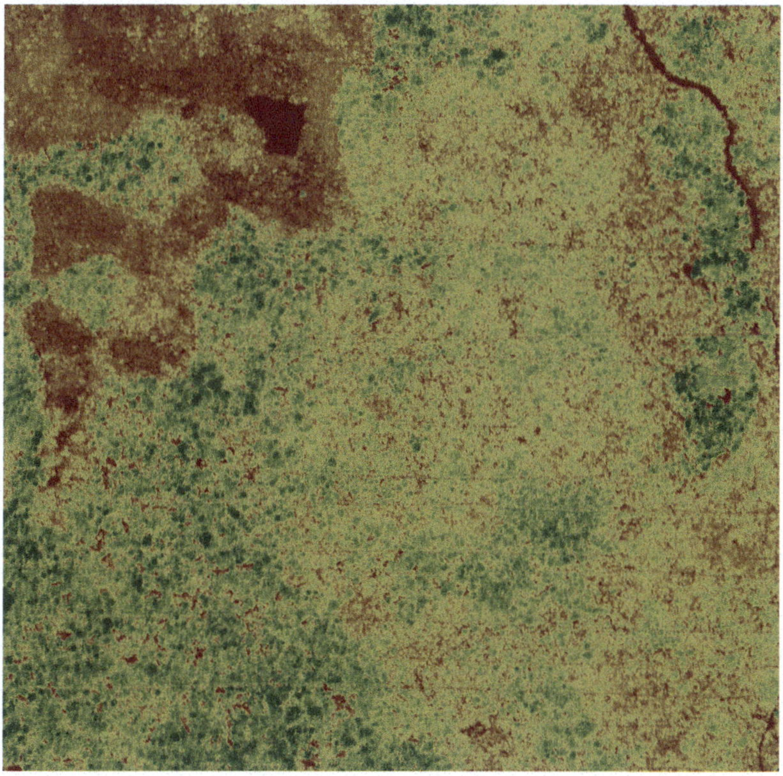

Fig. 12.16 El Pilar core area forest canopy cover variation

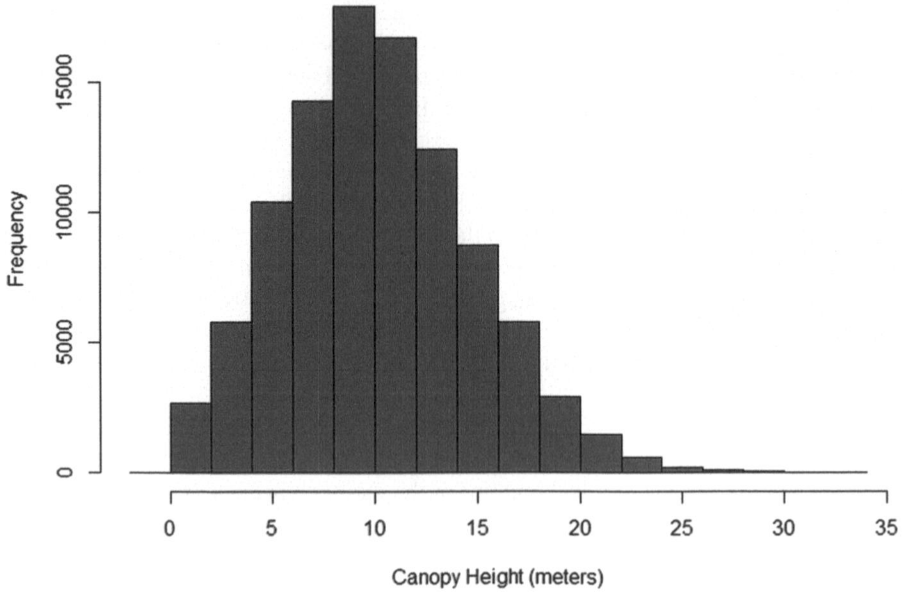

Fig. 12.17 Distribution of canopy height of El Pilar core

The third example comes from the Amatal uplands to the northwest of the El Pilar core area. Water flows from the southwest to the northeast, with the drainage flanked by two major architectural temple complexes built on overlooking ridges. Elevations are lower than those of the core area, beginning around 230 m and descending to about 180 m. Drainage is excellent and the presence of amate trees suggests availability of water. Settlements are large and residential unit density is 78 per km^2 around the monumental complex, equivalent to the site core. While forest cover is generally good, with an average above 15 m and the higher trees reaching above 30 m (Fig. 12.18), the variability demonstrates the impact of poor farming strategies in the area. The irregular blotches in at least seven areas represent the bracken fern that arrests succession where it takes hold. Large hardwoods are present across the area, and several large amate trees give the complex its name (Fig. 12.19).

Overview

The impact of human use in the past is evident with the density of residential units and the variety of other features recorded across El Pilar. Large and small structures are found throughout the area. Limestone quarries are found on most hillsides, and probable agricultural features (berms and terraces) are concentrated in specific zones toward the northeast. These are landscape modification features that relate to land use and land cover in the past.

In the example of vegetation at El Pilar, we see that the combination of Lidar classification and field assessment provides the means to understand the nature of past

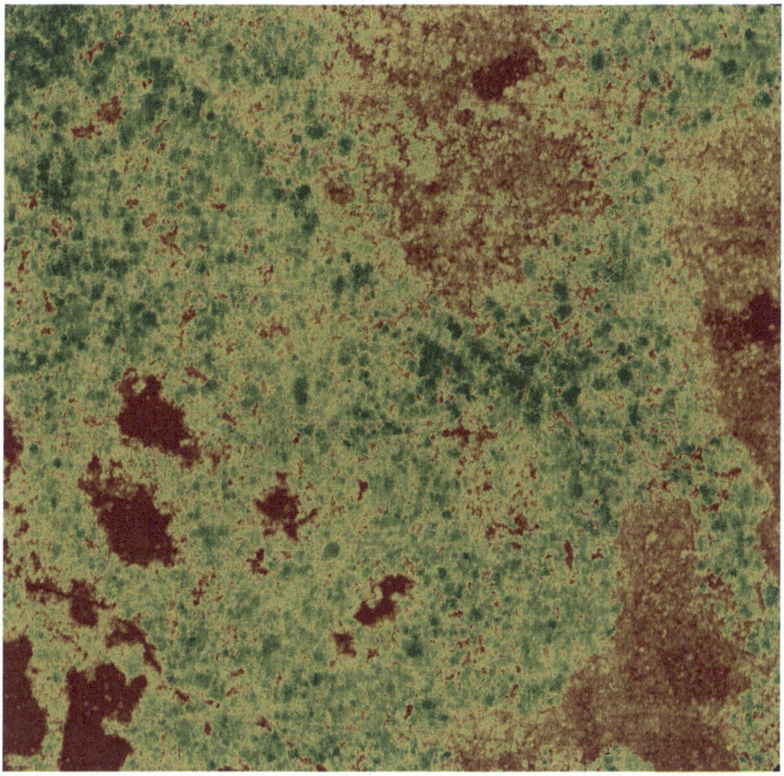

Fig. 12.18 Amatal forest canopy cover variation

influences on the forest species composition as well as historical and contemporary impacts of land use. The presence of the dominant plants of the Maya forest, all of diverse and relevant uses, indicates ancient Maya selection emphasizing important forest products. The establishment of the reserve in 1998 has led to the development and improvement of the forest cover within its boundaries while at the same time revealing agricultural pressures where private land use continues to expand.

Contemporary land use is evident in the vegetation characteristics and the state of the canopy (see Weishampel et al. 2012). Impacts of the agricultural frontier are evident in the east and have affected the vegetation more so here than in areas to the west. Yet throughout the reserve, the areas in the central monumental area and northeast Corozal exhibit the greatest impact. The presence of bracken fern, *Pteridium*, as a cosmopolitan plant that arrests succession where it is established, has inhibited forest recovery inside the reserve. Bracken is adapted to fire and grows fast on the well-drained hills and ridges, typically areas where the ancient Maya settled. Without substantial investment to recover these areas, we can expect the bracken fern will maintain or expand its coverage. Despite the prominent patches of bracken, the presence and persistence of the dominant plants of the Maya forest are evidence that the Maya influenced the landscape, directing biodiversity toward useful plants.

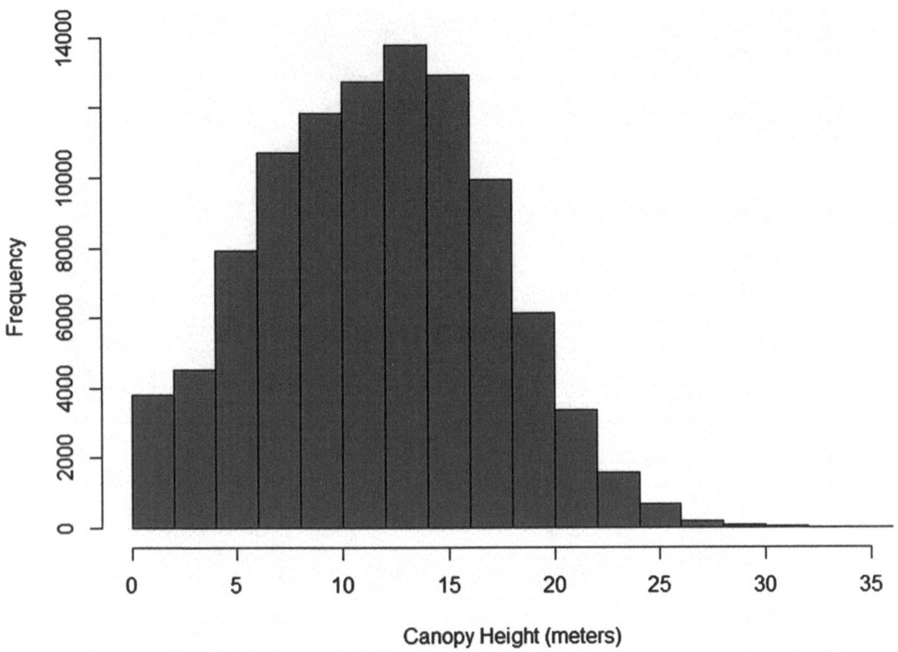

Fig. 12.19 Distribution of canopy height of Amatal

In review, the complexity of the Maya forest as exemplified by research at El Pilar provides a rich foundation upon which to build. To understand the past landscape, we need an appreciation of settlement patterns and how those patterns structured the environment. The study of surface visualizations indicates the promise of Lidar to reveal the secrets of ancient Maya land use. Residential units and many related features (chultuns, quarries, depressions and aguadas, terraces and berms) all point to manipulations that underwrote the successful adaptations of the Maya, past and present.

Conclusion

Fieldwork is the way forward for understanding visualizations created from Lidar data. The more field validation samples for archaeological features and vegetation character, the better the basis of extrapolation to the greater Lidar coverages for Belize, Guatemala, and Mexico. There is little doubt that Lidar is destined to become indispensable for ancient Maya settlement research, improving our understanding of the forested Maya Lowlands. Even as Lidar will have a role to play in the analyses of ancient settlement and the character, height, and density of the biomass captured in point clouds, there will *always* be a need for field reconnaissance and validation.

At the large scale of coverage that exists now, regionally defined topography and landforms alone have great potential for appreciating the nature of vegetation cover. This is the foundation for recognizing the role and influence of environmental factors on the distribution of ancient settlement and contemporary forest cover, which provides a call to action for scholars of land use and land cover to scrutinize sampling strategies and bring the value of Lidar to the detailed site scale.

Acknowledgments Our Lidar project owes much to Anfield Nickel and its associated company Mayaniquel. They have generously donated air photo and Lidar coverage for the El Pilar Archaeological Reserve. Our fieldwork was possible with the support of Exploring Solutions Past and numerous contributors through Experiment.com. Our field activities were accomplished with the support of Lic. Paulino Morales and Master Forest Gardener Narciso Torres. We gratefully thank the Institute of Archaeology Belize and the Instituto de Antropología e Historia Guatemala for their continued support of research and development at the El Pilar Archaeological Reserve for Maya Flora and Fauna.

Bibliography

Adams REW (1969) Maya Archaeology 1958-1968, a review. Lat Am Res Rev 4(2):3–45

Britanica (2019) Lidar scientific technique https://www.britannica.com/technology/lidar

Bullard WR Jr (1960) Maya settlement pattern in Northeastern Petén, Guatemala. Am Antiq 25(3):355–372

Bullard WR Jr (1964) Settlement pattern and social structure in the Southern Maya Lowlands during the classic period. XXXV Congreso Internacional de Americanistas. México City 1:279–287

Canuto MA, Estrada-Belli F, Garrison TG et al (2018) Ancient Lowland Maya complexity as revealed by Airborne Laser scanning of Northern Guatemala. Science forthcoming:1–12

Chase AF, Chase DZ, Weishampe JF et al (2011) Airborne LiDAR, archaeology, and the ancient Maya landscape at Caracol, Belize. J Archaeol Sci 38:387–398

Campbell DG, Ford A, Lowell K, Walker J, Lake JK, Ocampo-Raeder C, Townesmith A, Balick M (2006) The Feral Forests of the Eastern Petén. In: Balée W, Erickson C (eds) Time and Complexity in the Neotropical Lowlands: Studies in Historical Ecology. Columbia University Press, New York, pp 21–55

Chase A, Chase DZ, Awe JJ et al (2014) Ancient Maya regional settlement and inter-site analysis: the 2013 West-Central Belize LiDAR survey. Remote Sens 6(9):8671–8695

Chase ASZ, Chase DZ, Chase AF (2017) LiDAR for archaeological research and the study of historical landscapes. In: Masini N, Soldovieri F (eds) Sensing the Past. Springer International Publishing, New York, pp 89–100

Devereux BJ, Amable GS, Crow P et al (2005) The potential of Airborne Lidar for detection of archaeological features under Woodland Canopies. Antiquity 79(305):648–660

Dove M (1983) Theories of swidden agriculture, and the political economy of ignorance. Agric Syst 1:85–99

Dussol L, Elliott M, Michelet D et al (2017a) Ancient Maya sylviculture of breadnut (Brosimum alicastrum) and sapodilla (Manilkara zapota) at Naachtun (Guatemala): a reconstruction based on charcoal analysis. Quat Int 457:29–42

Dussol L, Elliott M, Théry-Parisot I (2017b) Experimental anthracology: evaluating the role of combustion processes in the representivity of archaeological charcoal records in tropical forests, a case study from the Maya Lowlands. J Archaeol Sci 12:480–490

Ebert CE, Hoggarth JA, Awe JJ (2016) Integrating quantitative Lidar analysis and settlement survey in the Belize river valley. Adv Archaeol Pract 4(3):284–300

Evans DH, Fletcher RJ, Pottier C et al (2013) Uncovering archaeological landscapes at Angkor using Lidar. Proc Natl Acad Sci 110(31):12595–12600

Fedick SL, Ford A (1990) The prehistoric agricultural landscape of the Central Maya Lowlands: an examination of local variability in a regional context. World Archaeol 22:18–33

Fedick SL, Clarke KC, Ford A (2016) Refining models of ancient Maya agricultural landscape archaeology in the Belize river area: initial results making use of LiDAR imagery. Res Rep Belizean Archaeol 13:121–128

Fernandez-Diaz WE, Carter RLS, Glennie CL (2014) Now you see it… now you don't: understanding Airborne mapping LiDAR collection and data product generation for archaeological research in Mesoamerica. Remote Sens 6(10):9951–10001

Ford A (1981) Conditions for the evolution of complex societies: the development of the Central Lowland Maya. Ph.D. dissertation, University of California, Santa Barbara

Ford A (1986) Population growth and social complexity: an examination of settlement and environment in the Central Maya Lowlands. Am Antiq 52(4):886–887

Ford A (2008) Dominant plants of the Maya forest and gardens of El Pilar: implications for paleoenvironmental reconstructions. J Ethnobiol 28(2):179–199

Ford A (1990) Maya settlement in the Belize river area: variations in residence patterns of the Central Maya Lowlands. In: Culbert TP, Rice DS (eds) Prehistoric population history in the Maya Lowlands. University of New Mexico Press, Albuquerque, pp p167–p181

Ford A (1991) Economic variation of ancient Maya residential settlement in the Upper Belize river area. Anc Mesoam 2:35–46

Ford A (2014) Using cutting-edge LiDAR technology at El Pilar, Belize-Guatemala, in discovering ancient Maya sites: there is still a need for archaeologists. Res Rep Belizean Archaeol 11:270–280

Ford A (in press) The Maya forest: a domesticated landscape. In: Hutson S, Arden T (eds) The Maya world. Routledge

Ford A, Bihr H, Morales P (2013) Usando Metodos Vanguardistas LiDAR en El Pilar, Guatemala-Belice: Cambiando la Arquelogia en la Selva Maya. *XXVII Simposio de Investigaciones Arqueologicas en Guatemala* 27(1):8

Ford A, Clarke KC (2019) Linking the past and present of the ancient Maya: Lowland use, population distribution, and density in the late classic. In: Isendahl C, Stump D (eds) Handbook of historical ecology and applied archaeology. Oxford University Press, Oxford, UK, pp 56–183

Ford A, Clarke KC, Raines G (2009) Modeling settlement patterns of the late classic Maya civilization with bayesian methods and geographic information systems. Ann Assoc Am Geogr 99(3):496–520. https://doi.org/10.1080/00045600902931785

Ford A, Crimmel T, Knudson C et al (2018) Using LiDAR at El Pilar. Popular archaeology https://popular-archaeology.com/article/using-lidar-at-el-pilar/(Spring)

Ford A, Fedick SL (1992) Prehistoric Maya settlement patterns in the Upper Belize river area: initial results of the Belize river archaeological settlement survey. J Field Archaeol 19:35–49

Ford A, Horn S (2017) El Pilar monuments retrospective & prospective: re-discovering El Pilar. Res Rep Belizean Archaeol 14:87–95

Ford A, Horn S (2018) Above and below the Maya forest: Advanced remote sensing technology raises questions about settlement and land use. Science 316:1313–1314

Ford A, Morales P, Lopez J (2015) Una Nueva Visiốn de Arqueologìa Bajo el Dosel: Beneficios de la Tecnologìa LiDAR. XXIX Simposio De Investigaciones Arqueològicas. Guatemala

Ford A, Nigh R (2015) The Maya forest garden: eight millennia of sustainable cultivation in the tropical Woodlands. Left Coast Press, Santa Rosa

Gaurav SG (2018) Brief history of LiDAR, its evolution and market definition http://blog.bccresearch.com/brief-history-of-lidar-evolution-and-market-definition. 2019

Hightower JN, Butterfield AC, Weishample JF (2014) Quantifying ancient Maya land use legacy effects on contemporary rainforest canopy structure. Remote Sens 6:10716–10732

Horn S, Ford A, Morales P. (2019) Lasers, lasers, everywhere – and all the trees did shrink: reliable methods and results from Lidar-guided survey at El Pilar. Res Rep Belizean Archaeol 16:143–155

Horn S, Ford A (in press) Beyond the Magic Wand: methodological developments and results from integrated Lidar survey at the ancient Maya Center El Pilar. Science and Technology in Archaeological Research

Hutson SR, Kidder B, Lamb C et al (2016) Small Budgets and Small Budgets Making Lidar Work in Northern Yucatan, Mexico. Adv Archaeol Pract 4(3):268–283

Johnson KM, Ouimet WB (2014) Rediscovering the lost archaeological landscape of Southern New England using Airborne Light Detection and Ranging (LiDAR). J Archaeol Sci 43:9–20

Magnoni A, Stanton TW, Barth N et al (2016) Detection thresholds of archaeological features in Airborne Lidar data from Central Yucatán. Adv Archaeol Pract 4(3):232–248

Mittermeier RA, Myers N, Mittermeier CG (2000) Hotspots: earth's biologically richest and most endangered terrestrial ecoregions. CEMEX, México City

NOAA (2019) What is Lidar? https://oceanservice.noaa.gov/facts/lidar.html

Parton PA, Clak G, Burley D (2018) The field of war: Lidar identification of earthwork defences on Tongatapu Island, Kingdom of Tonga. J Pac Archaeol 9(1):11–24

Pingel TJ, Clarke KC, Ford A (2015) Bonemapping: A LiDAR processing and visualization technique in support of archaeology under the canopy. Cartogr Geogr Inf Sci 42(S1):S18–S26

Prufer KM, Thompson AE, Kennett DJ (2015) Evaluating Airborne LiDAR for detecting settlements and modified landscapes in disturbed tropical environments at Uxbenka, Belize. J Archaeol Sci 57:1–13

Puleston DE (1973) Ancient Maya settlement patterns and environment at Tikal, Guatemala: implications for subsistence models. Ph.D. thesis, University of Pennsylvania

Puleston DE (1983) Tikal report No. 13: the settlement survey of Tikal. Philadelphia, Pennsylvania, The University Museum, University of Pennsylvania

Reese-Taylor K, Hernández AA, Esquivel FCAF et al (2016) Boots on the ground at Yaxnohcah ground-truthing Lidar in a complex tropical landscape. Adv Archaeol Pract 4(3):314–338

Rice DS (1976) The historical ecology of lakes Yaxhá and Sacnab, El Petén, Guatemala. Ph.D. dissertation, Pennsylvania State University

Rice DS (1978) Population growth and subsistence alternatives in a tropical Lacustrine environment. In: Harrison PD, Turner BL (eds) *Pre-hispanic Maya agriculture*. University of New Mexico Press, Albuquerque, pp 35–61

Rosenswig RM, Lopez-Torrijos R, Antonellli CE et al (2013) Lidar mapping and surface survey of the Izapa State on the tropical piedmont of Chiapas, México. J Archaeol Sci 40(2013):1493–1507

Ross NJ (2011) Modern tree species composition reflects ancient Maya "Forest Gardens" in Northwest Belize. Ecol Appl 21(1):75–84

Ross NJ, Rangel TF (2011) Ancient Maya agroforestry echoing through spatial relationships in the extant forest of NW Belize. Biotropica 43(2):141–148

Stephens JL (1969) Incidents of travel in Central America, Chiapas and Yucatán II. Dover Publications, New York

Stular BZK, Ostir K, Nuninger L (2012) Visualization of LIDAR-derived relief models for detection of archaeological features. J Archaeol Sci 39:3354–3360

Thompson KP, Hood A, Cavallaro D et al (2015) Connecting contemporary ecology and ethnobotany to ancient plant use practices of the Maya at Tikal. In: Lentz DL, Dunning NP, Scarborough VL (eds) Tikal: paleoecology of an ancient Maya City. Cambridge University Press, New York, pp 124–151

Turner BL II, Sabloff JA (2012) Classic period collapse of the Central Maya Lowlands: insights about human–environment relationships for sustainability. Proc Natl Acad Sci U S A 109(35):13908–13914

Weishampel JF, Hightower JN, Chase AF, Chase DZ (2012) Use of Airborne LiDAR to delineate canopy degradation and encroachment along the Guatemala-Belize Border. Trop Conserv Sci 5(1):12–24

Yaeger JK, Brown M, Cap B (2016) Locating and dating sites using Lidar survey in a mosaic landscape in Western Belize. Adv Archaeol Pract 4(3):339–356

Index

N. Torrescano-Valle et al. (eds.), *The Holocene and Anthropocene Environmental History of Mexico*, https://doi.org/10.1007/978-3-030-31719-5

D

E

F

The manufacturer's authorised representative in the EU is Springer Nature Customer Service Centre GmbH, Europaplatz 3, 69115 Heidelberg, Germany. If you have any concerns regarding our products, please contact ProductSafety@springernature.com

Printed and bound by CPI Group (UK) Ltd, Croydon, CR0 4YY
30/09/2026
02225657-0003